UNIVERSITY COLLEGE OF WALES
LIBRARY
ABERYSTWYTH

AF616127

BIOMEMBRANES

Volume 7

BIOMEMBRANES

A series edited by
Lionel A. Manson
The Wistar Institute
Philadelphia, Pennsylvania

1971 • Biomembranes • Volume 1
Articles by M. C. Glick, Paul M. Kraemer, Anthony Martonosi, Milton R. J. Salton, and Leonard Warren

1971 • Biomembranes • Volume 2
Proceedings of the Symposium on Membranes and the Coordination of Cellular Activities
Edited by Lionel A. Manson

1972 • Biomembranes • Volume 3
Passive Permeability of Cell Membranes
Edited by F. Kreuzer and J. F. G. Slegers

1974 • Biomembranes • Volume 4A
Intestinal Absorption
Edited by D. H. Smyth

1974 • Biomembranes • Volume 4B
Intestinal Absorption
Edited by D. H. Smyth

1974 • Biomembranes • Volume 5
Articles by Richard W. Hendler, Stuart A. Kauffman, Dale L. Oxender, Henry C. Pitot, David L. Rosenstreich, Alan S. Rosenthal, Thomas K. Shires, and Donald F. Hoelzl Wallach

1975 • Biomembranes • Volume 6
Bacterial Membranes in the Respiratory Cycle
By N. S. Gel'man, M. A. Lukoyanova, and D. N. Ostrovskii

1975 • Biomembranes • Volume 7
Aharon Katzir Memorial Volume
Edited by Henryk Eisenberg, Ephraim Katchalski-Katzir, and Lionel A. Manson

---

A Continuation Order Plan is available for this series. A continuation order will bring delivery of each new volume immediately upon publication. Volumes are billed only upon actual shipment. For further information please contact the publisher.

BIOMEMBRANES

Volume 7

Aharon Katzir Memorial Volume

Edited by

Henryk Eisenberg

and

Ephraim Katchalski-Katzir

The Weizmann Institute of Science
Rehovot, Israel

and

Lionel A. Manson

The Wistar Institute
Philadelphia, Pennsylvania

PLENUM PRESS · NEW YORK AND LONDON

Library of Congress Cataloging in Publication Data

Main entry under title:

Aharon Katzir memorial volume.

(Biomembranes; v. 7)
"Selected publications of Aharon Katzir-Katchalsky": p.
Includes bibliographies and index.
1. Biological transport. 2. Cell membranes. I. Katchalsky, Aharon. II. Eisenberg, Henryk. III. Katchalski-Katzir, Ephraim, 1916- IV. Manson, Lionel A. V. Series.
QH601.B53 vol. 7 [QH509] 574.8'75'08s
ISBN 0-306-39807-9 [574.8'75] 75-9934

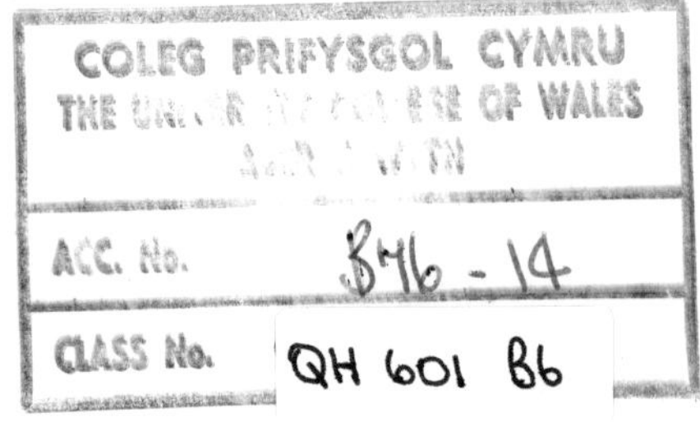

A Division of Plenum Publishing Corporation
227 West 17th Street, New York, N.Y. 10011

United Kingdom edition published by Plenum Press, London
A Division of Plenum Publishing Company, Ltd.
Davis House (4th Floor), 8 Scrubs Lane, Harlesden, London, NW10 6SE, England

Printed in the United States of America

CONTRIBUTORS

Margareta Baltscheffsky, Department of Biochemistry, Arrhenius Laboratory, University of Stockholm, Sweden

Ze'ev Barak, Department of Biochemical Sciences, Frick Chemical Laboratory, Princeton University, New Jersey

Britton Chance, Johnson Research Foundation, School of Medicine, University of Pennsylvania, Philadelphia

D. Chapman, Department of Chemistry, Sheffield University, Sheffield, England

W. W. Cheng, Johnson Research Foundation, School of Medicine, University of Pennsylvania, Philadelphia

Isidore S. Edelman, Departments of Biochemistry and Biophysics, University of California School of Medicine, San Francisco

E. B. Margareta Ekblad, Department of Medicine, University of California School of Medicine, San Francisco

Ayala Frenkel, Weizmann Institute of Science, Rehovot, Israel

Ilan Friedberg, Department of Electron Microscopy, Tel Aviv University, Israel

Charles Gilvarg, Department of Biochemical Sciences, Frick Chemical Laboratory, Princeton University, New Jersey

B. Z. Ginzburg, Botany Department, The Hebrew University, Jerusalem, Israel

M. Ginzburg, Botany Department, The Hebrew University, Jerusalem, Israel

Israel Goldberg, Department of Applied Microbiology, Hadassah Medical School, The Hebrew University, Jerusalem, Israel

David Nachmansohn, Department of Neurology and Biochemistry, Columbia, University, New York

Eberhard Neumann, Max-Planck-Institut of Biophysical Chemistry, Goettingen, Germany

Itzhak Ohad, Department of Biological Chemistry, The Hebrew University, Jerusalem, Israel

W. Wilbrandt, Department of Pharmacology, University of Bern, Switzerland

Preface

It was a warm, sunny morning in Rehovot. The sky was clear as it always is in June. As I walked to the Institute that morning, too many cars were passing by, too many people were hurrying onto the Institute's grounds. No one was smiling, acquaintances were recognized by a slight nod of the head. When I turned the corner, a few people already had gathered on the lawn in front of the Jacob Ziskind building. This number was to swell to thousands before the service was over. We were to be joined by the President of Israel, its first Prime Minister, many members of the cabinet, and other great, near-great, working colleagues, and residents of the town. The purpose of all this activity was written on everyone's face, and underlined by the casket that lay in the rotunda of the building. His wife was sitting there, his children, his brother, his students both past and present. One could hear the silence of the participants. I stood inside for a while, overlooking the rotunda. A long line of mourners filed by, offering their sympathies to the family. Suddenly a woman, dressed in black, fell to her knees in front of Rina, sobbing. It was the wife of the Japanese Ambassador to Israel. I moved to join the crowd outside. People were standing on the lawn, gingerly trying to avoid stepping on a flower. Aharon would have appreciated this, considering his great love of nature. Soon the eulogies were over, the procession formed and moved off to his final resting place. These are some of the remembrances I have of the funeral of Aharon Katzir-Katchalsky. He was one of 26 people senselessly murdered at Lod airport on May 30, 1972.

I left the Weizmann Institute on June 25, 1972, as my sabbatical leave had come to an end. Soon thereafter, with the help of Heini Eisenberg and Ephraim Katchalski, arrangements were made to publish a collection of scientific contributions by Aharon's colleagues and students. Invitations were sent to scientists in many countries. Not all who were contacted could participate.

The table of contents encompasses some of the scientific areas in which Aharon was involved. A more rounded picture is given in Heini Eisenberg's personal portrait of the man. It is not possible to describe this wonderful human being in words. Nor are pictures adequate. Those who met him and worked with him know that their lives have been enriched immeasurably. Others, like myself, who only had a few opportunities to speak with him or listen to his lectures, sincerely regret that we shall never get to know him.

Philadelphia
March 31, 1975

Lionel A. Manson

Contents

Chapter 3

Carotenoid and Merocyanine Probes in Chromatophore Membranes

Britton Chance and Margareta Baltscheffsky

Chapter 4

Effects of Sulfhydryl Reagents on Basal and Vasopressin-Stimulated Na^+ Transport in the Toad Bladder

Ayala Frenkel, E. B. Margareta Ekblad, and Isidore S. Edelman

Chapter 5

Biogenesis of Chloroplast Membranes in *Chlamydomonas Reinhardi*: Chloroplast-Controlled Transfer of Cytoplasmic Proteins As Visualized by Quantitative Radioautography

Israel Goldberg, Ilan Friedberg, and Itzhak Ohad

Chapter 6

Nerve Excitability—Toward an Integrating Concept

Eberhard Neumann and David Nachmansohn

Chapter 7

Peptide Transport

Ze'ev Barak and Charles Gilvarg

AHARON KATZIR-KATCHALSKY

Aharon Katzir-Katchalsky—An Appreciation

Aharon Katzir-Katchalsky was born in 1914 in Lodz, Poland, and came to Israel—then Palestine—in 1925, with his parents and his brother, Ephraim. He was educated and inspired in Jerusalem, where he obtained his Ph.D. from the Hebrew University in 1940. A brilliant career centered around the physical, biological, and humanistic aspects of science followed, dedicated to the service of his country and the world at large. At the peak of his creative ability, on May 30, 1972, his life was senselessly and brutally extinguished at the Lod airport massacre.

I first met Aharon Katzir in 1946 and our ways did not part until the day of his tragic death. He had just come back from a period of study in Basel, looking for tools to give reality to his exciting thoughts about the structure of the biological world. I had just finished five years of Army service in the War and was probing my way back into academic life, on the threshold of my scientific career. I will never forget that first meeting, a record of which is extant, buried somewhere in old papers. Aharon, with his deep and brilliant intuition, foresaw the explosive growth of molecular biology and its role in unraveling aspects of the structure and function of biological materials, many of them familiar even to high-school students today. In the days before DNA was recognized to be the carrier of genetic information and before protein structure was known, Aharon saw the link between biological function and the structure of large molecules, polymers, constructed by the joining of many atoms into long chains. He had studied biology, intent on understanding the processes of life, and had come across the idea that large molecules may play a vital role in terms of their basic structure. This refers to the fact that, in distinction to rigid, small molecules, from which relatively undeformable structures are built, macromolecules can assume a large number of conformations and form highly deformable structures, of which rubber is only one typical example. An engineering principle becomes immediately apparent in the creation of motion, for example: a principle spurned by the creators of artificial machines, but widely used in nature in many forms of biological motion and contractility.

Werner Kuhn had pioneered work in Basel on the properties of deformable macromolecules. To him Aharon went to acquire this knowledge. He realized that most macromolecules in nature carry electrical charges and therefore himself pioneered the field of polyelectrolytes, the study of macromolecules carrying electrically charged groups.

Our early exciting studies in one room on the second floor of the Physics Building on Mount Scopus in Jerusalem were soon interrupted by more pressing events. With the approach of the War of Independence, Aharon in 1947 was cocreator of Hemed, the forerunner of the Science Corps of the Israeli Defence Forces, Zahal. The full story of this venture has not been told, but Aharon fired our imagination and evolved the basic scheme whereby all our efforts in a soon to be besieged city were channeled toward survival in what appeared to be a hopeless situation. Books on pyrotechnics, aluminum powder in paint stores, sacks filled with potash, old rubber tires, and kerosene were marshalled to ensure our success. Many friendships were forged in those days and many of our friends have gone on to leadership in the nation today. We should also remember those who paid with their life for our own survival.

With the end of the war Aharon assembled a number of young novices and established the Polymer Department at the Weizmann Institute. Whereas elsewhere in the world great centers of learning tradition have taken decades and centuries to develop, here a miracle occurred. A center came into being which corresponded to the highest international standards. Recognition was immediate. In 1951 we had our first eminent scientific visitor (straight from Cairo, where he had visited a former student), who has become one of our staunch friends, and in April 1956 the first international scientific conference to be held in Israel devoted to polymer science was held in Rehovot on the initiative of Aharon Katzir. James Watson of DNA fame was one of many distinguished participants.

It is not possible in this brief space to analyze the achievements of Aharon Katzir in this fertile period of his scientific work. With his many collaborators, many of whom are now professors in their own right and head departments at the Weizmann Institute, he established the laws of polyelectrolyte behavior and found many useful applications both in biology and in practical technology. From the earliest days we worked on networks of charged polymers and fibers that could change chemical energy into mechanical work and, by way of conformational changes previously mentioned, lift weights and do work in analogy to natural system such as muscle, for instance.

Many aspects of polymer research have important applications in the field of plastics technology. In 1957 we were able, following Aharon's initiative, to establish the Plastics Laboratory, now grown into a full-fledged department, extending great help to Israeli industry in many varied aspects.

A basic contribution by Aharon and his colleagues was and still is the study of membranes and the thermodynamics of irreversible processes. There is hardly any process in life that is not based on structures in which membranes of different kinds are involved. In practical applications membranes are now studied in processes leading to the desalting of water, vital for the survival of mankind in this era of pollution of our natural stores. The Weizmann Institute in particular and other centers in Israel as well are deeply committed to research in this vital area. The basic approach leans heavily on the fundamental theoretical tools developed over many years by Aharon and his group.

Many other achievements can only briefly be mentioned. Aharon was deeply concerned with the origin of life, the problem of prebiotic synthesis, the way organized biological structure first came about, networks, hysteresis, time, memory, and many other applications and implications of his deep intuitive thoughts. Above all his thoughts were not restricted to Science in the narrow sense. Like Leonardo and not so many universal scientists thereafter, his mind roamed freely over all queries accessible to human thought. A true intellectual, he read widely and communicated with the leading thinkers of his time. The human aspects of science and pursuit of science in the service of mankind were dear to him. It is a tragic irony that a man with these unusual and outstanding qualities, wholly devoted to the service of mankind, was brutally murdered by a perverted group of maniacs claiming to be working toward the same goal.

So many thoughts come to mind that can only be crystallized briefly in these spaces. Aharon was a teacher and friend to all of us fortunate enough to experience his presence. There was no problem, large or small, on which he refused his help whenever he was approached, whether by someone he knew or by someone he had never before met; the warmth of his heart was unmatched. He was one of the most brilliant, erudite, stimulating, and entertaining lecturers I have ever encountered. He was widely in demand all over the world for the wisdom he could communicate by his presence. He was a member of many academies and international bodies, where his advice was eagerly sought. In social events he was always the center of attraction. He sparkled by his wit, erudition, and competence in fields as widely separated as science, philosophy, anthropology, art, literature, and music.

Aharon cannot be replaced, but we must continue his work. He has shown us the way and has taught us that the boundaries of what we can do are without limit. We must continue on the path he has charted and follow the example of his boundless devotion.

Henryk Eisenberg

Among the many awards and honors of Aharon Katzir-Katchalsky was the Presidency of the Israeli National Academy of Science and Humanities, as well as the International Union of Pure and Applied Biophysics, and Foreign Membership of the U. S. National Academy of Sciences. He was awarded the Israel Prize in Exact Sciences in 1961, and *Doctor Honoris Causa* in 1969 of the Free University of Brussels and the University of Bern. A selected list of publications follows. A full list is to be found in the Aharon Katzir-Katchalsky Memorial Issue of the *Israel J. Chem.* **11**(2–3):87–96 (1973).

Selected Publications of Aharon Katzir-Katchalsky

A. Katchalsky and P. Spitnik, Potentiometric titrations of polymethacrylic acid, *J. Polymer Sci.* **2**:432–446; 487 (1947).

W. Kuhn, O. Künzle, and A. Katchalsky, Verhalten Polyvalenter Fadenmolekelionen in Lösung, *Helv. Chim. Acta* **31**:1994–2037 (1948).

A. Katchalsky, Rapid swelling deswelling of reversible gels of polymeric acids by ionization, *Experientia*, **5**:319–320 (1949).

W. Kuhn, B. Hargitay, A. Katchalsky, and H. Eisenberg, Reversible dilation and contraction by changing the state of ionization of high-polymer acid networks, *Nature* **165**:514–517 (1950).

R. M. Fuoss, A. Katchalsky, and S. Lifson, The potential of an infinite rod-like molecule and the distribution of the counter ions, *Proc. Nat. Acad. Sci. U. S.*, **37**:579–589 (1951).

A. Katchalsky, Solutions of polyelectrolytes and mechanochemical systems, *J. Polymer Sci.* **7**:393–412 (1951).

A. Katchalsky and M. Paecht, Phosphate anhydrides of amino acids, *J. Am. Chem. Soc.* **76**:6042–6044 (1954).

P. Spitnik, A. Nevo, and A. Katchalsky, Interaction of polymeric acids with polymeric bases, *Bull. Res. Council Isr.* **4**:318–319 (1954).

A. Katchalsky and S. Lifson, Muscle as a machine, *Sci. Am.* **190**:72–75 (1954).

Z. Alterman and A. Katchalsky, Rate of burning of composite solid propellants, *Bull. Res. Council Isr.* **5A**:46–51 (1955).

J. Eliassaf, A. Silberberg, and A. Katchalsky, Negative thixotropy of aqueous solutions of polymethacrylic acid, *Nature* **176**:1119 (1955).

D. Vofsi and A. Katchalsky, Kinetics of polymerization of nitroethylene. II. Study of molecular weights (by cryoscopic method), *J. Polymer. Sci.* **26**:127–139 (1957).

Z. Alterman, U. Z. Littauer, and A. Katchalsky, Thermochemistry of composite propellants, *Bull. Res. Council Isr.* **7A**:165–170 (1958).

O. Kedem and A. Katchalsky, Thermodynamic analysis of the permeability of biological membranes to non-electrolytes, *Biochim. Biophys. Acta* **27**:229–246 (1958).

A. Katchalsky, D. Danon, A. Nevo, and A. De Vries, Interactions of basic polyelectrolytes with the red blood cell. II. Agglutination of red blood cells by polymeric bases, *Biochim. Biophys. Acta* **33**:120–138 (1959).

A. Katchalsky, S. Lifson, I. Michaeli, and M. Zwick, Elementary mechanochemical processes, *in* "Size and Shape of Changes of Contractile Polymers" (A. Wasserman, ed.), pp. 1–40, Pergamon Press, London, 1960.

O. Kedem and A. Katchalsky, A physical interpretation of the phenomenological coefficients of membrane permeability, *J. Gen. Physiol.* **45**:143–179 (1961).

A. Katchalsky and O. Kedem, Thermodynamics of flow processes in biological systems, *Biophys. J.* **2**:53–78 (1962).

A. Katchalsky, Polyelectrolytes and their biological interactions, *Biophys. J.* **4**:9–41 (1964).

A. Katchalsky and P. F. Curran, "Nonequilibrium Thermodynamics in Biophysics," Harvard University Press, Cambridge, Massachusetts, 1965.

I. Z. Steinberg, A. Oplatka, and A. Katchalsky, Mechanochemical engines, *Nature* **210**:568–571 (1966).

A. Katchalsky, Z. Alexandrowicz, and O. Kedem, Polyelectrolyte solutions, *in* "Chemical Physics of Ionic Solutions" (B. E. Conway and R. G. Barradas, eds.), pp. 295–346, Wiley, New York, 1966.

I. R. Miller and A. Katchalsky, The interaction of negatively charged polyacids with a positively charged mercury surface at different salt concentration, *in* "Physics and Physical Chemistry of Surface Active Substances" (J. Th. G. Overbeek, ed.), pp. 275–288, Gordon and Breach, New York, 1967.

A. Katchalsky, Membrane thermodynamics, *in* "The Neurosciences, A Study Program" (G. C. Quarton, T. Melnechuk, and F. O. Schmitt, eds.), pp. 326–343, The Rockefeller University Press, New York, 1967.

S. Reich, A. Katchalsky, and A. Oplatka, Dynamic-elastic investigation of the chemical denaturation of collagen fibers, *Biopolymers* **6**:1159–1168 (1968).

A. Katchalsky and R. Spangler, Dynamics of membrane processes, *Quart. Rev. Biophys.* **1**:127–175 (1968).

M. V. Sussman and A. Katchalsky, Mechanochemical turbine: A new power cycle, *Science* **167**:45–47 (1970).

M. Paecht-Horowitz, J. Berger, and A. Katchalsky, Prebiotic synthesis of polypeptides by heterogeneous polycondensation of amino acid adenylates, *Nature* **228**:636–639 (1970).

E. Neumann and A. Katchalsky, Thermodynamische Untersuchung der Hysterese im System Polyriboadenyl-Polyribouridylsäure—Modell einer Makromolekularen Gedächtnis-Aufzeichnung, *Ber. Bunsenges. Phys. Chem.* **74**:868–879 (1970).

A. Katchalsky, A thermodynamic consideration of active transport, *in* "Permeability and Function of Biological Membranes" (L. Bolis, A. Katchalsky, R. D. Keynes, W. R. Loewenstein, and B. A. Pethica, eds.), pp. 20–35, North-Holland, Amsterdam, 1970.

A. Katchalsky, Thermodynamics of flow and biological organization, *Zygon: J. Relig. Sci.* **6**:99–125 (1971).

G. Oster, A. Perelson, and A. Katchalsky, Network thermodynamics, *Nature* **234**:393–399 (1971).

A. Katchalsky and A. Oplatka, Mechano-chemical conversion, *in* "Handbook of Sensory Physiology," Vol. 1, "Principles of Receptor Physiology" (W. R. Loewenstein, ed.), pp. 1–17, Springer Verlag, Berlin, 1971.

A. Katchalsky, Thermodynamics and life, *in* "Proc. of the Int. Union of Physiological Sciences," Vol. VIII (1971).

A. Katchalsky and E. Neumann, Hysteresis and molecular memory record, *Int. J. Neurosci.* **3**:175–182 (1972).

E. Neumann and A. Katchalsky, Long-lived conformation changes induced by electric impulses in biopolymers, *Proc. Nat. Acad, Sci. U. S.* **69**:993–997 (1972).

R. A. Cox and A. Katchalsky, Hysteresis and conformational changes in ribosomal ribonucleic acid, *Biochem. J.* **126**:1039–1054 (1972).

A. Katzir-Katchalsky, Reflections on art and science, *Leonardo* **5**:249–253 (1972).

G. F. Oster, A. S. Perelson, and A. Katchalsky, Network thermodynamics: dynamic modelling of biophysical systems, *Quart. Rev. Biophys.* **6**:1–134 (1973).

A. Katchalsky, Prebiotic synthesis of biopolymers on inorganic templates, *Naturwiss.* **60**:215–220 (1973).

E. Neumann, D. Nachmansohn, and A. Katchalsky, An attempt at an integral interpretation of nerve excitability, *Proc. Nat. Acad. Sci. U. S.* **70**:727–731 (1973).

A. Katzir-Katchalsky, A. Silberberg, and A. Apelblat, A mathematical analysis of capillary-tissue fluid exchange, *Biorheology*, in press.

A. Katzir-Katchalsky, An Israeli scientist's approach to human values, *Bull. At. Sci.* **28**:19–24 (1972).

A complete bibliography can be found in *Israel J. Chem.* **11**(2–3):87–96 (1973).

BIOMEMBRANES

Volume 7

Chapter 1

Fluidity and Phase Transitions of Cell Membranes

D. Chapman

Department of Chemistry
Sheffield University
England, United Kingdom

I. INTRODUCTION

Recent years have seen the gap which exists between the different conceptual approaches to membranes, e.g., the "black box" approach of the physiologists and the "molecular structural" approach of the biochemist, begin to narrow. As basic studies of lipids have increased and model systems developed, so has our understanding increased of membrane structure and function.

In our laboratory we have for some time been studying the dynamics of lipids, their phase transitions and interactions with cholesterol, polypeptides, and proteins (Chapman and Wallach, 1968). Our aim in these studies has been to provide basic information relevant to the cell membrane structure and function. More recently we have also studied protein rotation in cell membranes and developed methods for extending such measurements (Naqvi *et al.*, 1973).

We hope that such studies will ultimately contribute to the situation where there will be a harmonious relationship among measurements of the rates of transport, their basic thermodynamics (studied and pioneered by Dr. A. Katchalsky), and the underlying detailed molecular mechanisms.

In this contribution we discuss aspects of fluidity and phase transitions of lipids and cell membranes. We shall emphasize the differing degrees of fluidity that occur in different cell membranes and at times within the same cell membrane.

II. PHASE TRANSITIONS AND FLUIDITY OF LIPID AND MEMBRANE SYSTEMS

For many years structural analyses of long-chain polar molecules were based upon the idea that even in lamellar and micelle form the hydrocarbon chains were rigid and fully extended. However, infrared spectroscopic studies of pure soap systems (Chapman, 1958) clearly revealed the nature of the fluidity of long-chain polar molecules when heated above a certain transition temperature. These studies showed that a polar lipid gives a "crystalline type" spectrum below this transition temperature and a "liquid type" spectrum above the transition temperature. The spectra above the transition temperature were interpreted in terms of a "melting" of the lipid chains involving a breakup of the all planar *trans* configuration of the chains by rotation of the methylene (CH_2) groups about their C—C bonds. In this way *gauche* isomers are formed, leading to a fluid state.

Many studies using a variety of physical techniques have since been carried out on soaps and phospholipids, which have confirmed and extended the details of this transition [for reviews see Luzzati (1968) and Chapman and Wallach (1968)]. Above the transition temperature the lipid chains contain *gauche* isomers as well as *trans* isomers, with increasing *gauche* isomers forming at higher temperatures. There is more molecular movement at the methyl end of the chain than occurs with the CH_2 groups near the polar end of the molecule. This gradient of molecular motion along the lipid chains was first pointed out with phospholipids, using NMR spectroscopy (Chapman and Salsbury, 1966), and later put on a more quantitative basis using spin labels (Hubbell and McConnell, 1971).

The cooperative nature of the lipid chain movements leading to the phase transition and in the liquid crystalline phase have also been emphasized (Whittington and Chapman, 1966; Rothman, 1973).

There are other aspects of molecular motion near the phase transition temperature which have also been studied. These include studies of the molecular motion of the polar groups of the lipids (Veksli *et al.*, 1969; Oldfield and Chapman, 1972). These show that there is a marked increase in motion of the lecithin polar groups prior to the main transition temperature. There also appears to be some reorganization of water around the polar group during this phase change.

The concept that lipid mobility was so great above the transition temperature that lipids could diffuse along the length of bilayer (Chapman *et al.*, 1967) has now been confirmed and diffusion coefficients measured in lipids and membranes by means of spin label methods ($D = 1.8 \times 10^{-8}$ cm^2/sec) (Kornberg and McConnell, 1971; Sackmann and Trauble, 1972; Devaux *et al.*, 1973). A more direct measurement, using NMR spin-echo methods, which does not involve strong perturbation effects has recently been applied to lipid diffusion studies (Lindblom *et al.*, 1975). The value of the diffusion coefficient obtained using this method is $D = 0.5 \times 10^{-7}$ cm^2/sec, although this is a provisional value at present.

Below the transition temperature, e.g., with lecithin–water systems, the lipid chains in the gel phase are fully extended and packed in a regular manner (Chapman *et al.*, 1967). With pure single lecithins, e.g., dipalmitoyl lecithin, the hydrocarbon chains are tilted to the basal planes and packed in a two-dimensional hexagonal lattice, while in lecithin mixtures (e.g., egg yolk lecithin) a phase can occur where the chains are vertical to the basal plane (Tardieu *et al.*, 1973).

Below the transition temperature in the gel phase the lipid chains have *less* molecular motion than in the liquid crystalline phase and hence diffusion processes will be much less. However the gel phase is a quasicrystalline phase, and there is *more* molecular motion of the molecule, e.g., of the lipid chains than occurs with the anhydrous *crystalline* material. This is indicated by NMR studies (Veksli *et al.*, 1969), where the linewidth at room temperature associated with the hydrocarbon chains is about 4 G at maximum water content. This can be compared with the rigid lattice value of 15 G at liquid N_2 temperatures for the anhydrous material.

The way in which the transition temperature shifts up or down dependent upon chain length, unsaturation, and polar groups has been well studied [see the review by Oldfield and Chapman (1972)]. Thus highly unsaturated lipids have a low transition temperature, and saturated lipids have a high transition temperature. These studies of lipid fludity and of phase transitions are important for cell membrane structure and function, since we know that many cell membranes contain regions of lipid bilayer. The importance of lipid fluidity for cell membranes has frequently been pointed out and related to permeability and metabolic characteristics (Chapman and Wallach, 1968). Thus membranes containing very unsaturated lipids, as in mitochondria, will have very fluid membranes and at body temperature are associated with considerable metabolic activity and easy permeability for water and some organic molecules.

It is important to appreciate that the fluidity of the lipids of cell membranes is not, however, related only to the lipid chain length or unsaturation. In membranes such as the myelin membranes, where the lipids are more saturated, the lipids would be rigid at body temperature except for the presence of cholesterol. In this membrane the cholesterol keeps the lipids in a fluid condition by interposing between the lipid chains to prevent chain crystallization from occurring (Ladbrooke *et al.*, 1968b). In fact cholesterol appears to have a dual effect on lipid systems in that at a temperature where the lipid would normally be in a gel condition, the presence of cholesterol causes the lipid to be in a fluid condition (Ladbrooke *et al.*, 1968a), whereas at a temperature when the lipid is in the fluid condition, the cholesterol inhibits some of the chain molecular motion *although fluidity characteristics are still retained* (Chapman and Penkett, 1966; Williams and Chapman, 1970; Chapman 1973).

The kinetics of water permeability through lipid systems has been studied and related to lipid fluidity (Bittman and Blau, 1972). Thus there is a marked increase in water permeability as the lipid chains become more unsaturated [this is also the case with nonelectrolytes such as glycerol, and erythritol (de Gier *et al.*, 1970)]. The self-diffusion rate of $^{22}Na^{+}$ through lecithin bilayers also shows a marked increase at the transition temperature to the liquid crystalline form (Papahadjopoulos *et al.*, 1973).

The effect of cholesterol on the lipid chains when the lipid is in the liquid crystalline condition is to decrease water permeability (Bittman and Blau, 1972). Black lipid membranes also show the same effect (Finkelstein and Cass, 1967). These results are consistent with a reduction in the fluidity characteristics of the lipid chains. On the other hand, the presence of cholesterol is to *enhance* the rate of water permeability of liposomes derived from saturated liposomes (Bittman and Blau, 1972).

The technique of NMR spectroscopy is a powerful one for showing this mobility or fluidity characteristics of cell membranes. Using this technique, we can compare the mobility of the lipid chains in two different cell membranes. When we compare the ^{13}C spectra of mitochondrial and erythrocyte membranes using this technique, we see that there is greater chain mobility of the mitochondrial membrane lipids (Keough *et al.*, 1973). Some inhibition of chain movement is occurring in the erythrocyte membranes. Some of this is a reflection of the different fatty acids present in these two membranes, but some is due to the effect of cholesterol and/or protein in the erythrocyte membrane system. The fluidity of the lipids can also be affected by interaction with polypeptides and proteins. Experiments with polypeptides, such as alamethicin, show that they can inhibit lipid chain mobility (Hauser *et al.*, 1970). Recently it has also been shown that rhodopsin, on incorporation into lipid bilayers, inhibits lipid chain segmental movement (Hong and Hubbell, 1972).

The concept that many membranes contain "fluid" lipid regions is now well accepted. It has indeed been shown that proteins can diffuse along the membrane systems (Frye and Edidin, 1970). This concept of movement of lipids and proteins in cell membranes is clearly important for any processes which may occur in the cell. In some cases lipid phase transitions can apparently cause protein aggregation to occur. Thus, with alveolar membranes grown at 28°C, particles as revealed by freeze–etch electron microscopy are randomly distributed. On the other hand, after chilling the cells to 5°C it is found that the particles appear to have aggregated (Speth and Wunderlich, 1973). Chapman and Urbina (1971) have already pointed out that proteins may be squeezed out of the fluid regions when lipid chain crystallization occurs.

The concept of heterogeneity of lipid packing in cell membranes has recently been emphasized, pointing out that some cell membranes can contain regions of fluid and also regions of rigid or stiff chains (Oldfield and Chapman, 1972). Model lipid–water systems have been examined of different chain lengths and the same lipid class and with the same chain length but with different lipid classes (Phillips *et al.*, 1970; Chapman *et al.*, 1974) and phase separation discussed.

Phase transitions similar to this have been observed with natural membranes. This is the case with the unsupplemented *A. choleplasma laidlawii* cells at their growth temperature, as well as cells when supplemented with fatty acids (Oldfield *et al.*, 1972; Steim *et al.*, 1969). It also occurs with supplemented *E. coli* systems (Esfahani *et al.*, 1971).

By varying the fatty acid in the media, we can shift the transition temperatures of the phase changes just as in the simple lipid–water systems. Thus different thermal phase changes associated with the phospholipids of membranes obtained by growing cells in oleic and linolenic acid-supplemented media have been observed (Esfahani *et al.*, 1971). The membranes of these cells also show differences in the Arrhenius plot of growth rates and of several membrane-associated functions. Fox and Tsukagoshi (1972) have carried out a number of studies of the *E. coli* system, shifting from growth in elaidic acid- to oleic acid-supplemented medium and vice versa. They show Arrhenius plots describing transport rates which are biphasic and sometimes triphasic in character, depending upon the temperature of incubation.

The lipids close to a protein in a cell membrane may have different fluidity than the remaining lipid. Thus heterogeneity of lipid packing appears to occur within liver microsomal membranes where a rather rigid halo of phospholipids in a semicrystalline form is considered to enclose a cytochrome enzyme while the bulk of the lipid is in a more fluid condition (Stier and Sackmann, 1973).

Although many cell membranes are fluid in character with regard to the lipid component, we should not ignore the fact that in some cell membranes the lipids and protein may not be particularly fluid or mobile and may be very well organized. This appears to be the case with the cell envelopes of *Halobacterium halobium* and *Halobacterium cutirubrum* (Blaurock and Stoeckenius, 1971), where X-ray studies show sharp reflections indicating a high degree of protein organization. Using spin labels, Esser and Lanyi (1973) have also shown that the lipid in the latter system is particularly immobilized and suggest that this occurs as a direct result of protein–lipid interactions.

In our laboratory recent studies on the protein mobility in cell membranes have shown with the purple membrane of the *Halobacterium halobium* (Naqvi *et al.*, 1973) that the protein appears to be particularly immobilized and has a rotational relaxation time of at least 20 msec. This can be contrasted with the rotational relaxation time of a similar protein, rhodopsin, which is reported to be some 20 μsec in rod outer segment membranes (Cone, 1972).

It may be that there are special reasons for the apparent rigidity of the envelopes of these halophilic cell envelope systems and there is a requirement to severely limit water permeability into the cell. Just as marked lipid fluidity can be related to marked water permeability, so limited fluidity may indicate very limited water permeability. (These cells require sodium chloride concentration of at least 3 M for growth and lyse when the salt concentration is lowered to less than 1 M).

The range of fluidity characteristics which can occur in cell membranes and envelopes appears to be quite great.

III. TRIGGERING MECHANISMS

A number of studies have shown that lipid phase transitions and hence fluidity and diffusion can be affected by interactions of the lipids with metal ions, polypeptides, and proteins (Chapman *et al.*, 1974). Thus interactions of divalent cations with the polar groups of the lipids raise the transition temperature. This behavior parallels the properties observed in monolayer systems. Electrostatic binding of proteins and basic polypeptides also shifts these transition temperatures. Some polypeptides, however, such as gramicidin A, cause a marked decrease in the energy associated with the lipid endothermic phase transition (Chapman *et al.*, 1974). This is analogous to the behavior of cholesterol and is consistent with the idea that this polypeptide is interdigitating among the lipid chains. A number of drug molecules also cause pronounced effects on this lipid transition, either shifting the transition or sometimes removing it completely (Cater *et al.*, 1974).

These model studies show that by a variety of processes, in principle, portions of membranes can be triggered to cause local changes of lipid fluidity and with it changes in permeability, transport processes, etc. These changes can be dramatic in character, ranging from gel to liquid crystal, or less dramatic in character, changing from a *fluid* condition to a *more fluid* condition or vice versa.

Many instances of cell stimulation in biology may have this change of fluidity as a first step in the subsequent chain of biochemical events—perhaps interactions of cell surfaces with concanavelin A and phytohemagglutinin. In the future many studies relating cell stimulation to changes of membrane fluidity can be expected, leading to increased understanding of membrane function.

IV. CONCLUSIONS

The concept of lipid fluidity of cell membranes is an important one. It can be related to transitions involving a "melting" of the lipid chains. Diffusional characteristics, protein movement, and enzyme function can all be affected by the lipid milieu.

The fluidity of the lipids can be changed in many ways, e.g., by varying chain length or degree of unsaturation, or by mixing different chain lengths or different lipid classes. Modulation of fluidity can also take place as a result of hydrophobic interactions with cholesterol, polypeptides, and proteins. Interactions with the polar groups by metal ions, polypeptides, and drugs can also affect lipid and membrane fluidity.

The fluidity of the lipids of cell membranes varies considerably, being very fluid with mitochondrial membranes, less fluid with erythrocyte membranes, and apparently particularly rigid with halophilic cell envelopes.

Trigger mechanisms affecting membrane fluidity may be important in certain cell stimulation processes, e.g., metal-ion interaction, drug action, and concanavalin A and lymphocyte effects.

Acknowledgments

I wish to acknowledge the financial support of our studies by the Royal Society and the Multiple Sclerosis Society.

REFERENCES

Bittman, R., and Blau, L., 1972, The phospholipid–cholesterol interaction. Kinetics of water permeability in liposomes, *Biochem.* **11**:4831–4839.

Blaurock, A. E., and Stoeckenius, W., 1971, Structure of the purple membrane, *Nature, New Biol.* **233**:152–155.

Cater, B., Chapman, D., Hawes S. M., and Saville, J., 1974, Lipid phase transitions and drug interactions, *Biochim. Biophys. Acta* **363**:54–69.

Chapman, D., 1958, An infrared spectroscopic examination of some anhydrous soaps, *J. Chem. Soc.* **1958**:784–789.

Chapman, D., and Penkett, S. A., 1966, Nuclear magnetic resonance spectroscopic studies of the interaction of phospholipids with cholesterol, *Nature* **211**:1304–1305.

Chapman, D., and Salsbury, N. J., 1966, Physical studies of phospholipids, Part 5, Proton magnetic resonance studies of molecular motion in some 2,3-diacyl DL phosphatidyl ethanolamines, *Trans. Farad. Soc.* **62**:2607–2621.

Chapman, D., and Urbina, J., 1971, Phase transitions and bilayer structures, *FEBS Lett.* **12**:169–172.

Chapman, D., and Wallach, F. H., 1968, *in* "Biological Membranes, Physical Fact and Function. Recent Physical Studies of Phospholipids and Natural Membranes" (D. Chapman, ed.) pp. 125–202, Academic Press, New York.

Chapman, D., Keough, K. M., and Urbina, J., 1974, Biomembrane phase transitions. Studies of lipid–water systems using differential scanning Calorimeter, *J. Biol. Chem.*, **249**:2512.

Chapman, D., Williams, R. M., and Ladbrooke, B. D., 1967, Physical studies of phospholipids. VI. Thermotropic and lytropic Mesomorphism of some 1,2 diacyl phosphatidylcholines (lecithins), *Chem. Phys. Lipids* **1**:445–475.

Chapman, D., 1973, *in* "Biological Membranes" Vol. II, Academic Press, New York.

Cone, R. A., 1972, Rotational diffusion of rhodopsin in the visual receptor membrane, *Nature, New Biol.* **236**:39–43.

Devaux, P., Scandella, C. J., and McConnell, H. M., 1973, Spin–spin interactions between spin labelled phospholipids incorporated into membranes, *J. Magn. Res.* **9**:474–485.

Esfahani, M., Limbrick, A. R., Knutton, S., Oka, T., and Wakil, S. J., 1971, The molecular organisation of lipids in the membrane of *Escherichia coli*, *Proc. Nat. Acad. Sci.* **68**: 3180–3184.

Esser, A. F., and Lanyi, J. K., 1973, Structure of the lipid phase in cell envelope vesicles from *Halobacterium cutirubrum*, *Biochem.* **12**:1933–1939.

Finkelstein, A., and Cass, A., 1967, Effect of cholesterol on the water permeability of thin lipid membranes, *Nature* **216**:717–718.

Fox, F. C., and Tsukagoshi, T., 1972, The influence of lipid phase transitions on membrane function and assembly, *in* "Membrane Research" (F. C. Fox, ed.), pp. 145–153, Academic Press, New York.

Frye, C. D., and Edidin, M., 1970, The rapid intermixing of cell surface antigens after formation of mouse–human hetero karyons, *J. Cell. Sci.* **7**:319–333.

de Gier, J., Haest, C. W. M., Mandersloot, J. G., and van Deenen, L. L. M., 1970, Valinomycin-induced permeation of 86 Rb^+ of liposomes with varying composition through the bilayers, *Biochim. Biophys. Acta* **211**:373–375.

Hauser, H., Finer, E. G., and Chapman, D., 1970, Nuclear magnetic resonance studies of the polypeptide alamethicin and its interaction with phospholipids, *J. Mol. Biol.* **53**:419–433.

Hong, K., and Hubbell, W. L., 1972, Preparation and properties of phospholipid bilayers containing rhodopsin, *Proc. Nat. Acad. Sci.* **69**:2617–2621.

Hubbell, W. L., and McConnell, H. M., 1971, Molecular motion in spin-labelled phospholipids and membranes, *J. Am. Chem. Soc.* **93**:314–326.

Keough, K. M., Oldfield, E., Chapman, D., and Beynon, P., 1973, Carbon-13 and proton nuclear magnetic resonance of unsonicated model and mitochondrial membranes, *Chem. Phys. Lipids* **10**:37–50.

Kornberg, R. D., and McConnell, H. M., 1971, Lateral diffusion of phospholipids in a vesicle membrane, *Proc. Nat. Acad. Sci.* **68**:2564–2568.

Ladbrooke, B. D., Jenkinson, T. J., Kamat, V. B., and Chapman, D., 1968a, Thermal analysis of myelin, *Biochim. Biophys. Acta* **164**:101–109.

Ladbrooke, B. D., Williams, R. M., and Chapman, D., 1968b, Studies on lecithin–cholesterol–water interactions by differential scanning calorimetry and X-ray diffraction, *Biochim. Biophys. Acta* **150**:333–340.

Lindblom, G., 1975, Diffusion in a phospholipid bilayer studied by pulsed NMR, to be published.

Luzzati, V., 1968, X-ray diffraction studies of lipid–water systems, *in* "Biological Membranes" (D. Chapman, ed.), Vol. I, pp. 71–123. Academic Press.

Naqvi, R. K., Gonzalez-Rodriguez, J., Cherry, R. J., and Chapman, D., 1973, A spectroscopic technique for studying protein rotation in membranes, *Nature*, in press.

Oldfield, E., and Chapman, D., 1972, Dynamics of lipids in membranes. Heterogeneity and the role of Cholesterol, *FEBS Lett.* **23**:285–297.

Oldfield, E., Chapman, D., and Derbyshire, W., 1972, Lipid mobility in *Acholeplasma* membranes using deuteron magnetic resonance, *Chem. Phys. Lipids* **9**:69–81.

Papahadjopoulos, D., Jacobson, K., Nir, S., and Isac, T., 1973, Phase transitions in phospholipid vesicles. Fluorescence polarisation and permeability measurements concerning the effect of temperature and cholesterol, *Biochim. Biophys. Acta* **311**: 330–348.

Phillips, M. C., Ladbrooke, B. D., and Chapman, D., 1970, Molecular interactions in mixed lecithin systems, *Biochim. Biophys. Acta* **196**:35–44.

Rothman, J., 1973, The molecular basis of mesomorphic phase transitions in phospholipid systems, *J. Theor. Biol.* **38**:1–16.

Sackmann, E., and Trauble, H., 1972, Studies of the crystalline–liquid crystalline phase transition of lipid model membranes. 11. Analysis of electron spin resonance spectra of steroid labels incorporated into lipid membranes, *J. Am. Chem. Soc.* **94**:4492–4498.

Speth, V., and Wunderlich, F., 1973, Membranes of tetrahymena. II. Direct visualisation of reversible transitions in biomembrane structure induced by temperature, *Biochim. Biophys. Acta* **291**:621–628.

Steim, J. M., Tourtellotte, M. E., Reinert, J. C., McElhaney, R. N., and Rader, R. L., 1969, Calorimetric evidence for the liquid crystalline state of lipids in a biomembrane, *Proc. Nat. Acad. Sci.* **63**:104–109.

Stier, A., and Sackmann, E., 1973, Spin labels as enzyme substrates. Heterogeneous lipid distribution in liver microsomal membranes, *Biochim. Biophys. Acta* **311**:400–408.

Tardieu, A., Luzzati, V., and Reman, F. C., 1973, Structure and polymorphism of the hydrocarbon chains of lipids. A study of lecithin–water phases, *J. Mol. Biol.* **75**: 711–733.

Veksli, Z., Salsbury, N. J., and Chapman, D., 1969, Physical studies of phospholipids. XII. Nuclear magnetic resonance studies of molecular motion in some pure lecithin–water systems, *Biochim. Biophys. Acta* **183**:434–446.

Whittington, S. G., and Chapman, D., 1966, Effect of density on configurational properties of long chain molecules using a Monte Carlo method, *Trans. Farad. Soc.* **62**:3319–3324.

Williams, R. M., and Chapman, D., 1970, Phospholipids, liquid crystals and cell membranes, *in* "Progress in the Chemistry of Fats and other Lipids" (R. T. Holman, ed.), Pergamon Press.

Chapter 2

Criteria in Carrier Transport

W. Wilbrandt

University of Bern
Department of Pharmacology
Bern, Switzerland

I. INTRODUCTION

Transport of matter across biological membranes is interpreted at present mainly on three principles of membrane passage: diffusion, carrier-mediated transport, and active transport. What is meant by these terms? Since biological membranes are complex molecular structures involving tight packing of molecules, it is obvious that processes identical with or closely approaching diffusion in free, diluted solution are unthinkable. The term "diffusion" in relation to membrane passage, therefore, should be defined. What is meant is movement of molecules under the exclusive influence of thermal molecular motion as driving force. In carrier-mediated transport an additional process assumed to be involved is the reaction of the penetrating molecules or ions with mobile components of the membrane (carriers) to form a transport complex which, in contradistinction to the free particles themselves, is capable of crossing the membrane in some way. This ability might be due to higher lipoid solubility of the complex compared to the free transport substrate or to configurational changes in proteins. Finally, "active transport" is characterized, according to the most widely accepted definition (Rosenberg, 1948; Wilbrandt, 1974), by the ability of operating against gradients of chemical activity or, in the case of ions, electrochemical activity ("uphill"). This requires the coupling to energy-yielding processes, for instance, chemical reactions. Additional involvement of carrier substrate reactions is frequently assumed. It is, however, not a logical consequence of the definition.

A critical comparison of criteria in current use for the recognition of these types of transfer appears useful. It will be convenient to discuss criteria for the three basic assumptions separately, but obviously the same criterion can be analyzed as to its relation to more than one of these basic assumptions. The discussion will focus on kinetic rather than thermodynamic considerations. The notation used will be that of previous publications of the author.

II. CRITERIA FOR DIFFUSION

One of the important properties of diffusion is the principle of independent diffusion streams. It implies that any movement by diffusion from one concentration S_1 to another S_2 can be considered as composed of two (or more) independent flows, e.g., two unidirectional movements in opposite directions, one from S_1 on side 1 to zero on side 2 and the other from S_2 on side 2 to zero on side 1. The rate of net movement, then, is the difference between these two unidirectional fluxes, which are proportional to S_1 and S_2, respectively.

A. Fick's Diffusion Law

The first consequence of this situation, which is in common use as a criterion for diffusion, is the postulate that the rate of net movement should be proportional to the difference of activities or concentrations on the two sides of the membrane, in other words, that the movement follows Fick's diffusion law:

$$v = P(a_1 - a_2) \tag{1}$$

In the case of ions the concentrations c or activities a have to be replaced by electrochemical activities $\bar{a}$ as defined by (Ussing, 1949)

$$RT \ln \bar{a} = RT \ln c + RT \ln f + zF\Psi \tag{1a}$$

In these equations is v is the transport rate, P is the permeability constant, a is the activity, $\bar{a}$ is the electrochemical activity, R is the gas constant, T is the absolute temperature, c is the concentration, f is the activity coefficient, F is the Faraday constant, z is the valence, and Ψ is the electrical potential. Two cautionary remarks appear appropriate with respect to this criterion. The first is that in carrier-mediated systems under conditions of low saturation of the carrier the unidirectional fluxes are practically independent of each other and the kinetics, therefore, is practically the same as in the case of diffusion. Thus in such cases the criterion may be misleading if it is not used over a sufficiently wide range of concentrations. Even in systems of active transport the criterion may be misleading if, of the two concentrations on the two sides of the membrane, only one is varied experimentally. This will become clear from later discussion.

B. The Flux Ratio

The second approach based on the independence of unidirectional fluxes in diffusion is the flux ratio criterion introduced by Ussing (1949). This criterion uses, instead of the difference between the unidirectional fluxes, their ratio. The advantage of this procedure is the elimination of proportionality factors which are not amenable to evaluation. Ussing has shown that the ratio of the two fluxes, in the case of diffusion, must be identical with the ratio of the activities (or electrochemical activities) on the two sides of the membrane:

$$M_1/M_2 = \bar{a}_1/\bar{a}_2 \tag{2}$$

Experimentally the availability of isotope-labeled compounds makes it possible to determine the two fluxes M_1 and M_2 separately and thereby obtain their ratio.

Since the flux ratio criterion is a consequence of independence of unidirectional fluxes, the same limitation applies as for linear kinetics: carrier-mediated transport under conditions of low saturation will result in a flux ratio practically indistinguishable from the case of diffusion. The important case of active transport in the sense of uphill movement due to energy coupling will be discussed later.

III. CARRIER-MEDIATED TRANSPORT

The carrier–substrate reaction assumed to be involved in mediated transport leads to a more complex relation between transport rate and substrate concentrations. Assuming rapid reaction between carrier and substrate and equal mobility of free and loaded carrier, the following rate equation, which is in frequent use, is readily derived (Stein, 1967; Widdas, 1952; Wilbrandt and Rosenberg, 1961):

$$v = v_{\max}\left(\frac{S_1}{S_1 + K_m} - \frac{S_2}{S_2 + K_m}\right) = v_{\max}\frac{K_m(S_1 - S_2)}{(S_1 + K_m)(S_2 + K_m)} \tag{3}$$

where v is the transport rate, $v_{\max}$ is the maximum transport rate, and K_m is the Michaelis constant (dissociation constant of the substrate–carrier complex).

A. Low Saturation

1. Linear Concentration Dependence

For conditions of low saturation ($S \ll K_m$) Eq. (3) reduces to

$$v = (v_{\max}/K_m)(S_1 - S_2) \tag{3a}$$

which is formally similar to Eq. (1). Thus, as mentioned before, in this case the kinetics is indistinguishable from diffusion kinetics.

2. Flux Ratio

The flux ratio under these conditions is

$$M_1/M_2 = S_1K_m/K_mS_2 = S_1/S_2 \tag{4}$$

In other words, the ratio will meet Ussing's criterion for diffusion.

B. General Case. Flux Ratio

If the condition of low saturation ($S \ll K_m$) is dropped, Eq. (4) changes to

$$M_1/M_2 = (S_1/S_2)(S_2 + K_m)/(S_1 + K_m) \tag{5}$$

indicating that (if $S_2 < S_1$) the flux ratio is smaller than the concentration ratio, resembling, as will be discussed later, the situation in active carrier transport systems operating downhill at high saturation.

C. High Saturation

1. Saturation Kinetics

a. Zero Trans Concentration, Lineweaver–Burk Plot. If $S_2 = 0$, then, under conditions of high saturation ($S \gg K_m$), the transport rate approaches a maximum ($v \to v_{max}$). The criterion most frequently used to test saturation kinetics is the double reciprocal plot of transport rate against concentration according to the method of Lineweaver and Burk (1934). The plot is linear for carrier transport and saturation is indicated by a finite, positive ordinate intercept.

Although saturation kinetics certainly is a consequence of carrier mediation, it is not the best criterion for differentiating among transport forms, for more than one reason. First, saturation phenomena can occur due to reasons of limited space without involvement of chemical reactions. This has been pointed out by Zierler (1961), who devised an instructive mechanical model for diffusion across porous channels, not involving binding. This model showed distinct saturation experimentally. Figure 1 gives Zierler's data plotted normally and according to the Lineweaver–Burk approach. Secondly, saturation phenomena can be mimicked in systems operating exclusively with diffusion but displaying osmometer properties. This can be shown as follows. Introducing an equation for osmotic equilibrium,

$$(m + m_n)/V = N \tag{6}$$

(in which V is the volume, m and m_n are the intracellular amounts of penetrating and nonpenetrating substances, respectively, and N is the external concentration of nonpenetrating substances), and inserting it into the following equation for diffusion from an intracellular concentration $S_1 = m/V$ into a medium with $S_2 = 0$:

$$v = Dm/V \tag{7}$$

we obtain the relation

$$v = DNm/(m + m_n) \tag{8}$$

This equation is formally similar to Eq. (3) (with $S_2 = 0$). In other words, in a diffusion system involving osmotic volume changes the translocation kinetics resembles carrier kinetics closely. Experiments on glycerol penetration in beef red cells have corroborated this conclusion (Wilbrandt *et al.*, unpublished).

b. Finite Trans Concentration, Biphasic Concentration Dependence. It can be shown from Eq. (3) that, if $S_2 \neq 0$ but $S_2/S_1 = r = \text{const}$, the transport rate passes through a maximum at the concentration $S_1 = K_m(1/r)^{1/2}$ [or $(S_1/K_m)(S_2/K_m) = 1$] and, at higher substrate concentrations, decreases again. Figure 2 shows theoretical and experimental (Wilbrandt, 1972a) curves. This rate decrease at high concentrations can serve as a criterion for carrier mediation.

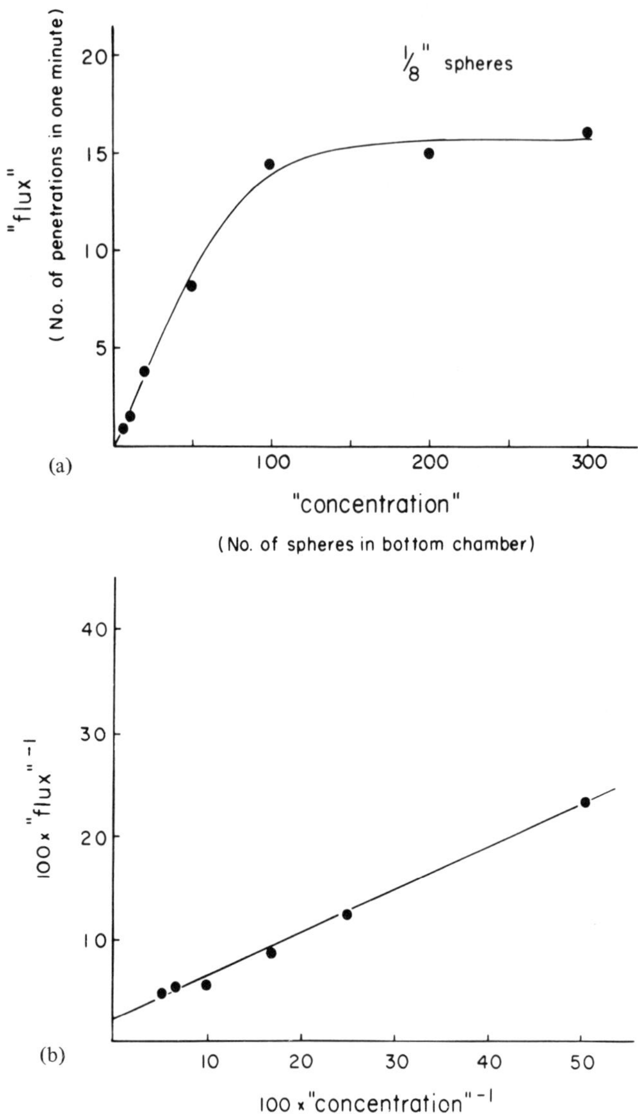

Fig. 1. Data indicating "saturation" in experiments with a mechanical model imitating diffusion through a porous channel. In the model, macroscopic spheres agitated mechanically in a bottom chamber have to pass through a hole in a separating wall to reach an upper chamber. The numbers of spheres in the two chambers represent "concentrations" on the two sides of the membrane. (a) "Rate" versus "concentration" in a normal plot, (b) in a double reciprocal plot corresponding to the procedure of Lineweaver and Burk. The curves shown have distinct saturation character.

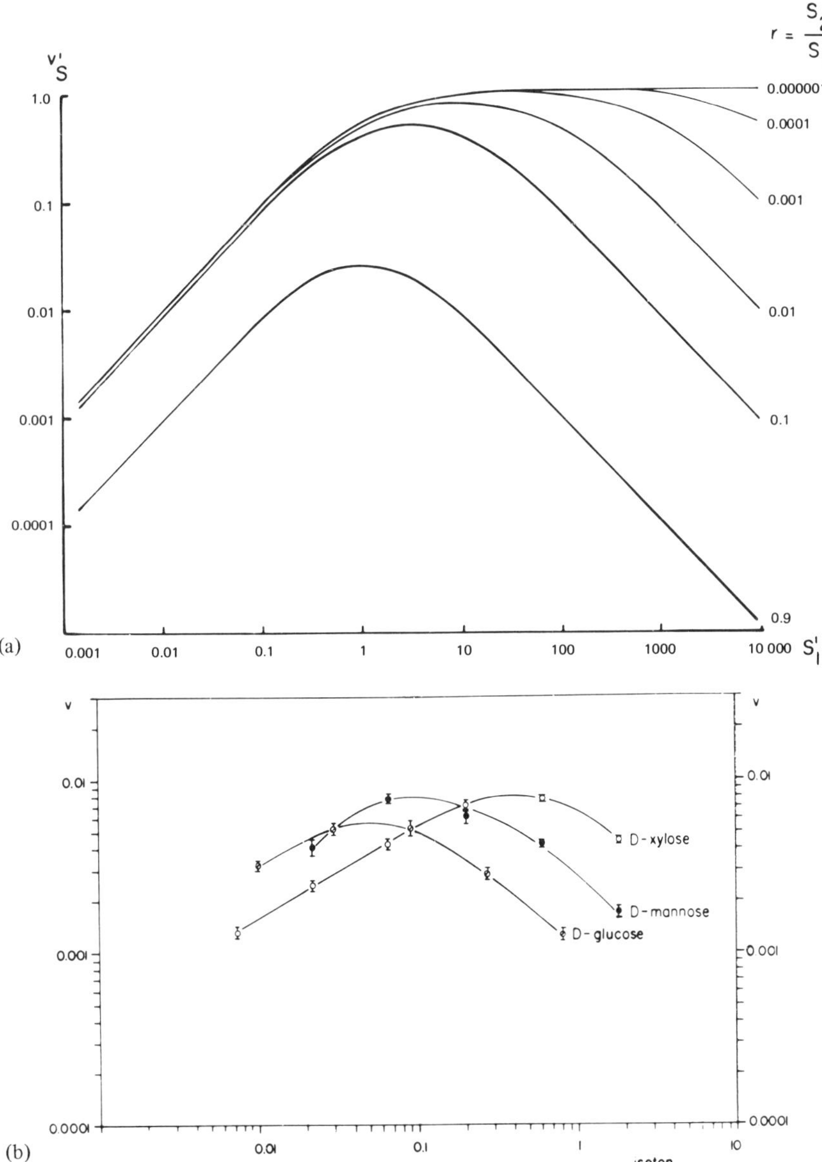

Fig. 2. Dependence of penetration rate on substrate concentration in a carrier system under conditions of constant ratio between the two concentrations on the two sides of the membrane. (a) Calculated curves. Ordinate: rate of transport for the substrate S; abscissa: normalized concentration of the substrate S on side 1; parameter: concentration ratio $r = S_2/S_1$. With $S_2 \rightarrow 0$ the curves approach saturation in the sense of constant rate at high concentration. With rising concentration S_2 the curves assume biphasic form. (b) Experimental verification of biphasic concentration dependence as shown in (a) in experiments involving the transport of various sugars across the human red cell membrane. The difference in the position of the maxima of the curves reflects the difference in affinity of the three sugars to the system (glucose > mannose > xylose, in accordance with other experimental data). The concentration ratio $r = S_2/S_1$ was 0.333. Temperature 10°C.

2. *Competition*

a. Introduction. Besides saturation kinetics, the involvement of a chemical binding reaction in carrier-mediated transport has the second consequence of possible competition, i.e., decrease of transport rate v_S in the presence of a second substrate R with affinity for the same binding site. The most convenient way to treat the kinetics under these conditions in a quantitative manner is given by the following equation:

$$v_S = v_{\max}\left(\frac{S_1'}{S_1' + R_1' + 1} - \frac{S_2'}{S_2' + R_2' + 1}\right) \tag{9}$$

$$= v_{\max}\frac{S_1'(R_2' + 1) - S_2'(R_1' + 1)}{(S_1' + R_1' + 1)(S_2' + R_2' + 1)} \tag{9a}$$

in which S' and R' are "normalized" concentrations related to the half-saturation concentrations ($S' = S/K_{CS}$, $R' = R/K_{CR}$).

b. The Equilibrium Exchange Test. Since obviously the quantitative effect of competition on transport rate depends on the ratio of the two substrate concentrations S and R, the highest degree of sensitivity for a competition test can be reached when $S' \ll R'$. This can be realized if S is isotope-labeled and used in trace concentrations in the presence of high concentrations of R. Then, if R is in equilibrium ($R_1 = R_2 = R$), Eq. (9) reduces to

$$v_S = \frac{v_{\max}}{K_{CS}}\frac{S_1 - S_2}{(R' + 1)} = K_p(S_1 - S_2) \tag{10}$$

indicating that the transport kinetics of S in this case is linear with a proportionality factor equal to $K_p = v_{\max}/(R' + 1)K_{CS}$, enabling the dissociation constant K_{CR} to be evaluated graphically from a plot of the reciprocal proportionality factor against the concentration R according to the equation

$$\frac{1}{K_p} = \frac{K_{CS}}{v_{\max}}\left(\frac{R}{K_{CR}} + 1\right) \tag{11}$$

in which K_p is the proportionality factor. In this plot K_{CR} is given by the negative abscissa intercept.

This test can be successfully applied even in cases of rather low affinities (high values of K_{CR}). The substrates R and S may be identical or different molecular species ("homogeneous" or "heterogeneous" arrangement). Since the movement of isotope-labeled substrate occurs by exchange against cold substrate, the test has been termed the "equilibrium exchange test" (Stein, 1967) or "isotope exchange test."

D. Counter Transport

(Rosenberg and Wilbrandt, 1957; Widdas, 1952; Wilbrandt 1969b)

The test which, as discussed below, probably is the most specific criterion for a carrier mechanism is the demonstration of "counter transport." If, of two substrates present, S and R, S is in equilibrium, then a movement of R due to an existing gradient (i.e., $R_1 \neq R_2$) will induce a movement of S out of equilibrium, in other words, against an increasing gradient. This is what was originally termed counter transport (Rosenberg and Wilbrandt, 1957). Later the term was used more loosely as will be discussed later. It is useful, both for the sake of simplicity of conditions and for the sake of clarity of concept, to maintain the definition given here for the following discussion. Counter transport, then, is a movement of one substrate out of equilibrium and against rising gradients, induced by the simultaneous movement of a second substrate with affinity for the same binding site of the same carrier.

Kinetically, counter transport is easily derived from Eq. (9), assuming $S_1 = S_2 = S$. Then, as long as $R_1 \neq R_2$, implying a downhill movement of R, S will move ($v_S \neq 0$) even though $S_1 = S_2$:

$$v_S = v_{\max} \frac{S'(R_2' - R_1')}{(S_1' + R_1' + 1)(S_2' + R_2' + 1)} \tag{12}$$

In the absence of R, however, there is no movement of S as long as $S_1 = S_2$.

Conceptually, counter transport can be understood as a consequence of asymmetric inhibition of the reaction between S and carrier on the two sides of the membrane due to the difference between R_1 and R_2.

The specificity for mobile carriers of this criterion can be demonstrated by a comparison between a carrier system with mobile carrier and a system with fixed binding sites located along the walls of a porous channel. In the latter system the substrate, in order to pass the membrane, would have to jump successively from one binding site to the next. It can be shown that in such a system the transport kinetics of a substrate S in the absence of other substrates is identical with that of a true mobile carrier system. However, the system shows no counter transport (Rosenberg and Wilbrandt, 1957).

Let the channel have n successive binding sites between the two sides of the membrane. Then, in the steady state, the rate of net movement from the Kth binding site to the neighboring, $(K + 1)$th site will be the same for all values of K. It can be calculated as the difference between the rates of exchange from site K to site $K + 1$ (direction 1) and from site $K + 1$ to site K (direction 2).

In direction 1 the rate will be proportional to the fraction of occupied sites at site K and to the fraction of free sites at site $K + 1$, and in the opposite direction 2, the rate will be proportional to the fraction of occupied sites at site $K+1$ and to the fraction of free sites at site K. Denoting the fraction of the occupied sites as f and, therefore, that of the free sites as $1-f$, the net rate of movement from K to $K+1$ will be

$$v = K_e[f_K(1-f_{K+1}) - f_{K+1}(1-f_K)] = K_e(f_K - f_{K+1}) \tag{13}$$

with K_e the constant of exchange rate.

Adding up all equations of the type of Eq. (13) for the $n+1$ successive similar steps, and considering that in the steady state v is identical in all these equations, we obtain

$$v = [K_e/(n-1)](f_1 - f_n) \tag{14}$$

Since site 1 is in equilibrium with S_1 and site n with S_2, Eq. (14) can be written as

$$v = \frac{K_e}{n-1}\left(\frac{S_1'}{S_1'+1} - \frac{S_2'}{S_2'+1}\right) \tag{15}$$

indicating a concentration dependence identical with that of a simplified case with only two fixed binding sites ($n=2$).

Equation (15) is formally similar to Eq. (3). Thus under these conditions the kinetics of a system involving fixed binding sites and a system involving mobile binding sites are similar.

After the introduction of a second substrate R this no longer holds. It was shown above that for $S_1 = S_2$ but $R_1 \neq R_2$ in a mobile carrier system a net movement of S ensues. Treating the fixed binding site system for the sake of simplicity only for the special case $n=2$ (although a rigorous treatment is not bound to this simplifying assumption), and following the same argument as above, we find the rate of transport for S as

$$v = K_e\left\{\frac{S_1'}{S_1'+R_1'+1}\,\frac{1}{S_2'+R_2'+1} - \frac{1}{S_1'+R_1'+1}\,\frac{S_2'}{S_2'+R_2'+1}\right\} \tag{16}$$

$$= K_e\frac{S_1'-S_2'}{(S'+R'+1)(S'+R'+1)} \tag{16a}$$

Equation (16a) shows that for $S_1 = S_2$ the rate of transport of S will be zero in spite of the presence of R. In other words, there is no counter transport.

E. Other "Exchange" Phenomena

In the kinetic treatment of counter transport so far we have held the assumption $S_1 = S_2$. If this assumption is dropped the mutual interdependence between the movements of S and of R persists but then the effect of a movement of R on the simultaneous movement of S consists in a change of rate of movement which, according to conditions, may be an acceleration or a retardation. The basic mechanism, however, will still be the same: asymmetric competitive inhibition by the second substrate on the two sides of the membrane.

The presence of a second substrate R can, however, produce changes in rate of movement of S also by a different basic mechanism. If the above-mentioned condition of equal mobility for free and loaded carrier is dropped, then, if there is a net movement of S from side 1 to side 2, the rate of this movement becomes dependent on whether the back transport of carrier (which is necessary for the maintenance of the steady state) occurs in the form of free carrier C or of CR. The fractions of back transport by C and by CR, respectively, depend on the concentration and affinity of the second substrate R.

This creates a second type of interdependence between the movements of R and S which differs fundamentally in the basic mechanism from counter transport. This type of interdependence was first demonstrated by Heinz and Walsh (1958) for the case of amino acid uptake in tumor cells. They found that preloading the cells with a second amino acid affects the rate of uptake, in most cases in the accelerating direction. Similar observations were made with respect to exit of transport substrates from red cells and its dependence on the external concentration of other substrates (Levine *et al.*, 1965). The two types of interdependence have in common that exchange of two different types of loaded carrier is involved. They are both, therefore, occasionally loosely referred to as "exchange." It may not always be easy to differentiate between them experimentally. It should be kept in mind, however, that conceptually they are quite different although both are consequences of the carrier concept.

An extreme variant is what is usually referred to as "exchange diffusion." The concept, introduced by Ussing (1947), is that of a carrier which is not able to move in unloaded form at all. In this case no net transport of substrate is possible without simultaneous counter movement of a second substrate.

F. Comparative Considerations

It is useful to compare the criteria discussed so far with respect to reliability, specificity, and sensitivity. It was pointed out that the criterion for saturation kinetics is somewhat unsatisfactory with respect to reliability.

With respect to sensitivity it is obvious that in cases of low affinity it may be difficult or impossible to reach concentration levels showing saturation.

The equilibrium exchange test, in our experience, has the highest sensitivity in the sense that even in cases of low affinity it may give reliable and satisfactory indications of binding. However, it does not differentiate between binding to fixed and mobile sites. Thus it may be considered as particularly sensitive but not as specific as counter transport. Counter transport, as shown above, does differentiate between the two types of binding sites. However, it depends strongly on saturation conditions both for the inducing and for the counter transported molecular species. This feature warrants some discussion.

In an analysis of this dependence two points have to be considered: the maximum of asymmetric distribution of the counter transported substrate and the rate of the counter transport movement.

From Eq. (9a), in which S is the counter transported (passive) substrate and R the inducing substrate, it emerges that the rate of counter transport v_S will become zero when

$$S_1/S_2 = (R_1' + 1)/(R_2' + 1) \tag{17}$$

This equation shows that *maximum inequality of distribution* for S depends on carrier saturation with R but not with S. Thus low affinity for the carrier of R but not of S may become limiting.

On the other hand, the *rate* of counter transport depends on the affinity of both substrates. The dependence is such that both low saturation and high saturation reduces the rate, as can be seen by closer inspection of Eq. (12). The equation shows that the normalized concentrations S' and R' are unimportant in the denominator as long as they are smaller than 1. In this range the counter transport rate is proportional to the normalized concentrations of both substrates [in the enumerator of Eq. (12)]. When they rise above the value 1, however, they become important in the denominator to the second power, implying that now increase of both substrate concentrations will reduce the rate of counter transport. Thus only in the concentration range around the half saturation concentration can counter transport be expected to be readily demonstrated experimentally.

Figure 3 may serve to illustrate this situation. The figure shows computed curves of counter transport for a variety of saturation conditions characterized by different values of K_{CR} and K_{CS}. Both for very low and very high values of the Michaelis constants the rate of counter transport becomes very low (Wilbrandt, 1972b).

Figure 4 shows computed and experimental curves of iso-counter transport for a number of different values of K_{CR} (Wilbrandt, 1972a). In the case of low affinity for R the maximum of unequal distribution of S is so

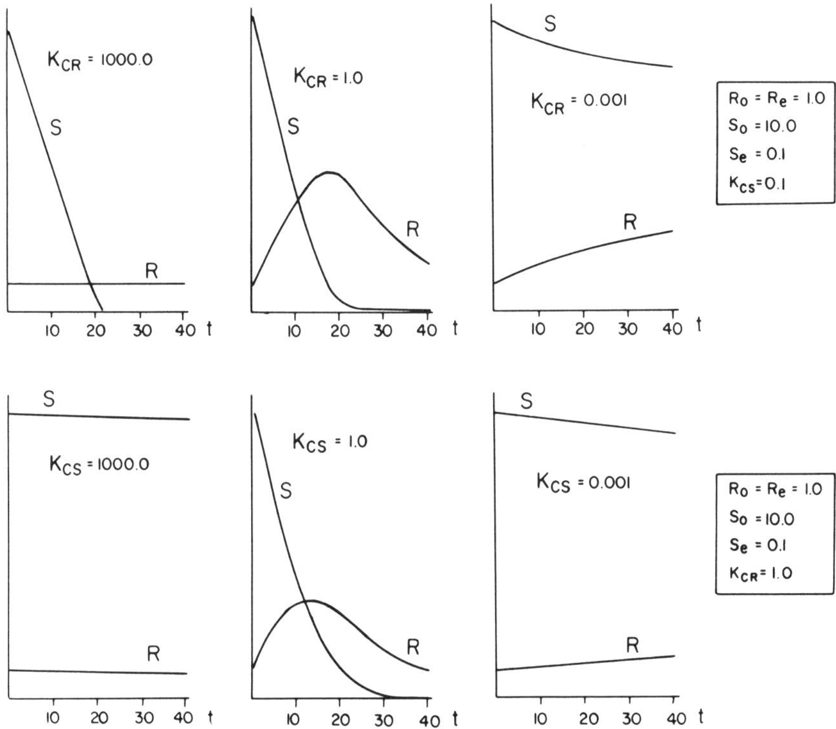

Fig. 3. Computed curves of counter transport of a substrate *R* induced by the downhill movement of substrate *S*. The curves were computed using Eq. (9) with the Michaelis constants indicated. They show the influence of saturation both of the inducing substrate *S* and the induced substrate *R*. The middle curves show the counter transport in the region around half-saturation. In the curves to the left (low affinity of *R* in the upper and of *S* in the lower curve) no counter transport can be recognized. In the curves to the right (with high affinity of *R* in the upper and of *S* in the lower curve) counter transport is recognizable but extremely slow. Thus both very low and very high saturations are unfavorable for the demonstration of counter transport.

close to 1 that experimental verification may become difficult. The experimental curves demonstrate this for a series of sugars in the system of transport across the human red cell membrane. With decreasing affinity of the inducing sugar the extent of counter transport decreases more and more and finally the demonstration of counter movement becomes doubtful.

Another case of transport across the mammalian red cell membrane which is instructive has been observed recently (Wilbrandt and Becker, unpublished). It was found that urea and related compounds display mutual inhibition, indicating that a carrier system might be involved. Equilibrium

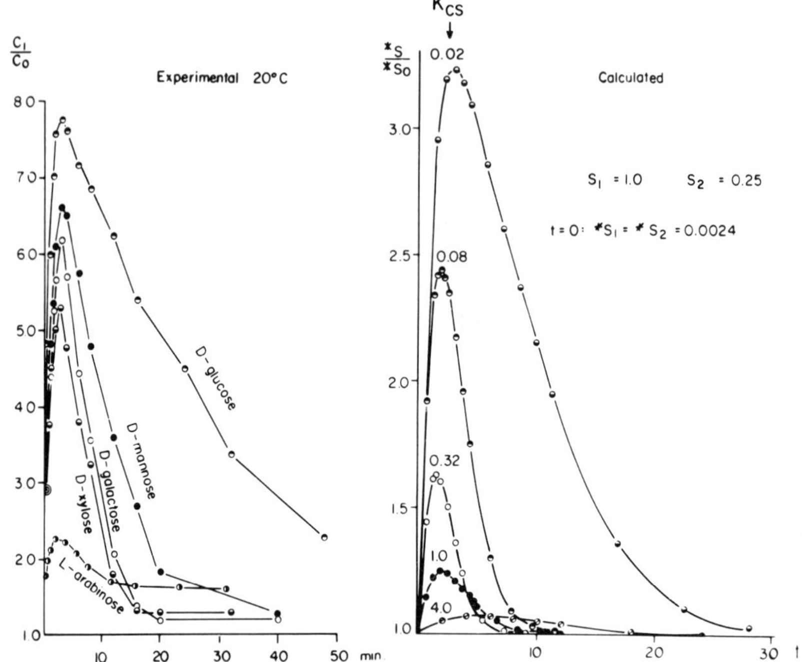

Fig. 4. Iso-counter transport curves for substrates of varying affinity. Left: experimental curves with sugar transport across the human red cell membrane. Ordinate: ratio of inside to outside concentrations. Abscissa: time. Right: computed curves for the indicated values of Michaelis constants.

exchange experiments with thiourea as inhibitor yielded a rectilinear plot with a slope distinctly different from zero. From the abscissa intercept a Michaelis constant for thiourea of about 2.7 M was obtained, indicating a very low affinity. The attempt to demonstrate iso-counter transport with thiourea, however, failed completely. Thus the sensitivity of the equilibrium exchange test proved to be sufficient for this low affinity, while for the counter transport test the affinity appeared to be too low.

IV. ACTIVE TRANSPORT

A. Uphill Operation

Since the definition of active transport = uphill transport has been adopted here, the first criterion obviously is movement against a gradient.

In the case of electrically neutral substrates it is necessary to know the activities or at least concentrations of the substrate on the two sides of the membrane. In the case of ions, in addition, the electrical potentials must be known, in order to obtain the difference in electrochemical activity according to Eq. (1a).

Demonstration of transport against electrochemical potential is a valid criterion for active transport, independent of the mechanism, resting on the definition.

B. Dependence on Metabolism

Metabolic energy is certainly the form of energy most frequently used in active transport systems. If the energy of uphill operation is compensated by the energy of chemical metabolic reactions, then suppression of these reactions of course must suppress the "activity" of the transport. Criteria on this basis are in frequent use. Means to inhibit metabolism include: lack of metabolic substrate, lack of oxygen, the effect of metabolic inhibitors like cyanide, azide, dinitrophenol, iodoacetic acid, and others. It should be kept in mind, however, that metabolic inhibitors or other changes of metabolic conditions may well have an effect on passive permeability. For instance, the frequently used inhibitor dinitrophenol has been shown to increase the hydrogen ion permeability of mitochondria (Mitchell, 1961). Metabolic effects on the structure of the membrane are difficult to exclude. Thus the criterion of metabolic dependence is of limited reliability.

C. Short-Circuit Current

An important criterion introduced by Ussing (1954) is the short-circuit criterion for ion transport. The basic consideration is the following: In diffusion the ions move under the influence of two driving forces—osmotic forces and electrical forces. If the osmotic force across a membrane is abolished by the use of identical solutions on the two sides of the membrane and if the two media adjacent to the membrane are electrically short-circuited, both osmotic and electrical driving forces are eliminated. If, under these conditions, a movement of ions does take place, other driving forces must be operative and the transfer occurs against increasing gradients, meeting the definition of active transport adopted here. The short-circuit criterion has been used extensively in many cases, particularly in the case of the frog skin in Ussing's own and many other experiments (Ussing, 1954) as well as in the intestinal wall (Schultz and Zalusky, 1964). It has been extremely valuable for the recognition of active transport. Combination with the use of unidirectional fluxes of labeled ions has furthermore made it possible to identify the movement of specific ions by comparing the amount of electricity carried by the electrical current and by the ion fluxes, respectively.

D. Asymmetric Michaelis Constants

In Eq. (1) the Michaelis constants in the two saturation terms in the parentheses are numerically identical. It can easily be shown that only under these conditions is the steady state characterized by equality of the concentrations S_1 and S_2. If numerically different values are inserted for K_m in the two saturation terms, say K_1 and K_2, the steady state condition $v = 0$ is only compatible with unequal distribution in the steady state ($S_1 \neq S_2$ or, more specifically $S_2{:}S_1 = K_2{:}K_1$). Since unequal distribution in the steady state implies active transport, the demonstration of $K_1 \neq K_2$ also constitutes a criterion for active transport. It has been used, for instance, in the case of bacterial permeases. It must be pointed out, however, that in this case K_1 and K_2 can of course not be true thermodynamic constants (dissociation constants), since the same complex is involved at the two sides of the membrane. The "apparent" Michaelis constants K_1 and K_2 in such a case are complex terms containing information about the type of energy coupling involved (Wilbrandt, 1969a).

This can be illustrated for two cases, the simple case of counter transport discussed above and the case of coupling to metabolic reactions treated kinetically by Rosenberg in so-called *CZ* systems (Rosenberg and Wilbrandt, 1963).

In the case of counter transport, Eq. (9) can be written in the form

$$v_S = v_{\max}\left[\frac{S_1}{S_1 + K_{CS}(1 + R_1')} - \frac{S_2}{S_2 + K_{CS}(1 + R_2')}\right] \qquad (18)$$

In this case, thus, $K_1 = K_{CS}(1 + R_1')$ and $K_2 = K_{CS}(1 + R_2')$. The "apparent Michaelis constants" K_1 and K_2 thus contain, besides the true Michaelis constants, the two normalized concentrations of R which are involved in the energy coupling. The energy used for the uphill operation is the flow energy of the inducing substrate R.

In Rosenberg's treatment of active transport driven by metabolic energy two forms of carrier are assumed, capable of interconversion: a low affinity form Z and a high affinity form C. If the interconversion is asymmetric, e.g., spontaneous on one side of the membrane, say the external side, but catalyzed due to coupling with metabolic reactions on the other side of the membrane, then the ratio $Z/C \equiv \Phi$ may differ on the two sides of the membrane. The rate of transport has been shown to be given by the equation

$$v_S = v_{\max}\left(\frac{S_1}{S_1 + \theta_1} - \frac{S_2}{S_2 + \theta_2}\right) \tag{19}$$

in which

$$\theta = K_{ZS}(1 + \Phi)/(q + \Phi) \tag{20}$$

Unequal distribution in the steady state ($S_2 \neq S_1$ due to $\theta_1 \neq \theta_2$) results under the condition that $\Phi_1 \neq \Phi_2$ and $K_{ZS} \neq K_{CS}$. Again the terms θ_1 and θ_2 (analogous to K_1 and K_2) are complex terms containing information about the type of coupling involved.

E. The Flux Ratio Criterion of Ussing

Coming back now to the flux ratio criterion discussed above, the demonstration of a flux ratio not identical with the ratio of activities or electrochemical activities is one of the criteria most frequently used for active transport at present. An understanding of its meaning and its limitations is therefore of practical interest.

In Ussing's original derivation of the flux ratio criterion no assumption was made with respect to the mechanism involved in the uphill operation. In fact the condition of uphill operation was not mentioned in the derivation. The derivation was given for diffusion and therefore, without further specification, the criterion is originally one for diffusion versus nondiffusion rather than for active transport. It has, however, become current use to take it as indicating active transport in the direction $S_1 \rightarrow S_2$, if $M_1/M_2 > S_1/S_2$. This situation will, in the following discussion, be referred to as the "positive flux ratio criterion." Consequently, the criterion will be termed "negative" if $M_1/M_2 < S_1/S_2$, as in the case of uncoupled carrier operation [Eq. (5)].

In the steady state, regardless of the mechanism, the flux ratio criterion is obviously "positive" since the fluxes are identical and their ratio therefore equals one, while $S_1/S_2 < 1$ (if the transport is "active" in the sense of the definition adopted here). However, whether and how far outside of the steady state the validity of the criterion is maintained has to be analyzed in terms of a specific model. Of particular interest is the question of whether the criterion provides information as to the nature of the system ("pump" versus uncoupled mediated transport) or as to the actual operation (uphill or downhill). A "pump" may, due to existing concentrations ($S_2 < S_1$), operate downhill, although it would be capable of uphill operation (resembling an automobile that moves downhill with the motor running). Can the criterion recognize a pump when it operates downhill?

Since in most cases of active transport carrier criteria are positive and carrier mediation is assumed, it appears useful to analyze a carrier model. This will now be done in terms of the kinetic model discussed in the previous section involving two numerically different values of apparent Michaelis constants K_1 and K_2, and using, for the sake of simplicity, concentrations rather than activities or electrochemical activities. As long as the mobilities of free and loaded carrier are equal the unidirectional fluxes M_1 and M_2 of the model will be $v_{\max} S_1/(S_1 + K_1)$ and $v_{\max} S_2/(S_2 + K_2)$, respectively, yielding a flux ratio of

$$\frac{M_1}{M_2} = \frac{S_1}{S_2}\frac{S_2 + K_2}{S_1 + K_1} = \frac{S_1}{S_2} F_M \tag{21}$$

This ratio differs from the concentration ratio by the factor $F_M = (S_2 + K_2)/(S_1 + K_1)$, which, depending on saturation conditions, ranges between S_2/S_1 and K_2/K_1. It is useful to consider these two limiting cases, $F_M = K_2/K_1$ for low saturation, and $F_M = S_2/S_1$ for high saturation, more closely. Since for uphill carrier transport in the direction $1 \rightarrow 2$ we always have $K_2/K_1 > 1$, the flux criterion under conditions of low saturation ($F_M = K_2/K_1$) will always be "positive" regardless of whether $S_1 > S_2$ (actual movement downhill) or $S_1 < S_2$ (actual movement uphill). On the other hand, in the case of high saturation, with $F_M = S_2/S_1$, obviously only for uphill movement ($S_1 < S_2$) will the criterion be "positive," while for downhill movement ($S_1 > S_2$), we always have $F_M < 1$. Under these conditions, then, the flux ratio will differ from the concentration ratio in the same way as in the case of uncoupled carrier transport (not operating uphill; $K_1 = K_2$). Thus in the case of coupled carrier transport there appear to be two opposite trends: one, due to energy coupling, tending to raise F_M, i.e., to render the flux ratio criterion "positive"; the other, due to carrier operation, tending to lower F_M and to render the flux ratio criterion "negative" as in uncoupled carrier transport.

Figure 5 gives a graphic presentation of the dependence of F_M on saturation which complements the above discussion (Wilbrandt, 1972a). The basic assumption is a living cell, accumulating the substrate S inside (side 2) due to active transport from a constant external concentration S_1, with a maximum concentration ratio $S_2/S_1 = Q$ in the steady state. The progress of penetration is indicated by a, the fraction of the total increase of S_2 (from zero to QS_1) at a given time. Thus at any time

$$S_2 = aS_1Q \tag{22}$$

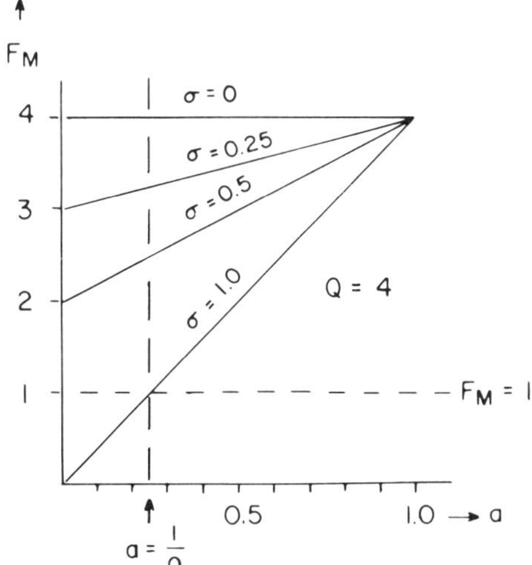

Fig. 5. Flux ratio coefficient calculated for coupled carrier systems according to Eqs. (21) and (24). $M_1/M_2 = F_M a_1/a_2$. For details and discussion see text.

Therefore, considering $K_2/K_1 = Q$ and Eq. (20) and (21), we have

$$F_M = \frac{aS_1Q + K_1Q}{S_1 + K_1} \tag{23}$$

$$= Q\left(\frac{S_1}{S_1 + K_1}a + \frac{K_1}{S_1 + K_1}\right) \tag{23a}$$

Introducing the saturation fraction $\sigma = S_1/(S_1 + K_1)$ and, therefore, $K_1/(K_1 + S_1) = 1 - \sigma$, we see that Eq. (23a) becomes

$$F_M = Q(\sigma a + 1 - \sigma) = Q[\sigma(1 - a) + 1] \tag{24}$$

with the special cases

$$F_M = \begin{cases} Q & \text{for } a = 1 \text{ (steady state) regardless of } \sigma \\ Q & \text{for } \sigma = 0 \text{ (zero saturation) regardless of } a \\ Qa & \text{for } \sigma = 1 \text{ (complete saturation)} \end{cases}$$

and $\sigma = 1 - (1/Q)$ for $a = 0$, and $F_M = 1$.

Figure 5 shows that for all saturation degrees higher than $\sigma = 1 - (1/Q)$, we always have $F_M > 1$, i.e., then the flux ratio is "positive" regardless of whether the system operates uphill or downhill. At higher saturation there appears a range of values for a and thereby for S_2 for which $F_M < 1$ ("negative" flux ratio criterion). If $\sigma \to 1$, this range extends to $S_2 = S_1$, i.e., then the criterion is "positive" as long as the actual movement is uphill and "negative" as long as it is downhill. The flux ratio criterion, under these conditions, characterizes the actual operation rather than the nature of the system.

One more remark should be made with respect to this situation. For the evaluation of the flux ratio test the electrochemical activities $\bar{a}_1$ and $\bar{a}_2$ must be known. With this knowledge, however, the question as to whether the transport actually is uphill or downhill can be decided by the mere determination of net flux. The use of the two unidirectional fluxes, then, would appear to be a procedure of unnecessary sophistication.

V. CONCLUSION

A review of the most common transport criteria in current use has been given. The purpose of the review was to show that the criteria can be compared from different points of view, that they differ considerably in reliability, specificity, and sensitivity, and that these properties depend strongly on saturation conditions. In general it appears advisable, therefore, to take into account as far as possible the probable range of saturation in the observations to be analyzed and, whereever possible, not to rely on the result of one single criterion.

REFERENCES

Heinz, E., and Walsh, P. M., 1958, Exchange diffusion, transport, and intracellular level of amino acids in Ehrlich carcinoma cells, *J. Biol. Chem.* **233**:1488.

Levine, M., Oxender, D. L., and Stein, W. D., 1965, Substrate-facilitated transport of the glucose carrier across the human erythrocyte membrane, *Biochim. Biophys. Acta* **109**: 151–163.

Lineweaver, H., and Burk, D., 1934, The determination of enzyme dissociation constants, *J. Am. Chem. Soc.* **56**:658–666.

Mitchell, P., 1961, Chemiosmotic coupling in oxidative and photosynthetic phosphorylation, *Bioch. J.* **79**:23P–24P.

Rosenberg, Th., 1948, On accumulation and active transport in biological systems, *Acta Chem. Scand.* **2**:14–33.

Rosenberg, Th., and Wilbrandt, W., 1957, Uphill transport induced by counterflow, *J. Gen. Physiol.* **41**:289–296.

Rosenberg, Th., and Wilbrandt, W., 1963, Carrier transport uphill, *J. Theor. Biol.* **5**: 288–305.

Schultz, St. G., and Zalusky, R., 1964, Ion transport in isolated rabbit ileum. I. Short-circuit current and Na-fluxes, *J. Gen. Physiol.* **47**:567–584.

Stein, W. D., 1967, "The Movement of Molecules across Cell Membranes," Academic, Press, New York and London.

Ussing, H. H., 1947, Interpretation of the exchange of radiosodium in isolated muscle *Nature* **160**:262–263.

Ussing, H. H., 1949, The distinction by means of tracers between active transport and diffusion, *Acta Physiol. Scand.* **19**:43–56.

Ussing, H. H., 1954, Active transport of inorganic ions, *Symp. Soc. Exp. Biol.* **8**:407–422.

Widdas, W. F., 1952, Inability of diffusion to account for placental glucose transfer in the sheep and consideration of the kinetics of a possible carrier transfer, *J. Physiol., Lond.* **118**:23–39.

Wilbrandt, W., 1969a, Carrier systems in biological transport, *in* "Proc. 4th Int. Congr. Pharmacology, Basel 1969," Vol. 4, pp. 391–401, Karger, Basel.

Wilbrandt, W., 1969b, Specific transport mechanism in the erythrocyte, *Experientia* **25**:673–677.

Wilbrandt, W., 1972a, The bearing of membrane and transport studies for pharmacology, *in* "Pharmacology and the Future of Man" (Fifth Int. Congr. Pharmacology, San Francisco, 1972), Vol. 4, pp. 375–389.

Wilbrandt, W., 1972b, Carrier diffusion, *in* "Passive Permeability" (F. Kreuzer, ed.), In press.

Wilbrandt, W., 1974, Active transport, *in* "MTP International Review of Science," in press.

Wilbrandt, W., and Becker, Ch., unpublished.

Wilbrandt, W., and Rosenberg, Th., 1961, The concept of carrier transport and its corollaries in pharmacology, *Pharmacol. Rev.* **13**:109–183.

Wilbrandt, W., Fuhrmann, G. F., and Touabi, M., 1975, in preparation.

Zierler, K. L., 1961, A model of a poorly-permeable membrane as an alternative to the carrier hypothesis of cell membrane penetration, *Bull. Johns Hopkins Hospital* **109**: 35–48.

Chapter 3

Carotenoid and Merocyanine Probes in Chromatophore Membranes

Britton Chance

Johnson Research Foundation, School of Medicine
University of Pennsylvania
Philadelphia, Pennsylvania

and

Margareta Baltscheffsky

Department of Biochemistry, Arrhenius Laboratory
University of Stockholm
Stockholm, Sweden

I. INTRODUCTION

Increasing awareness of the vast amount of information that can be obtained from intrinsic and extrinsic probes of membrane phenomena has led to a wide-ranging survey of the properties of both types of probe. Intrinsic probes of photosynthetic membranes are provided by the carotenoid pigments. The carotenoids are assumed to be located in the lipid phase of the membrane close to chlorophyll, since they act as accessory light-harvesting pigments. Isolated reaction center preparations from photosynthetic bacteria usually contain tightly bound carotenoids, which further supports the concept of a close association between them and the photosynthetic unit. The spectral and kinetic properties of these pigments respond to several parameters: light, oxygen, ATP, and inorganic pyrophosphate (PP_i). In single-flash illumination experiments, the rate of decay of the membrane charge is assumed to depict the degree of coupling of the chromatophore or chloroplast membrane. The origin of the carotenoid absorbance change has been much debated and recently it has been shown to coincide with the formation of a diffusion potential in chromatophore membranes from *Rhodopseudomonas spheroides* (Jackson and Crofts, 1969, 1971; Crofts, 1974). In plant chloroplasts, the corresponding absorbance change at 518 nm has been stated by Witt (1972) to indicate the formation of a transmembrane electric field. When induced in the dark by energizing conditions, or by illumination, the steady state response of the pigments is abolished by uncouplers but the response to a single flash is unaltered in rise time and extent, even though the decay time is accelerated.

While ideally, the intrinsic probe has advantages because it affords information on the native state of the membrane, it often lacks those very properties that make a probe most useful: (1) occupancy of a characteristic site identified with a single function of the membrane; (2) signaling a response to a single phenomenon; (3) affording signals of optimal clarity. For these reasons, extrinsic probes added to the membrane system may be selected to afford highly specific locations, responses to single functions of the system, and, most important, a clearly identifiable signal. They may have the disadvantage that the system is altered by the probe, but appropriate selective tests of the gamut of functions of the membrane system can be used to show that the altered system functions as well as the natural system. It is on this basis that we have been studying fluorescence probes in membranes for some time (Azzi *et al.*, 1969; Chance, 1974).

While the efficacy, localization, clear readout, and single-function response have made ANS (1-anilino-8-naphthalene sulfonic acid) and its homologues extremely valuable for the study of submitochondrial particles and membrane fragments derived from mitochondria, their applicability to chromatophores and chloroplasts has been comparatively limited. For example, the maximal fluorescence changes of ANS upon energization are limited to 60% in photosynthetic membranes (Vainio *et al.*, 1972); the emission maximum may be subject to spectroscopic interferences, particularly those due to chlorophyll and/or carotenoids.

The remarkable results obtained with merocyanine probes in squid axons (Davila, 1973; Cohen *et al.*, 1975), much better than with ANS-like compounds, together with the wide choice of excitation and emission wavelengths, has suggested their application as electrochromic probes to membrane fragments from mitochondria and from photosynthetic bacteria, as indicators of local or transmembrane charge separation. The use of related cyanine dyes in red blood cell membranes has been reported (Laris and Hoffman, 1973).

Since the wavelengths of the fluorescence and absorption maxima for the merocyanine dyes can be adjusted by varying the number of conjugated double bonds between the end groups (Brooker *et al.*, 1951), those merocyanine dyes showing absorbance changes and emission characteristics in the region between 610 and 720 nm appear attractive for the study of chromatophores or bacteriochlorophyll, since this band lies between the 590- and 860-nm maxima of bacteriochlorophyll and beyond the wavelengths at which carotenoids absorb. Such a probe is afforded by merocyanine-V and its homologues (Platt, 1956; Platt, 1961).

The specific objective of this chapter is to compare the responses of the extrinsic merocyanine probe with those of the intrinsic carotenoid probe, and thus to distinguish whether they are measuring a common transmembrane potential or different localized charge separations in different patches of lipid.

II. MATERIALS AND METHODS

A. Experimental Methods

Wavelength scanning spectrophotometry of the merocyanine probes was carried out with a dual wavelength spectrophotometer specially adapted for the scanning procedure (Chance and Graham, 1971). A motor drive was applied to one of the monochromators, which operated a synchronized *X–Y* chart recorder to provide a precise wavelength scale. A baseline correction was afforded by the simple procedure of recording, by means of an A/D converter and a digital memory, the corrections that needed to be applied to the photomultiplier dynode voltage in order to maintain the output corresponding to the scanned wavelength exactly equal to that for the nonscanned wavelength. Such voltages were applied to the dynode of the photomultiplier via a controlled high-voltage power supply in the form of pulses, each appropriate to the time when the measuring wavelength flashes through the sample. On subsequent scans, this voltage was read out from the memory via a D/A converter to control the dynode voltage appropriately throughout the scans. Under these conditions, a very small absorbance change due to addition of the dye, illumination of the suspension, or addition of ATP or pyrophosphate was easily detected.

Dual wavelength spectroscopy of the red shift (Chance, 1973) of the merocyanine dye was satisfactorily measured by the absorbance difference between 620 and 630 nm in a Johnson Foundation dual wavelength spectrophotometer. The light-induced absorbance changes of the carotenoid pigments of *Rhodospirillum rubrum* were measured by the dual wavelength technique at 530 and 508 nm. Fluorescence changes in the merocyanine dye were appropriately excited at 577 nm, somewhat displaced from the excitation maximum at 620 nm, and emission was measured at 660 nm, also somewhat displaced from the emission maximum at 640 nm. The reason for employing displaced excitation and emission wavelengths was not only to obtain a good exclusion of excitation energy from the emission-measuring photomultiplier, but also to allow the use of the dual wavelength spectrophotometer to measure the absorbance changes due to the red shift of the probe between the excitation and emission wavelengths.

The actinic excitation was provided by a tungsten lamp using an 860-nm interference filter. The light transmitted by the sample was measured directly by a silicon diode detector employing a similar filter. Thus the actinic light absorbed by the chromatophore suspension was precisely measured.

Fluorescence excitation and emission spectra were obtained with a Perkin-Elmer MP-4F fluorescence spectrophotometer. Red Corning filters were used to afford additional protection for the emission-measuring photomultiplier.

For scans of most emission spectra and for some excitation spectra as well, a filter cutting absorbance below 600 nm was used (CS-2-62), while for other excitation spectra a red filter absorbing below 650 nm was used (CS-2-64). In order to obtain the excitation and emission spectra, the slits of the excitation and emission monochromators were set at approximately 20 nm; this spectral resolution was adequate for the probe spectra. The instrument was operated at its maximal sensitivity.

Recordings of excitation and emission show that an optimal wavelength for excitation of the probe fluorescence is at 625 nm and for emission measurements is at 635 nm. The proximity of the excitation and emission optima makes it desirable to separate the wavelengths actually employed by a significant amount in order to avoid the transmission of excitation light through the emission filter. For this reason, the mercury line at 577 nm, a wavelength at which the excitation efficiency is about half that of the maximum, is appropriate. Measurements of the fluorescence from 620 nm to the cutoff point of an S-10 photomultiplier capture most of the emission.

B. Preparations

Chromatophores were prepared by the method of Baltscheffsky (1967) and chlorophyll was determined at 880 nm, based upon an *in vivo* extinction coefficient at 860 nm of 140 cm^{-1} mM^{-1} (Clayton, 1963). The chromatophores were suspended in 5 mM Tris-sulfate, brought to *p*H 7.2 by NaOH unless otherwise noted. The antibiotics were obtained through the kindness of Dr. C. P. Lee. Other chemicals employed were reagent grade.

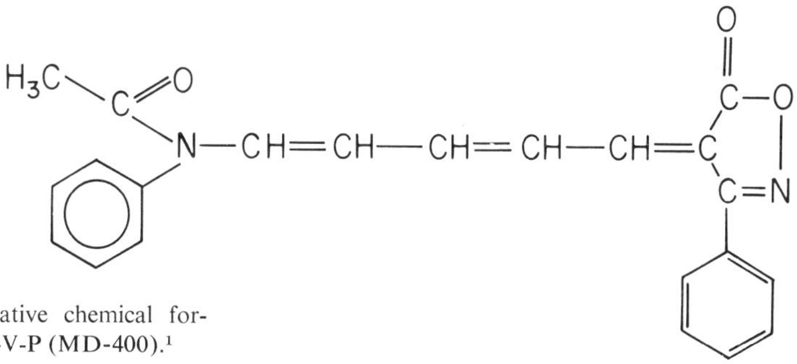

Fig. 1. Tentative chemical formula of MC-V-P (MD-400).[1]

C. Merocyanine Probes

The dicarbocyanine dye MC-I[2] used in voltage-clamp and real-time experiments in the squid axon (Davila, 1973; Cohen *et al.*, 1975) was found to be highly satisfactory in experiments with submitochondrial particles (Chance, 1973) but unsatisfactory for chromatophores. For this reason, two other merocyanine dyes, which show significant responses in SMP and chromatophores, have been used. They are MC-V-P, whose formula is shown in Fig. 1, and another of very closely related but not as yet exactly determined structure, MC-V, on which experiments are reported here. The former, MC-V-P, is termed "R-6" in the work of Brooker *et al.* (1951), and the latter occurs in the fraction from which MC-V-P, or R-6, has been purified. MC-V was examined by thin-layer chromatography and showed only one component. These two probes differ from the MC-I used by Cohen *et al.* (1975) in that they lack the negative charge on the sulfonic acid and the long aliphatic chains. Thus neither of the probes employed here contains an isolated charge group; however, they are highly lipophilic.

An approximation to the structure is based upon work of Dr. Barry Cooperman and his colleagues (see Acknowledgment) (pers. comm.) and is indicated in Fig. 1; however, the exact structure may be a tautomer (Dr. L. M. Loew, pers. comm.). Also, charge separation across the molecule with one of the nitrogens being negatively charged and the other being positively charged is to be considered with corresponding rearrangements in the conjugated double-bond system (Brooker *et al.*, 1951; Platt, 1956, 1961). The absorption bands of the probe, in addition to being affected by the number of conjugated double bonds, are also dependent upon the environment in which the two ends of the molecule (b_R and b_L) are located and by

[1] Further information on figures may be obtained through this and following reference numbers.

[2] Eastman Organic Chemicals Merocyanine 540.

the potential difference between them, described by Platt's equation (1961) as

$$E^2 = E_I^2 + (b_R - b_L + V_{RL})^2 \quad (1)$$

where E is the energy of the absorption band, V_{RL} is the potential difference between the end groups, and b is a basicity generally related to the dielectric constant of the medium. The equation indicates that E will vary with V_{RL} in both linear and quadratic dependence, as is further discussed in the appendix. Bücher *et al.* (1969) have also explored the relationship between the absorption maximum and the potential both experimentally and theoretically; the appendix considers their conclusions as well. It is important to note that whereas a simplistic interpretation of Platt's basic formula (1961) suggests that the quadratic response is either a blue or a red shift of the probe with increasing potential, the equation given in the appendix suggests that a red shift is to be expected.

By combining the probe in a fatty acid multilayer, evaporating aluminum electrodes across it, and applying alternating potentials, Bücher *et al.* (1969) find that the light transmitted by the multilayer is varying not only at the same frequency but also at twice the frequency. By measuring the linear wavelength dependence at the same frequency and the quadratic dependence at twice the frequency, they obtained a blue shift and a red shift, respectively, at both room and low temperatures. The relationships between the potential and the magnitude of the shift are not determined in detail, but the spectra are clearly delineated at the two temperatures. As yet, no theoretical or experimental relationships have been developed between the fluorescence emission and the red shift of the absorption band.

D. Binding Determinations

Washed chromatophores were suspended in 10 ml of 5 mM Tris-SO_4, *p*H 7.4, supplemented with a given concentration of merocyanine. After centrifugation for 1 hr, 50,000*g*, the merocyanine concentration in the supernatant was determined spectrophotometrically at 600 nm. The difference between the absorbance before and after centrifugation was taken as bound merocyanine. In some experiments, the merocyanine-loaded chromatophores were resuspended in fresh buffer without merocyanine and recentrifuged. Subsequent determination of merocyanine in the supernatant showed that usually less than 5% of the chromatophore-bound merocyanine was released from the chromatophores.

E. Energy Transfer between Chlorophyll, Carotenoid, and Merocyanine Probes

Two types of experiment were undertaken to determine whether energy transfer conditions were appropriate between the probe and the two principal pigments of the chromatophores. The first was designed to determine whether illumination of the probe-labeled chromatophores would increase the quantum yield for cytochrome *b* reduction measured at 430 nm; here, the excitation wavelength was changed to 530 nm in order to ensure that probe excitation rather than chlorophyll excitation was involved. Also, the light levels were reduced to the point where the response of cytochrome *b* required 20 sec to reach the steady state, with a half-time of 7 sec. With a sufficient probe concentration to cause a large membrane occupancy, cytochrome *b* exhibited the same time to reach the steady state, and a measured half-time of 6 sec.

The second experiment was intended to determine whether or not excitation of chlorophyll at 425 nm would cause excitation of probe fluorescence at 600 nm. No detectable excitation of the probe fluorescence occurred, although a slight quenching of excitation of the chlorophyll fluorescence was observed. In addition, there was no excitation of the probe fluorescence in the region where carotenoids absorb.

III. EXPERIMENTAL RESULTS

A. Responses to Steady State Illumination

1. Absorbance Changes

Figure 2 illustrates the technique of wavelength scanning of the probe-labeled chromatophore suspension. Traces 1 and 2 correspond to the absorption spectrum of the chromatophore suspension prior to the addition of the probe. Trace 1 is flat because the absorption of the chromatophore suspension had previously been read into the memory of the computer controlling the spectrophotometer, and this trace shows only the error of the spectrophotometer in suppressing the chlorophyll and carotenoid spectra of the chromatophore suspension. Under these conditions, illumination of the suspension with 680-nm light causes only small effects because the wavelength scan does not include the region in which the light-induced carotenoid changes are seen, and the chlorophyll absorption changes are not significant when recorded at this sensitivity.

Following trace 2, the light is turned off and 5.2 μM of the probe MC-V is added, and after a few minutes traces 3–5 are recorded, with 44 sec between scans. These traces are not superimposed, because the occupancy of the membrane by the probe, which causes a decrease in absorption in the region of 600 nm, has not been completed. However, illumination of the

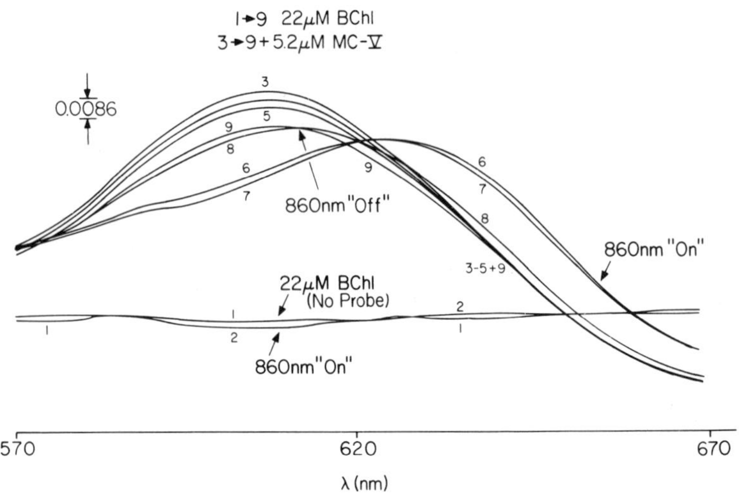

Fig. 2. Experimental chart of dual wavelength scanning spectrophotometer records. In traces 1–9, *R. rubrum* chromatophores containing 22 μM BChl were suspended in 5 mM Tris-sulfate at *p*H 7.2. In traces 3–9, 8.8 μM MC-V was also present. The time interval between scans is 44 sec. The labels "On" and "Off" refer to the actinic illumination. The wavelength scale is linear between the markings ($1061\text{-}3^{V}$).

probe-labeled chromatophore membrane at this point causes a decrease of absorbance at 600 nm and an increase of absorbance at 640 nm. It is seen that the apparent shift of the long-wave edge of the absorption band amounts to approximately 10 nm. On cessation of illumination, the apparent red shift disappears (traces 8, 9) and the absorbance at 600 nm returns to a value appropriate to the increased occupancy of the membrane by the probe that has occurred in the time between trace 5 and trace 9. At the same time, fluorescence decreases occur which have an excitation maximum near 600 nm and an emission maximum near 640 nm (see below).[3]

2. Dark Titrations

By employing a dual wavelength spectrophotometer and fluorometer as described in Section IIA, it is possible simultaneously to measure both the

[3] *Note Added in Proof:* It is important to note that the intensity of the absorption band and its shape remain largely unaffected. Thus, neither altered binding nor "stacking," i.e. polymerization (Ross, W. N., Salzberg, B. M., Cohen, L. B., Davila, H. V., Waggoner, A. S., and Wang, C. H., 1975, A large change in axon absorption during the action potential, *Biophys. Soc. Abs.* **15**: 133*a*) seem to explain the results, which is consistent with an electrochromic shift due to localized or transmembrane potentials.

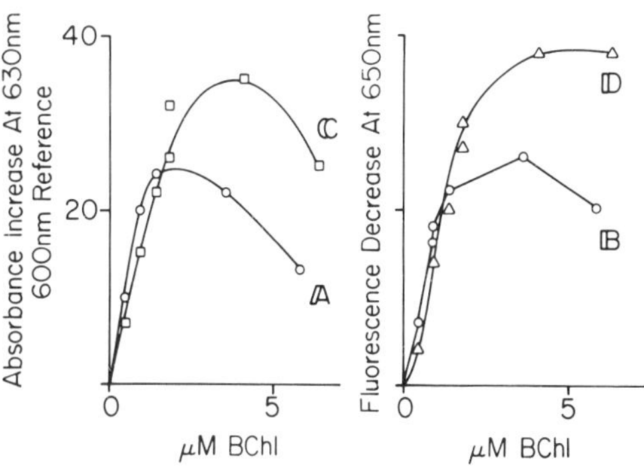

Fig. 3. Titration of the MC-V probe with chromatophores. Traces A and B, 2.7 μM MC-V; traces C and D, 8.8 μM MC-V. The concentration of BChl in μmol/liter is given along the abscissa, and the corresponding absorbance and fluorescence changes are given along the ordinate. Traces A and C represent absorbance changes measured at 630 nm with respect to 620 nm; traces B and D refer to fluorescence decreases measured at 650 nm with excitation at 577 nm. The reaction medium was 5 mM Tris-sulfate, *p*H 7.2 (1058-CV).

absorbance and the fluorescence changes as two given amounts of probe are titrated with *R. rubrum* chromatophores, as shown in Fig. 3. Plotting the absorbance and fluorescence increments (absorbance increase at 630 nm vs. 620 nm due to the red shift of the probe, and fluorescence decrease at 650 nm with respect to that of the solution of the probe in water), we obtain a linear relationship for both quantities with respect to chromatophore concentration. This slope represents the line of maximal occupancy of the chromatophore membrane by the probe. The traces then break sharply and pass through a maximum. The slope of the line of maximum occupancy has the dimensions of a molecular extinction coefficient, and has a value of 0.8 cm^{-1} mM^{-1}, indicative of the ability of the membrane, as represented by its chlorophyll content, to bind the probe. Converted to a lipid basis, by the ratio of the chlorophyll to phospholipid content of the chromatophore, the value is 0.2 cm^{-1} mM^{-1}. The value for maximal occupancy in terms of chlorophyll to merocyanine is obtained from the intersection of the maxima of the curve with the slope of maximum occupancy. These can be estimated satisfactorily where the plateau is relatively broad, as in the case of three of the curves, and the average value for the two probe concentrations is between two and three merocyanines per chlorophyll.

One of several explanations for the existence of maxima in the curves is that the less effective sites are occupied as the ratio of chlorophyll to probe decreases. Another explanation is that the concentration of the probe in the membrane reaches a level where quenching occurs. The latter view is supported by centrifugation studies, which indicate a very efficient binding of the probe to the chromatophores. The binding increases linearly with increasing probe concentration and reaches probe to chlorophyll ratios of 2 or more (Table I).

Table I

Merocyanine Binding to *R. rubrum* Chromatophores as Studied by Centrifugation[a]

BChl, μM	Merocyanine bound, μM				
	Added:[b] 5	10	20	30	40
3	2.1	3.4	—	—	—
10	3.7	7.6	15.7	24.6	—
20	3.6	8.5	17.9	26.9	35.9

[a] For experimental details see Section IIA.
[b] Merocyanine added, μM.

3. Light Titrations

Figure 4 illustrates a light titration of a chromatophore suspension in which the 860-nm light absorbed at each illumination intensity is measured and the spectrum corresponding to the absorption difference is recorded. In this case, trace A is the result of the suppression of both chromatophore and probe absorbance, corresponding closely to trace 9 of Fig. 2, but under experimental conditions where the bacteriochlorophyll concentration is 44 μM and the probe concentration is 2.6 μM. Absorption at 622 nm

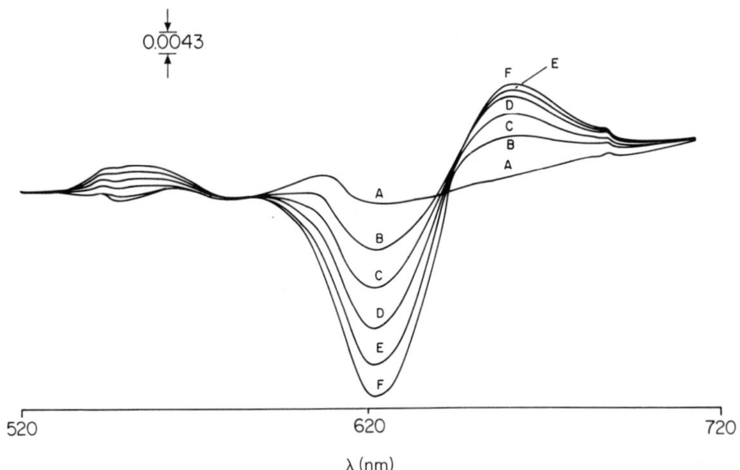

Fig. 4. Response of an *R. rubrum* chromatophore suspension to illumination at 860 nm. The suspension contains 44 μM BChl; 4.4 μM MC-V in 5 mM Tris-sulfate, *p*H 7.92. The scans are taken at an increasing light intensity (A–F); the time interval between scans is 44 sec. The experiment is similar to that of Fig. 2, except that the baseline for this experiment corresponds to trace 5 of Fig. 2 (V1062-7).

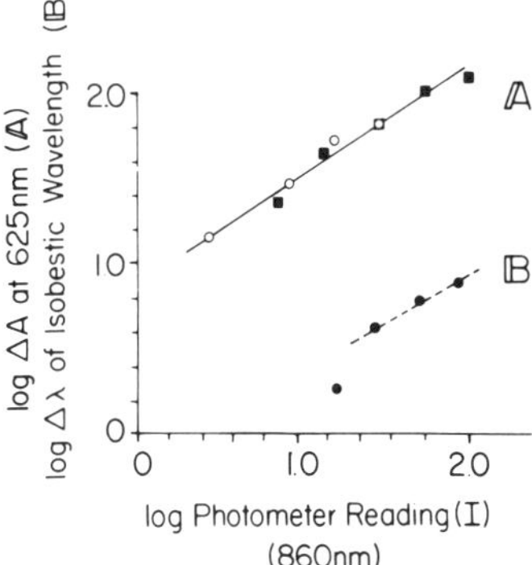

Fig. 5. A plot of the absorbance changes of Fig. 4 as a function of the intensity of the actinic light (abscissa). Trace A corresponds to the absorbance changes of Fig. 4, while trace B corresponds to the shift of the wavelength of the isosbestic point (1062[V]).

decreases and absorption at 640 nm increases regularly with increasing light intensity, although it appears that saturation occurs somewhat earlier at 640 nm than at 620 nm. It is also observed that the isosbestic point shifts to a longer wavelength with each progressive increase of light, suggesting a true shift of the existing absorption band rather than the appearance of a new absorption band.

Since the actinic light transmitted by the sample is measured by a silicon diode photometer, it is appropriate to plot light intensity vs. absorbance change and shift of isosbestic wavelength, as indicated in Fig. 5 for the experiment of Fig. 4 and for another, similar experiment which covered a somewhat higher intensity range. The data points for the absorbance decrease at 625 nm (trace A) fall upon the line of slope 0.5 over the range of the log-log plot. The points for the wavelength of the isosbestic point (trace B) are obviously less precise but are consistent with the idea of a square-root dependence between the intensity of the exciting light and the magnitude of the absorbance change and the spectral shift.

4. *Fluorescence Changes: Excitation and Emission Spectra*

Figure 6A shows the excitation spectra of the membrane-bound probe, trace *a* without magnesium and trace *b* in the presence of 6.7 mM magnesium sulfate, both with an excitation maximum at 625 nm. The ratio of the excitation efficiency at 625 nm to that at 577 nm is approximately 2:1.

Figure 6B shows, in trace *a*, that the fluorescence emission of the chromatophores alone (17 μM BChl), excited at 570 nm, decreases sharply above 600 nm. In trace *b*, the fluorescence of the chromatophores supplemented with 11 μM merocyanine shows a prominent emission band peaking in the region between 620 and 640 nm. In the presence of 6.7 mM magnesium sulfate (trace *c*) the emission spectrum is greatly intensified and has a broad apparent peak between 620 and 650 nm. Since these spectra are not corrected

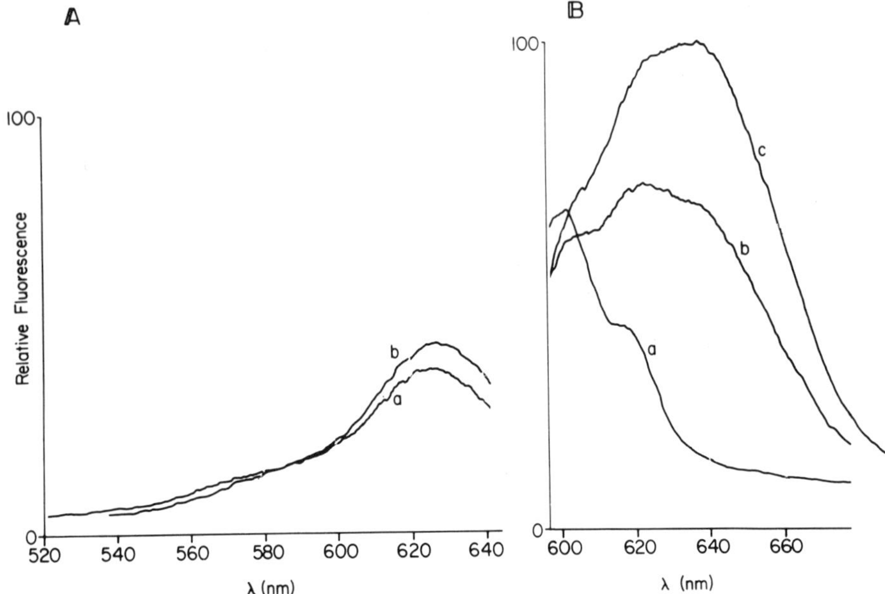

Fig. 6. Excitation and emission spectra of the merocyanine probe bound to chromatophores of *R. rubrum*, 17 μM BChl, supplemented with 18.6 μM merocyanine in (A) and curves *b* and *c* of (B). (A) Excitation spectra (a) without magnesium and (b) in the presence of 7.7 μM magnesium sulfate. (B) Emission spectra (a), without merocyanine, (b) in the presence of 11 μM merocyanine; (c) in the presence of 11 μM merocyanine and 6.7 μM magnesium sulfate (V-1132).

for the optical properties of the spectrofluorometer, which decline above 630 nm, optimal sensitivity for emission spectra is obtained with a red filter and the cutoff of the S-10 characteristics of the EMI 99592B photomultiplier or that of a Corning infrared absorbing filter (CS-1-37). Other experiments indicate that the probe bound to the chromatophore membrane has a quantum yield roughly three times that of the probe in water under these conditions, and thus the free probe does not contribute significantly to these spectra. The emission maximum for the probe in alcohol is at 635 nm, and the intensity is sixfold greater than that in the chromatophore membranes.

5. *Effect of Ion Fluxes on the Probe Response*

The experiment of Fig. 7 was carried out in order to determine the absorption and fluorescence changes of the MC-V probe in *R. rubrum*

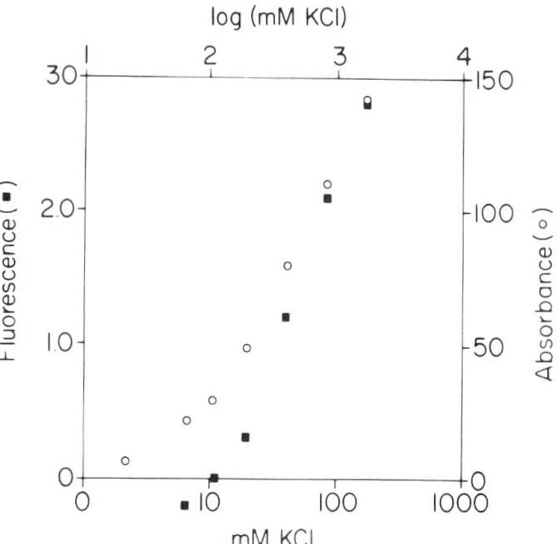

Fig. 7. Effect of valinomycin-induced potassium ion flux upon the fluorescence and absorbance of MC-V in *R. rubrum* chromatophores. The abscissa (logarithmic coordinates) represents the concentration of KCl added, while the ordinates represent relative amplitudes of fluorescence and absorbance change. The fluorescence is excited at 577 nm and measured above 640 nm. The absorbance is measured at 620 nm with a reference to 630 nm. 15 μM BChl, 5 mM Tris-sulfate, *p*H 7.5, 5 μM valinomycin, 16.4 μM MC-V, temperature 23° (1128V).

chromatophores in the presence of a fixed concentration of valinomycin and variable amounts of potassium. Empirically, it has been found that the fluorescence of the probe is larger in the presence of magnesium, and for this reason 2.5 mM Mg^{2+} is included. The data are plotted in semilogarithmic fashion since the logarithm of the added K^+ concentration is related to the Nernst potential, providing the valinomycin has afforded a pathway for diffusion of K^+ across the membrane. The usefulness of the scale furthermore depends upon the fact that the K^+ inside the chromatophores has not accumulated significantly during the experiment.

The experimental data show observations of absorption changes at 620–630 nm (circles) and fluorescence changes excited at 577 nm and measured at 640 nm (squares). The values of the abscissas are plotted on linear but arbitrary scales which are nearly matched at the highest concentrations of KCl employed. The two traces soon diverge, the fluorescence following an approximately linear profile and the absorption deviating from linearity at lower concentrations of KCl. According to the formulas above and in the appendix, the absorption would be expected to be linearly or quadratically related to the diffusion potential. No further attempt has been made to analyze these data for proof of linear or quadratic responses.

B. Kinetic Responses

Since the previous data indicate binding of the probe to chromatophores and light-induced red shifts of the probe, the kinetics of the response to

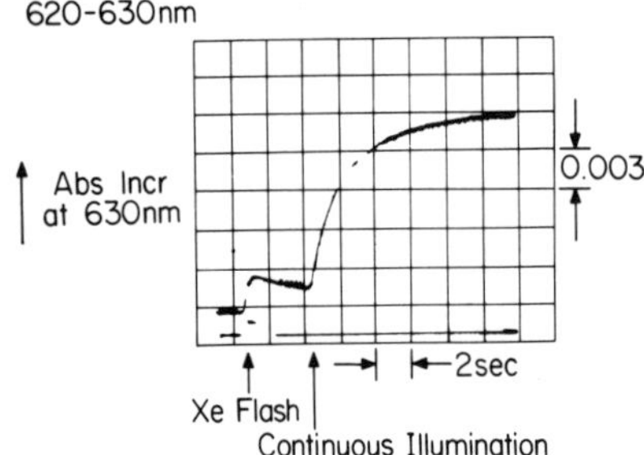

Fig. 8. Comparison of flash and steady state responses of the absorbance bands of MC-V. Time proceeds from left to right, and the arrows mark the moments of flash and continuous illumination at 860 nm, both at saturation levels. 15.2 μM BChl, 16.4 μM MC-V (V-1131).

pulsed illumination may well identify the stage in energy conversion that is indicated by the merocyanine probe.

1. Comparison of Flash and Steady State Illumination

The fraction of the steady state absorbance change of the probe obtained in a single flash is shown in Fig. 8. Both the flash and steady state intensities were saturating. On the slow time scale employed here, the response to the flash and steady state illuminations show a ratio of 1:5, or 20%. In other experiments where both fluorescence and absorbance were recorded, both were approximately 15% of the steady state response. In this respect, the single turnover response of the probe is a small fraction of the steady state response, as is also the case for the carotenoid; in fact, this probe gives an even smaller response relative to the steady state level.

2. Rise Time of Absorbance and Fluorescence

In Figure 9, expansion of the time scale shows the rise time of the red shift of the merocyanine in the *R. rubrum* chromatophore membranes. Here the time scale is 50 msec per division, and the flash artifact obscures

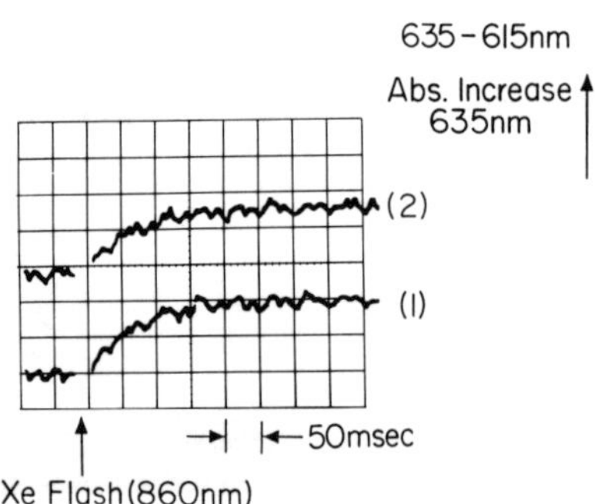

Fig. 9. Time-resolved absorbance changes of the merocyanine probe response in *R. rubrum* chromatophores. 45 μM BChl, 4.6 μM MC-V. Saturating xenon flash illumination in two successive flashes, first one, then two. The displacement of the traces upward indicates summation of the response (1059-6[V]).

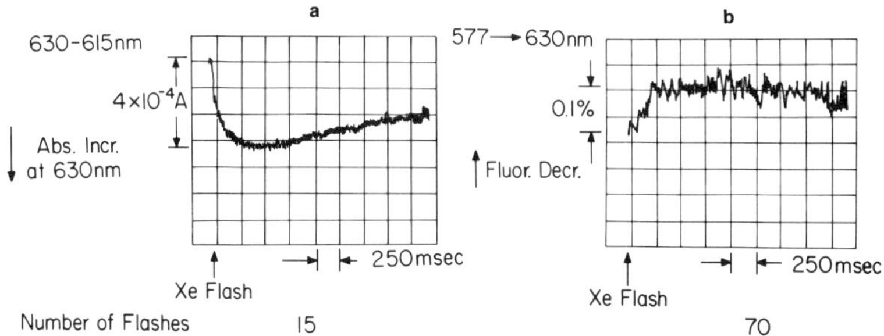

Fig. 10. Computer averaging of the flash response of fluorescence and absorbance changes of the merocyanine probe in *R. rubrum* chromatophores. 5.5 μM BChl, 40 μM MC-V, 6.7 mM Mg^{2+}. (a) Average of 15 flashes, one every minute, at 615–635 nm; (b) average of 70 flashes, one every minute, of fluorescence excitation at 577 nm, emission at 630 nm. Calibrations obtained from single traces (V-1126).

the first 10 msec of the trace. Thereafter, the half-time for the red shift of the probe is 50 msec in both traces, with two successive flashes following one upon the other within 3 sec, so that the amplitude of the second summates with that of the first. The amplitude of the second response is slightly smaller than that of the first.

Under the same conditions, computer averaging permits a comparison of the rise time of the red shift with the fluorescence change of the merocyanine probe in *R. rubrum* chromatophore membranes. The top trace of Fig. 10 illustrates, by a downward deflection, the absorbance change as a result of 15 computer averages on a time scale of 200 msec per division. Shown as an upward deflection is the pulsed-light-induced fluorescence decrease as an average of 70 flashes. It is seen that the rise times of the fluorescence and absorbance are comparable, so that on a kinetic basis they are responding to the same phenomenon or to simultaneous different phenomena.

3. *Effect of Ionophores on the Flash Response*

Figure 11 illustrates the sensitivity of the pulsed light response in terms of both absorbance and fluorescence changes to the addition of valinomycin, nigericin, and K^+. In Fig. 11A, the top trace represents a fluorescence decrease following a light pulse, and the bottom trace represents an absorbance increase. The rise time of the spectrophotometric trace portrays faithfully the rise time of the probe absorbance increase. However, since in the absence of computer averaging of the fluorometric data the response time of the amplifier was necessarily increased, the rise time is limited by

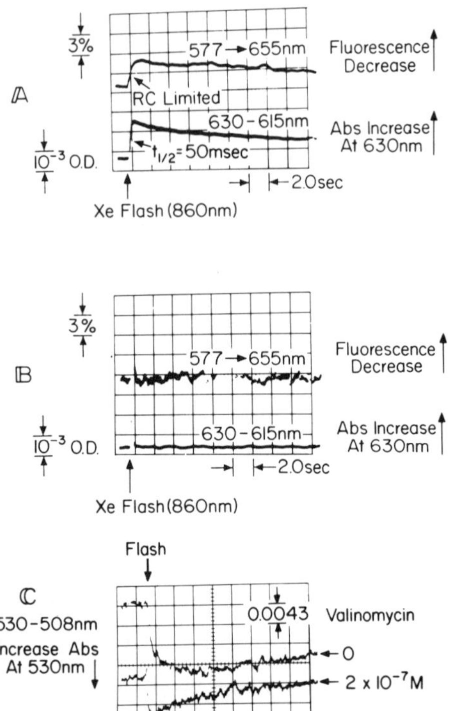

Fig. 11. Effect of ionophores on the flash response. Conditions generally similar to those of Fig. 10, showing pulsed changes of absorbance and fluorescence in response to a xenon flash, but without computer averaging. 18 μM BChl, 8.6 μM MC-V. In trace (B), where supplements of 6 mM K_2SO_4, 0.8 μg/ml of nigericin and 0.15 μg/ml of valinomycin have been made, there is no probe response. Part (C) illustrates the flash response of the carotenoid signal of *R. rubrum*. The upper trace is a control experiment. In the lower trace 2×10^{-7} M valinomycin is added and the experiment is repeated. The two traces are stored on an oscilloscope and photographed together. Absorbance and time scales are included in the photograph. *R. rubrum* chromatophores, 34 μM BChl, 5 mM Tris-sulfate, *p*H 7.4, 5 mM K_2SO_4 (MB-75).

the time constant *RC* as indicated on the figure. Nevertheless, both traces rise to a plateau within a fraction of a second. Thereafter, they decay, as has been observed in previous studies of the response of the carotenoid of *R. rubrum* (Baltscheffsky, 1974).

Figure 11B shows, however, that supplements of K^+, nigericin, and valinomycin completely eliminate both the absorbance and fluorescence responses of the probe. The decrease in amplitude is not due to an increase of the rate of decay to a value exceeding the rate of rise of the probe response; instead, the amplitude of the rise of the trace decreases and no "spike" remains in the "uncoupled" system, as in the case of the carotenoid response. Therefore the merocyanine response is completely dependent on energy coupling. The characteristic carotenoid response, an accelerated decay following the rapid rise, is shown in Fig. 11C.

Since the bulk of the work on carotenoid responses in photosynthetic bacteria has been done with chromatophores from *Rps. spheroides*, it is appropriate to illustrate, in Fig. 11C, the response of the carotenoid band shift of *R. rubrum* chromatophores to pulsed illumination in the presence and absence of valinomycin, K^+ being present. The absorbance increases are measured at 530 nm with 508 nm as a reference wavelength. The top trace shows the response to flash illumination without valinomycin, and the general characteristics are seen to be similar to those of the carotenoid band shift of *Rps. spheroides*. However, the slow phase of the absorbance increase at 530 nm is a distinct second segment following the initial fast rise. The half-time for this second phase is not precisely determinable since it merges with the kinetics of the decreased absorbance, which has a half-time under these conditions of approximately 1.5 sec, as mentioned below. Upon addition of 2×10^{-7} M valinomycin (6 mM K^+ already being present), the slow phase is no longer identifiable and the decay time has diminished to approximately 100 msec, in an effect similar to that shown in *Rps. spheroides* in Fig. 12.

4. *Comparison of Carotenoid and Merocyanine Responses in* Rps. spheroides

Figures 11 and 12 justify our saying that the pulsed light responses of the carotenoids of *R. rubrum* and *Rps. spheroides* chromatophores are similar. In addition, in Fig. 11 the merocyanine and carotenoid responses are measured simultaneously, using the method described for previous experiments except that the wavelengths of the spectrophotometer are shifted to those appropriate to either the carotenoid or the merocyanine. The flash responses, when stored on the oscilloscope trace, are photographed to give the composites shown in traces A, B, and C. Trace A shows the vast difference in the speed of the response of the carotenoid and of the probe; the half-time for the probe response is about 50 msec, while that of the carotenoid is much less than the rise time of the instrument (0.1 sec) (Jackson and Crofts, 1969). Furthermore, the completion of the rise of the merocyanine continues as the decay of the carotenoid occurs. These relationships are true only for *Rps. spheroides*; in *R. rubrum*, the half-time of the carotenoid decay is 1.5 sec (see Fig. 11). The decay of the carotenoid is biphasic and the slow phase is not completed on the time scale of the recording shown here. Addition of 1.4 mM K_2SO_4 and 0.03 μg of valinomycin per ml diminishes the amplitude of the response of both probes to approximately one-half. The merocyanine probe has approximately the same half-time of roughly 50 msec in the presence of K^+ and valinomycin, while the decay rate of the carotenoid is more rapid and no slow phase is seen. Doubling of the valinomycin concentration results in a scarcely detectable merocyanine response; the amplitude is scarcely 20 % of the original. Nevertheless, the amplitude of the carotenoid change under these conditions appears to be very little diminished.

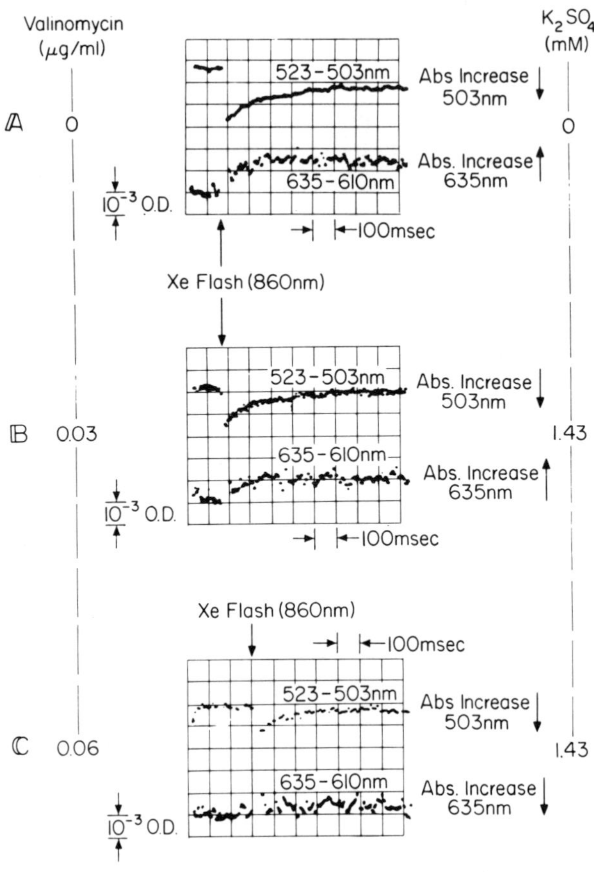

Fig. 12. Comparison of carotenoid and merocyanine responses in *Rps. spheroides* chromatophores. 20 μM BChl, 11 μM MC-V, 5 mM Tris-sulfate, *p*H 7.5 (1079-30, 31, 32^V).

IV. DISCUSSION

A. Location of the Probes

Studies on the location of the merocyanine probe have been pursued from a number of standpoints in several types of membranes. The environment in which the probe is located is hydrophobic, and the fluorescence is several-fold enhanced over that in water but not so intense as that observed in ethanol, where the peak is shifted to 635 nm. In the chromatophores, the emission peak is broader, ranging from 620 to 640 nm.

Occupancy of the membrane by merocyanine and by carotenoid is approximately equal to that by bacteriochlorophyll. Both are electrochromic, both respond to light and to ionophores, uncouplers, etc. Thus a comparison of those properties is of the greatest interest in order to compare their responses quantitatively.

Evidence for the localization of the probe in the lipid phase of a variety of membranes arises from the collisional quenching of the probe fluorescence by oxidized ubiquinone, as observed in submitochondrial particles (Azzi *et al.*, 1969; Chance, 1974). Thus contact between oxidized ubiquinone and the probe is possible in SMP membranes. In addition, titration of the membrane with bovine serum albumin does not result in the enhancement of probe fluorescence characteristic of the binding of the ANS probe to BSA, indicating that no merocyanine probe is extracted from the membrane by BSA.

B. Energy Transfer between Chlorophyll, Carotenoid, and Merocyanine Probes

Further studies to locate the probe in chromatophore membranes by energy transfer were negative, as described in the experimental portion; neither was energy transfer from the probe to chlorophyll or from chlorophyll to the probe feasible.

It is apparent that the overlap integrals between the chlorophyll, carotenoid, and merocyanine absorption bands are insufficient to give significant energy transfer within the available fluorometric sensitivities of the Perkin-Elmer MP-4F.

C. Response of Merocyanine to Valinomycin-Stimulated K^+ Diffusion

Both the fluorescence and the absorption of the MC-V probe are altered as the K^+ gradient is increased in the valinomycin-supplemented *R. rubrum* chromatophores. The relationship between the probe absorption, the probe fluorescence, and the diffusion potential is nonlinear for the absorption and roughly linear for the fluorescence. While the exact nature of the function for absorption is not precisely obtainable from these data, it is apparent that the probe absorption and fluorescence are sensitive indicators of the diffusion potential.

D. Responses to Flash Illumination of Chlorophyll, Carotenoid, and Merocyanine

The amplitude of the light response in *R. rubrum* chromatophores to a single flash is approximately 30% of the saturating value for the carotenoid, and about 10% for the merocyanine absorbance change. This suggests that the fraction of the total response obtained in a single flash with the merocyanine probe is smaller than that for the carotenoid.

Another difference between the two responses is that of rate. The carotenoid response occurs in less than 20 nsec (Jackson and Crofts, 1969), beyond the limit of measurement of most techniques, while the absorbance responses of the merocyanine probe have a half-time of approximately 50 msec. The half-time of the fluorescence change is similar, but is much less accurately measured.

Lastly, and perhaps most important, is the fact that ionophores completely eliminate the merocyanine response but leave a characteristic spike response for the carotenoid (Jackson and Crofts, 1969). Thus the merocyanine probe response is completely dependent on energy coupling.

E. Electrochromic Responses

The overall conclusion of this communication is that the kinetics of the absorbance changes of merocyanine and carotenoid differ in both coupled and uncoupled chromatophores.

The nature of the carotenoid shift in *Rps. spheroides* has been reported in detail by Reich and Schmidt (1972), Schmidt and Reich (1972a, b) and Schmidt *et al.* (1972), who show that the electrochromic shift in lutein, as measured *in vitro* by a system similar to that used by Bücher *et al.* (1969), has absorption peaks at the nodes of the light-induced spectrum for the carotenoids of *Rps. spheroides*. The sole carotenoid component of the *Rps. spheroides* mutant G-1-C is neurosporene, which has a light-induced spectrum similar to that observed in the wild type (Crofts, 1974). Thus the lack of correlation between the electrically induced shift in lutein and the light-induced shift in neurosporene lacks an obvious explanation. A further lack of correlation is based upon the observation that lutein shows only a quadratic electrochromic shift, while the neurosporene responds linearly to a KCl-induced diffusion potential in the presence of valinomycin. A number of explanations for this discrepancy may be considered, beginning with the possibility that the neurosporene of *Rps. spheroides* is showing an electrochromic response not to a transmembrane potential but rather to a local charge separation. This is supported by the hypothesis put forward by Schmidt and Reich (1972b) to explain why the quadratic electrochromic response of lutein *in vitro* is converted to a linear electrochromic response of neurosporene *in situ*; these authors postulate that a local fixed charge greatly in excess of the light-induced charge displaces the carotenoid response *in situ* to a small and approximately linear segment of the quadratic response. Whatever the validity of this explanation, it emphasizes that the localized charges are much greater than the transmembrane charges and, according to our view, only a small, light-induced variation of the "fixed" charges would be required to explain the response of the carotenoid in the membrane. However, in order to continue a balanced discussion of the two approaches, we will retain both hypotheses—of local and transmembrane charges—as not mutually exclusive, and attempt to indicate an appropriate assignment of the various events following a light flash to these two types of charge distribution.

The membrane model that is identified with this discussion is one in which the carotenoid, or at least a part of it, is intimately associated with the reaction center chlorophyll. The merocyanine probe is located elsewhere, presumably beyond energy transfer distance.

The initial event following a light flash corresponds most probably to a rapid local charge separation across the reaction center chlorophyll, causing a corresponding absorbance change of the reaction center carotenoid. Subsequently, this charge disappears at rates which depend upon the activation of electron transfer to other parts of the system and the movement of ions through the membrane. The decay phase of the carotenoid response, at least in *Rps. spheroides* chromatophores, corresponds to the rising phase of the absorption changes of the merocyanine.

The merocyanine absorption and fluorescence changes are completely dependent upon energy conservation, i.e., they are eliminated by valinomycin, K^+, and nigericin, whereas a rapid rise and decay of the carotenoid response is observed under these conditions (Jackson and Crofts, 1969, 1971; Crofts, 1974; Witt, 1972; Baltscheffsky, 1975).

Returning now to the two alternatives given above, it appears difficult to support the view that either probe is principally responsive to transmembrane potentials; the experimental data seem more consistent with the view that both respond to local charge redistributions. Upon illumination of the chromatophores a structure or charge change in the reaction center causes an instant absorbance change in the carotenoid, but not in the merocyanine, which is located too far away to sense the structure or charge change. As electron transfer occurs in the reaction center and charge redistributions due to electron transfer related to energy coupling ensue, the merocyanine responds with a characteristic red shift and a fluorescence decrease. The time for the decay of the carotenoid absorption shift, the rise of the merocyanine absorption shift, and the establishment of energy coupling are thus closely related. In this sense, the merocyanine serves as a probe of local charge redistributions characteristic of the energized state.

ACKNOWLEDGMENTS

The authors gratefully acknowledge the advice of Drs. Gary Strichartz, Sol Harrison, and J. K. Blasie in these probe studies. Dr. J. M. Vanderkooi, Mr. Guy Salama, and Ms. J. H. Owen have also assisted in preliminary experiments. Purification and characterization of the probe has been carried out by Dr. W. D. Phillips and his colleagues at DuPont Experimental Station and by Dr. Barry Cooperman and Mssrs. Paul Russ, Greg Stern, and Chris Denny. Studies on electrochromic calibration have been carried out by Dr. J. Smith in collaboration with Dr. Paul Mueller and Gene Ching. Information and advice on MC-I and MC-II from Drs. Waggoner, L. B. Cohen, and B. Salzberg is gratefully acknowledged. The research has been supported by USPHS GM-12202, NCHD O-6274, NINDS 10939, and the Swedish Natural Science Research Council.

REFERENCES[4]

Azzi, A., Chance, B., Radda, G. K., and Lee, C. P., 1969, A fluorescence probe for energy dependent structural changes in fragmented membranes, *Proc. Natl. Acad. Sci.* **62**:612–619.

Baltscheffsky, M., 1967, Inorganic pyrophosphate and ATP as energy donors in chromatophores from *R. rubrum*, *Nature* **216**:241–243.

Baltscheffsky, M., 1974, Reversible energization in photosynthesis as measured with endogenous carotenoid, *in* "Dynamics of Energy Transducing Membranes." (L. Ernster, R. W. Estabrook, and E. C. Slater, eds.), pp. 365–376, BBA Library, Elsevier, Amsterdam.

Brooker, L. G. S., Keyes, G. H., Sprague, R. H., Van Dyke, R. H., Van Lare, E., Van Zandt, G., White, F. L., Cressman, H. W. J., and Dent, S. G., 1951, Color and constitution. X. Absorption of the merocyanines. *J. Am. Chem. Soc.* **73**:5332.

Bücher, H., Wiegand, J., Snavely, B. B., Beck, K. H., and Kuhn, H., 1969, Electric field induced changes in the optical absorption of a merocyanine dye, *Chem. Phys. Lett.* **3**:508–511.

Chance, B., 1973, Electrochromic responses of merocyanine probes in energy coupling responses of submitochondrial particles (smp), *Fed. Proc. Abs.* **32**:669.

Chance, B. (1974), Deep and shallow probes of natural and artificial membranes, *in* "4th Int. Biophys. Congress—Moscow" (L. Kayushin, ed.), pp. 911–923, USSR Academy of Sciences, Moscow.

[4] Manuscript submitted June, 1974.

Chance, B., and Graham, N., 1971, A rapid-scanning dual wavelength spectrophotometer, *Rev. Sci. Instr.* **42**:941–945.

Clayton, R. K., 1963, Absorption spectra of photosynthetic bacteria and their chlorophylls, *in* "Bacterial Photosynthesis" (H. Gest, A. San Pietro, and L. P. Vernon, eds.) pp. 495–500. The Antioch Press, Yellow Springs.

Cohen, L. B., Salzberg, B. M., Davila, H. V., Ross, W. N., Landowne, D., Waggoner, A. S., and Wang, C. H., 1974, Changes in axon fluorescence during activity: A search for useful probes, *J. Memb. Biol.* **19**:1–36.

Crofts, A. R., 1975, The electron transport system as a H^+ pump in photosynthetic bacteria, *in* "Perspectives in Membrane Biology Symposium—Oaxaca, Mexico" (C. Gitler, ed.).

Davila, H. V., Salzberg, B. M., and Cohen, L. B., 1973, A large change in axon fluorescence that provides a promising method for measuring membrane potential, *Nature, New Biol.* **241**:159–160.

Jackson, J. B., and Crofts, A. R., 1969, The high energy state in chromatophores from Rhodopseudomonas spheroides, *FEBS Lett.* **4**:185–189.

Jackson, J. B., and Crofts, A. R., 1971, The kinetics of light induced carotenoid changes in Rhodopseudomonas spheroides and their relation to electrical field generation across the chromatophore membrane, *Eur. J. Biochem.* **18**:120–130.

Laris, P. C., and Hoffman, J. F., 1973, Membrane potentials in human red blood cells determined using a fluorescent probe, *Fed. Proc. Abs.* **32**:326.

Platt, J. R., 1956, Wavelength formulae and configuration interaction in Brooker dyes and chain molecules, *J. Chem. Phys.* **25**:80–105.

Platt, J. R., 1961, Electrochromism, a possible change in color producible in dyes by an electric field, *J. Chem. Phys.* **34**:862–863.

Reich, R., and Schmidt, S., 1972, Über den Einfluß elektrischer Felder auf das Absorptionsspektrum von Farbstoffmolekülen in Lipidschnichten. I. Theorie, *Ber. Bunsenges. Physik. Chem.* **76**:589–598.

Schmidt, S., and Reich, R., 1972a, Über den Einfluß elektrischer Felder auf das Absorptionsspektrum von Farbstoffmolekülen in Lipidschnichten. II. Messungen an Rhodamin B, *Ber. Bunsenges. Physik. Chem.* **76**:599–602.

Schmidt, S., and Reich, R., 1972b, Über den Einfluß elektrischer Felder auf das Absorptionsspektrum von Farbstoffmolekülen in Lipidschnichten. III. Elektrochemie eines Carotinoids (lutein), *Ber. Bunsenges. Physik. Chem.* **76**:1202–1208.

Schmidt, S., Reich, R., and Witt, H. J., 1972, Electrochromic measurements in vitro as a test for the interpretation of field indicating absorption changes in photosynthesis, *in* "Proc. 2nd Int. Congr. on Photosynthesis Research" (G. Forti, M. Avron, and A. Melandri, eds.), p. 1087, Dr. W. Junk N. W. Publishers, The Hague.

Vainio, H., Baltscheffsky, M., Baltscheffsky, H., and Azzi, A., 1972, Energy-dependent changes in membranes of Rhodospirillum rubrum chromatophores as measured by 8-anilino-naphthalene-1-sulfonic acid, *Eur. J. Biochem.* **30**:301–306.

Witt, H. T., 1972, Energy transduction in the functional membrane of photosynthesis, *Bioenergetics* **3**:47–54.

Appendix

Electrochromic Properties of Membrane Probes

W. K. Cheng

Johnson Research Foundation, School of Medicine
University of Pennsylvania, Philadelphia

The purpose of this appendix is to continue the discussion of "electrochromism" as implied by the following two equations:[1]

$$E^2 = E_I{}^2 + (b_R - b_L + V_{RL})^2 \tag{A0}$$

and by

$$h\,\Delta\nu = -\Delta\boldsymbol{\mu}\cdot\mathbf{F} - \tfrac{1}{2}\,\Delta\alpha\,F^2 \tag{A1}$$

(Electrochromism refers specifically to the induced spectral shift in the absorption or emission spectrum of a given molecule due to the presence of an external electric field **F**; this phenomenon is also known as the Stark effect.) It should be noted that Eq. (A0) involves terms quadratically in energies, whereas Eq. (A1) involves energies only linearly; as we shall see below, this difference has an important implication for the spectral shift, which is quadratic in **F**.

(I) We shall begin with Eq. (A1). As is well known, as far as the interaction between the dye molecule and the *local* electric field **F** (as seen by the dye molecule) is concerned, the dye molecule can be simply treated as an electric dipole having a dipole moment **m** located at the center of the molecule. This electric dipole in the field **F** then is governed by an interaction Hamiltonian given by

$$H_{\text{int}} = -\mathbf{m}\cdot\mathbf{F}$$

The standard second-order perturbation theory of quantum mechanics then predicts that, treating H_{int} as a small perturbation, the new energy level E_i' of the ith quantum state of the dye molecule is given by the expression

$$E_i' = E_i - \langle\psi_i \mid \mathbf{m}\cdot\mathbf{F} \mid \psi_i\rangle - \sum_{j\neq i} \frac{|\langle\psi_j \mid \mathbf{m}\cdot\mathbf{F} \mid \psi_i\rangle|^2}{E_j - E_i} \tag{A2}$$

where E_i and E_j are the energy levels of the ith and jth quantum states, and ψ_i and ψ_j are the corresponding wave functions in the absence ($F = 0$) of

[1] Eq. (A0) is Eq. (1) in the text.

the field. In terms of classically more familiar quantities $\boldsymbol{\mu}$, the permanent electric dipole moment, and the electric polarizability α, the above equation can be recast into the form

$$E_i' = E_i - \boldsymbol{\mu}_i \cdot \mathbf{F} - \tfrac{1}{2}\alpha_i F^2 \tag{A3}$$

with

$$\boldsymbol{\mu}_i \equiv \langle \psi_i \mid \mathbf{m} \mid \psi_i \rangle \tag{A4}$$

and

$$\alpha_i \equiv 2 \sum_{j \neq i} \frac{|\langle \psi_j \mid m \cos \sigma \mid \psi_i \rangle|^2}{E_j - E_i} \tag{A5}$$

(σ is the angle between **m** and **F**). Note that $\alpha\mathbf{F}$ is the induced electric dipole moment. Applying now Eq. (A3) to the first excited state and the ground state of the dye molecule and then taking their difference, we obtain, via Bohr's condition, the spectral shift in frequency given by

$$h\,\Delta\nu = -\Delta\boldsymbol{\mu} \cdot \mathbf{F} - \tfrac{1}{2}\,\Delta\alpha\,F^2 \tag{A6}$$

$$(\Delta\boldsymbol{\mu} \equiv \boldsymbol{\mu}_e - \boldsymbol{\mu}_g\,;\ \Delta\alpha \equiv \alpha_e - \alpha_g)$$

which is just Eq. (A1). [More explicitly, $h\,\Delta\nu \equiv (E_e' - E_g') - (E_e - E_g)$, where the subscripts e and g refer to the excited and ground states, respectively.] It should be pointed out explicitly that Eq. (A6) is equally applicable to the frequency shift in fluorescence spectra. The ith and jth quantum states involved in the dipolar transitions in the case of fluorescence spectra differ from those in absorption spectra, hence $\boldsymbol{\mu}_i$ and α_i as well as $\Delta\boldsymbol{\mu}$ and $\Delta\alpha$ [Eqs. (A4) and (A5)] also differ from their counterparts in absorption spectra. The above remark also implies that a relationship between the frequency shift in absorption spectra and that in fluorescence spectra may be obtained only if detailed information about the electronic structure of the dye molecule in the absence and presence of the local electric field **F** is available.

It is clear from Eq. (A1) that the spectral shift linear in F is inherently associated with the permanent dipole moment $\boldsymbol{\mu}$. The vanishing of this linear effect may arise from the following situations.

(a) If the dye molecule possesses a center of symmetry in both the excited and ground states, then it follows from Eq. (A4) that $\boldsymbol{\mu}_e = \boldsymbol{\mu}_g = 0$ identically. Hence, $\Delta\boldsymbol{\mu} = 0$.

(b) If environmental conditions are such that $\boldsymbol{\mu}_e = \boldsymbol{\mu}_g \neq 0$, then again $\Delta\boldsymbol{\mu} = 0$.

(c) If $\boldsymbol{\mu}_e$ and $\boldsymbol{\mu}_g$ are randomly oriented, then the net dipole moment is zero, so $\Delta\boldsymbol{\mu} = 0$.

(d) If $\Delta\boldsymbol{\mu}$ is a vector perpendicular to **F**, then $\Delta\boldsymbol{\mu} \cdot \mathbf{F} = 0$.

Therefore, if none of the situations (a)–(d) is realized, a linear spectral shift is to be expected. Moreover, (i) depending on the relative orientation of $\Delta \boldsymbol{\mu}$ and **F**, the spectral shift can be a blue shift ($\Delta\nu > 0$) or a red shift ($\Delta\nu < 0$), and (ii) a reverse of the sense of **F** ($\mathbf{F} \to -\mathbf{F}$) will result in changing the shift from red to blue, and vice versa.

We next discuss the spectral shift quadratic in F. From Eq. (A6), it is easily seen that this effect is connected with the induced dipole moment and is independent of the sense of F. Similar to the linear effect, the quadratic effect can be a blue or red shift, depending on the algebraic sign of $\Delta\alpha$. However, in the case when the transition moment matrix elements $\langle \psi_j \mid \mathbf{m} \mid \psi_e \rangle$ and $\langle \psi_j \mid \mathbf{m} \mid \psi_g \rangle$ (j refers to higher excited states of the dye molecule) are quantities of the same order of magnitude, then it can be argued on the basis of the energy difference denominators appearing in Eq. (A5) that $\Delta\alpha > 0$, implying a red shift. For convenience, we summarize the above discussion in Table I.

We also note that since the permanent and induced dipole moments $\boldsymbol{\mu}$ and $\alpha\mathbf{F}$ are vectors essentially oriented along the chain axis of the dye molecule, therefore, to serve as a useful probe of *trans* membrane potential, the dye must be bound to the membrane with its chain axis not in parallel or near parallel to the membrane wall.

(II) We now discuss Eq. (A0), the Platt equation. It should be immediately pointed out that this equation was proposed by Platt (1961) from a generalization of the following relation [i.e., $V_{RL} = 0$ in Eq. (A0)]:

$$E^2 = E_I^2 + (b_R - b_L)^2 \qquad \text{(A7)}$$

This equation was formulated to account for the main empirical conclusions concerning the transition energies of Brooker's dyes (1951) in terms of the "isoenergetic" energy E_I and the difference in basicities of the right and left end groups of the dye molecule. The latter quantity ($b_{RL} \equiv b_R - b_L$) is a measure

Table I

Properties of the Equation

Dipole moment	Spectral shift		F dependence of $\Delta\nu$
Permanent ($\mu \neq 0$)	Blue	$\Delta\nu > 0$	Linear
	Red	$\Delta\nu < 0$	
Induced ($\alpha F \neq 0$)	Blue	$\Delta\nu > 0$ ($\Delta\alpha < 0$)	Quadratic
	Red	$\Delta\nu < 0$ ($\Delta\alpha > 0$)	

of the stabilization of one of the two extreme resonance structures relative to the other. (For example, referring to the chemical formula of MC-V-P given in Fig. 1, the two extreme resonance structures are, schematically, Ⓛ−C=C−C=C−C=Ⓡ and Ⓛ=C−C=C−C=C−Ⓡ.) However, it must also be pointed out that the actual quantum mechanical derivation of Eq. (1) has not been given in the literature.

We now proceed to obtain an explicit expression giving the spectral shift. Regarding the transition energy E as a function of the potential difference V_{RL} between the right and left end groups, we take the differentials of both sides of Eq. (A0), obtaining

$$h\,\Delta\nu = \Delta E = [(b_R - b_L + V_{RL})/E]\,\Delta V_{RL} \tag{A8}$$

Observe that V_{RL}, on the other hand, can be considered as the additional interaction energy between the two extreme resonance structures due to the presence of the field **F**, so ΔV_{RL} is simply V_{RL} itself. Thus we have

$$h\,\Delta\nu = \frac{b_R - b_L}{E} V_{RL} + \frac{1}{E} V_{RL}^2 \tag{A9}$$

which is to be considered as the counterpart of Eq. (A1). The following statements can be made on the basis of the above equation:

(i) The linear effect vanishes if (a) $b_R = b_L$; this is true for the symmetric dyes or for an arbitrary dye at its isoenergetic point; and (b) $V_{RL} = 0$; this may happen in a very rigid medium. Moreover, the direction of the spectral shift (blue vs. red) depends, in the present case, on the relative sign between b_{RL} and V_{RL}.

(ii) The quadratic effect is a blue shift. This prediction follows uniquely from the particular form of Eq. (A0), which involves only *squares* of the energies. It may be mentioned in this connection that Eq. (A7) was formulated specifically to account for what is now known as Brooker's deviation—namely the absorption is blue-shifted by an amount equal to Brooker's deviation, which involves quadratically the difference in energy of the two extreme resonance structures.

From a theoretical point of view, Platt's generalization, Eq. (A0), of Eq. (A7) to the case when an external electric field is present by no means represents a unique one. We may, for instance, regard the unperturbed energies E_i and E_j of Eq. (A2) as being related by Eq. (A7), and use the standard second-order perturbation theory to treat the perturbation due to the electric field **F**. We would then obtain the same expression as Eq. (A1) for the present case, taking Brooker's empirical results explicitly into account, but obtaining a different result from that of Platt. Note that the remark just made is particularly relevant in view of the recent merocyanine experiments of Bücher *et al.* (1969), where a red shift quadratic in the field strength F was experimentally observed.

Finally, it should be emphasized that, as already discussed, unless certain specific conditions are realized in a given experimental setup, the observed spectral shift is expected to be a sum of linear and quadratic effects.

(III) Finally, we remark that the orientation of the dye molecule bound to the membrane can be determined by nuclear magnetic resonance, X-ray, and other techniques. On this basis, then, appropriate dyes can be chosen which have, additionally, the property that the asymmetry of the left and right end groups is minimal in order that b_{RL} is minimal. Optimally, dyes with identical end groups ($b_{RL} = 0$) are better yet. Such dyes may prove to be very sensitive probes of membrane potential.

NOTE ADDED TO MANUSCRIPT

After the completion of this appendix, we learned that in a series of papers Schmidt *et al.* (1972a, b) had given a similar discussion of Eq. (A1) and applied the theory to the interpretation of electrochromism observed with (i) rhodamin B in lipid layers and (ii) lutein in a thin capacitor. In addition, they also gave an interesting discussion of the corresponding induced change in the extinction coefficient ($\Delta\epsilon$) and in the absorption (ΔA).

REFERENCES

Brooker, L. G. S., Keyes, G. H., Sprague, R. H., Van Dyke, R. H., Van Lare, E., Van Zandt, G., White, F. L., Cressman, H. W. J., and Dent, S. G., 1951, Color and constitution. X. Absorption of the merocyanines, *J. of Amer. Chem. Soc.* **73**:5332.

Bücher, H., Wiegand, J., Snavely, B. B., Beck, K. H., and Kuhn, H., 1969, Electric field induced changes in the optical absorption of a merocyanine dye, *Chem. Phys. Lett.* **3**:508–511.

Platt, J. R., 1961, Electrochromism, a possible change in color producible in dyes by an electric field, *J. Chem. Phys.* **34**:862–863.

Schmidt, S., and Reich, R., 1972a, Über den Einfluß elektrischer Felder auf das Absorptionsspektrum von Farbstoffmolekülen in Lipidschnichten. II. Messungen an Rhodamin B, *Ber. Bunsenges. Physik. Chem.* **76**:599–602.

Schmidt, S., and Reich, R., 1972b, Über den Einfluß elektrischer Felder auf das Absorptionsspektrum von Farbstoffmolekülen in Lipidschnichten. III. Elektrochemie eines Carotinoids (lutein), *Ber. Bunsenges. Physik. Chem.* **76**:1202–1208.

Chapter 4

Effects of Sulfhydryl Reagents on Basal and Vasopressin-Stimulated Na^+ Transport in the Toad Bladder[1,2]

Ayala Frenkel,[3] E. B. Margareta Ekblad,[4] and Isidore S. Edelman

Cardiovascular Research Institute
and
Departments of Medicine and Biochemistry and Biophysics
University of California School of Medicine
San Francisco, California

I. INTRODUCTION

The two-barrier, in series, model of Koefoed-Johnsen and Ussing (1958) has been widely accepted as representative of the basic mechanism of transepithelial active Na^+ transport. The first step in the process is presumed to be passive penetration of Na^+ across the apical plasma membrane driven by the electrochemical gradient, and the second step is active extrusion of Na^+ across the basal–lateral plasma membrane. The Na^+ pump in the basal–lateral membrane apparently derives its energy from the hydrolysis of ATP (Skou, 1965).

Our studies concern the role of SH groups in the apical plasma membrane in the regulation of Na^+ transport across toad bladder epithelium. The apical

[1] This paper is dedicated to Aharon Katzir-Katchalsky; we cherish the memory of his warm friendship and honor him for his distinguished contributions to our field.

[2] Financial support was provided by U. S. PHS National Institute of Arthritis, Metabolism and Digestive Diseases Grant No. AM-13659.

[3] During the tenure of a National Kidney Foundation Fellowship. Present address: Department of Organic Chemistry, Weizmann Institute of Science, Rehovot, Israel.

[4] During the tenure of a postdoctoral traineeship provided by U. S. PHS National Heart and Lung Institute Grant No. HL-05725.

boundary was chosen for study because of its accessibility to reagents, the assignment of a regulatory role in the response to hormones to this boundary, and the evidence of a saturable component in Na^+ permeation, implying the participation of functional groups in this process (Frazier *et al.*, 1962, Civan and Frazier, 1968).

Sulfhydryl reagents have been useful probes in the analysis of a variety of membrane functions. The location of a protein in the plasma membrane and the local site of the SH group within the membrane protein, including the ionic and lipid environment, determine the accessibility of the group to titration (Rothstein, 1970). With the aid of a variety of SH reagents, differential roles of SH groups have been described in active and passive ion transport, sugar transport, membrane ATPase activity, and surface antigenicity (Rega *et al.*, 1967; van Steveninck *et al.*,1965; Banerjee and Sen, 1969; Green, 1967).

The objectives of this study were (1) to define the conditions needed to limit the SH reactions to the apical plasma membrane of toad bladder epithelium, (2) to characterize the effects of titration of apical membrane SH groups on basal and hormone-stimulated active Na^+ transport, and (3) to assess the effect of vasopressin or cyclic AMP on the accessibility of SH groups in the apical plasma membrane to titration with SH reagents.

II. MATERIALS

Sodium *para*-chloromercuriphenyl-sulfonate (PCMPS) and adenosine 3′5′-cyclic monophosphoric acid (cyclic AMP) were obtained from Sigma Chemical Co., sodium *para*-chloromercuribenzoate (PCMB) from Mann Research Lab., 5,5′-dithiobis(2-nitrobenzoic acid) (DTNB) from Aldrich Chem. Co., *N*-ethylmaleimide (NEM) from CalBioChem., vasopressin (Pitressin) from Parke, Davis & Co., and theophylline (U.S.P.) from Nutritional Biochem. Corp. Amiloride was generously provided as a gift by Merck, Sharp & Dohme Research Lab. 2-Nitro-5-thiocyanato-benzoic acid (NTCB) was synthesized by the method of Degani and Patchornik (1971). All of the conventional chemical reagents were analytical grade.

III. METHODS

A. Short-Circuit Current Experiments

Female *Bufo marinus*, indigenous to Colombia, were obtained from Tarpon Zoo, Florida. The toads were stored at room temperature with continuous access to tap water, without food, for about one week prior to use. After the toads were double-pithed, the urinary bladders were removed, divided in half, and transferred to aerated frog-Ringers solution at room temperature. The composition of the frog-Ringers solution was NaCl = 111, KCl = 3.4, $KHCO_3$ = 2.4, $CaCl_2$ − 2.7, all in mM; osmolarity was 224 mosmol/liter; *p*H was adjusted to 7.5 or 8.2. The hemibladders were mounted as diaphragms in chambers, and active Na^+ transport was measured by the short-circuit current (SCC) method of Ussing and Zerahn (1951). Two chamber systems were used; single-glass chambers with a capacity of 20 ml and an orifice area of 2.54 cm^2, or double-barreled Lucite chambers with a 10 ml capacity and orifice areas, on each side, of 2.52 cm^2. The hemibladders were supported by nylon mesh from the serosal side. The hemibladders were usually preincubated in frog-Ringers solution for 2–3 hr before the addition of the reagents. Four groups of SCC experiments were performed, to determine (1) the response to the addition of various concentrations of the SH reagents to the luminal (apical side) medium, (2) the response to vasopressin, or to cyclic AMP + theophylline, added to the serosal medium (with or without glucose present) about 30 min after addition of the SH reagent, (3) the effect of simultaneous addition of cyclic AMP + theophylline to the serosal medium and the SH reagent to the mucosal medium, and (4) the effect of the addition of vasopressin to the serosal medium 5–10 min before the addition of the SH reagent to the luminal medium. The details of these protocols are given with the results.

B. Assay of Soluble SH Compounds

The degree of exclusion of the SH reagents from the interior of the epithelial cells (i.e., limitation of the reaction to the apical boundary) was assessed by assaying for soluble, unreacted SH groups in high-speed supernatants after homogenization in TCA. Hemibladders, freed of blood by prior perfusion with frog-Ringers solution, were mounted in glass chambers. Care was taken not to stretch the bladders. Hydrostatic pressure (~1 cm of H_2O) was applied from the serosal side to create standard sacs with reproducible areas of exposure of the luminal surface to the SH reagents. In each experiment, four pairs of hemibladders were used: (1) The SCC response to the reagents, after preincubation for 2 hr in frog-Ringers solution, was measured in one pair. The reagents were added to the luminal medium of one hemibladder and the diluent to the control. (2) The remaining three pairs of hemibladders were mounted and preincubated under the same conditions but without monitoring of the SCC. The SH reagents were added to the luminal medium of one of each pair and the diluent to that of the other. At various times after the addition of the reagents, the reaction was terminated by removing all the media and washing twice with equal volumes of fresh frog-Ringers solution. The hemibladders were immediately transferred to ice-cold Lucite blocks, luminal side up, and the epithelial cells were scraped off with the edge of a glass slide. The scrapings of the four experimental and four control hemibladders were pooled separately and put into TCA. These pools were homogenized in a glass–Teflon homogenizer (Talboy Instr. Corp.) with 20 strokes at maximum speed. The homogenates were transferred into Eppendorf vials (1.5 ml) inserted in an ice-bath and allowed to stand for 30 min. The homogenates were then centrifuged at 19,000 $\times$ *g* for 4 min in a model 3200 Eppendorf centrifuge. The protein-free supernatants were assayed for soluble SH content by the DTNB method (Ellman and Lysko, 1967). The reaction of DTNB with SH groups (equimolar) results in the formation of 5-mercapto-2-nitrobenzoic acid (TNB), which has an absorption maximum at 412 nm. One milliliter of 5×10^{-4} M DTNB in 0.5 M Na_2HPO_4–5 mM EDTA buffer (*p*H 8) was mixed with a 0.5-ml aliquot of the supernatant by agitation in a Vortex Mixer (Scientific Products, Inc.) and allowed to stand for 15 min, and the absorbance was read at 412 nm in the Zeiss spectrophotometer. The protein content of the TCA precipitate was determined by the method of Lowry *et al.* (1951).

C. Effects of Amiloride on Titratable SH Groups

Paired hemibladders, freed of blood by *in vivo* cardiac perfusion with frog-Ringers solutions, were mounted in Lucite double chambers in 4.5 ml (each side) of frog-Ringers solution and preincubated for 2 hr. Amiloride (final concentration 10^{-4} M) was added to the luminal media of one of each quarter-bladder and the diluent to the control. When the SCC reached zero in the treated quarter-bladder, DTNB (final concentration 5×10^{-4} M) was added to the luminal media of all chambers. Periodically, 0.5-ml aliquots were removed from the luminal media, absorbance at 412 nm was determined, and the samples were returned to the chambers.

D. Effects of Vasopressin on Titratable SH Groups

The effect of vasopressin on the number of SH ligands in the apical plasma membrane and on the leakage of SH-containing compounds from the epithelial intracellular compartment was also assayed by titration with DTNB. Paired hemibladders were mounted in Lucite double chambers and preincubated in 4.5 ml of frog-Ringers solution on each side (luminal and serosal) for 2 hr. DTNB was added to the luminal side, to a final concentration of 5×10^{-4} M, of all of the quarter-bladders and at the same time vasopressin was added to the serosal sides, to a final concentration of 30 mU/ml, to one of each quarter-bladder and the diluent to the control. Aliquots of the luminal media of 0.5 ml were withdrawn periodically; the absorbance was read at 412 nm and the sample was returned to the same chambers. In some cases, the bathing media were increased to 7 ml and 0.5-ml aliquots, which were not returned to the chambers, were withdrawn simultaneously from both the luminal and serosal media to maintain equal hydrostatic pressures across the wall of the toad bladder. To assess the effects of vasopressin on the leakage of SH-containing compounds from the interior of the epithelial cells, pairs of hemibladders were mounted in Lucite double chambers in frog-Ringers solution (7 ml on each side) and preincubated for 2 hr and then vasopressin (final concentration 20 mU/ml) or the diluent was added to the serosal media. Periodically, 0.5 ml of luminal and serosal media were removed simultaneously and assayed for SH content with DTNB as described above.

IV. RESULTS

A. Penetration of SH Reagents into Toad Bladder Epithelium

To determine whether a particular SH reagent titrated only SH groups in the plasma membrane, we assayed TCA extracts of toad bladder epithelium for free SH content, on the assumption that intracellular penetration by the reagent would result in titration of free SH groups in small molecules, e.g., glutathione. The results in Table I indicate that the TCA-soluble SH content varied from 4.6 to 35.5×10^{-6} M/mg of extracted protein in the control population. On a paired basis, however, reproducibility was excellent. Thus, PCMB (10^{-4} M), PCMPS (5×10^{-4} M), NTCB (10^{-3} M), and DTNB (5×10^{-4} M) in the luminal media had no significant effect on TCA-soluble SH content of the epithelium. In addition, PCMPS (5×10^{-4} M) apparently failed to penetrate from the serosal side either. In contrast, PCMB (10^{-6} M) on the serosal side and NEM (10^{-4} M) on the luminal side reduced TCA-soluble SH content by 45% and 90%, respectively. To determine whether stimulation of Na^+ transport by vasopressin would affect the penetration of SH reagents (excluded under basal conditions) into the epithelial cell, the assays were carried out either after 5 min of exposure to vasopressin, followed by 60 min of exposure to NTCB (luminal), or after 60 min of exposure to

Table I

Effect of Reagents on TCA-Soluble SH Groups

Reagent	Number of pairs	Concentration, M	Time of contact, min	SH content[a]		
				Expt.	Control	Δ
PCMB	8	10^{-4} (luminal)	30	8.7	8.1	+0.6
	4		90	8.1	10.3	−2.2
	4		200	9.6	10.0	−0.4
PCMB[b]	4	10^{-4} (luminal)	90	5.3	4.6	+0.7
PCMB	4	10^{-6} (serosal)	90	4.6	8.2	−3.6
	4		120	5.5	10.0	−4.5
PCMPS	8	5×10^{-4} (luminal)	25	7.2	6.6	+0.6
	8		60	5.8	5.8	0
PCMPS	4	10^{-4} (serosal)	90	6.4	5.5	+0.9
	4	5×10^{-4}	90	7.0	8.1	−1.1
NTCB[c]	4	10^{-3} (luminal)	60	35.5	35.5	0
	4		60	18.2	20.2	−2.0
DTNB	16	5×10^{-4} (luminal)	90	5.3	6.1	−0.8
NEM	4	10^{-6} (luminal)	30	18.8	18.0	+0.8
	4	10	30	17.2	20.0	−2.8
	4	5×10^{-5}	30	18.6	30.0	−11.4
	4	10^{-4}	30	3.0	28.0	−25.0
Vasopressin	16	20 mU/ml (serosal)	5–10	7.3	7.1	+0.2

[a] SH content is given in units of $\times 10^{-6}$ M/mg of protein extracted with TCA.
[b] Treated with vasopressin (20 mU/ml) for 60 min before addition of PCMB
[c] Treated with vasopressin (20 mU/ml) for 5 min before addition of NTCB.

vasopressin followed by 90 min to PCMB (luminal). In neither case did pretreatment with vasopressin facilitate the entry of the SH reagent into the cell. These results imply that PCMB, PCMPS, NTCB, and DTNB do not cross the apical membrane barrier either in the presence or absence of vasopressin and can be used as reagents that act only on the apical membrane when added to the luminal media.

Farah *et al.* (1969) reported a significant correlation between vasopressin-induced increases in SCC and decreases in protein-bound disulfide groups, although he found no significant change in protein-bound SH groups in toad bladder epithelium. We extended these observations by assessing the possibility of an effect of vasopressin on TCA-soluble SH content of the epithelium and found no such effect (Table I).

B. Effects of SH Reagents on SCC

The time-dependent effects of titrating apical membrane SH groups on SCC were determined with PCMPS, NTCB, and DTNB. Four issues were considered: (1) effects on basal SCC, (2) effects of pretreatment with SH reagents on the SCC response to vasopressin or cyclic AMP, (3) effects of pretreatment with vasopressin or cyclic AMP on the SCC response to the SH reagents, and (4) modifying effects of exogenous glucose on the response to SH reagents followed by addition of hormone.

1. *PCMPS*

Addition of PCMPS to the luminal medium evokes an increase in SCC within 5 min; the magnitude and duration of this effect are concentration dependent. Two time points, 15 and 60 min, were chosen to characterize these effects (Fig. 1). At the 15-min time point, the increase in SCC was

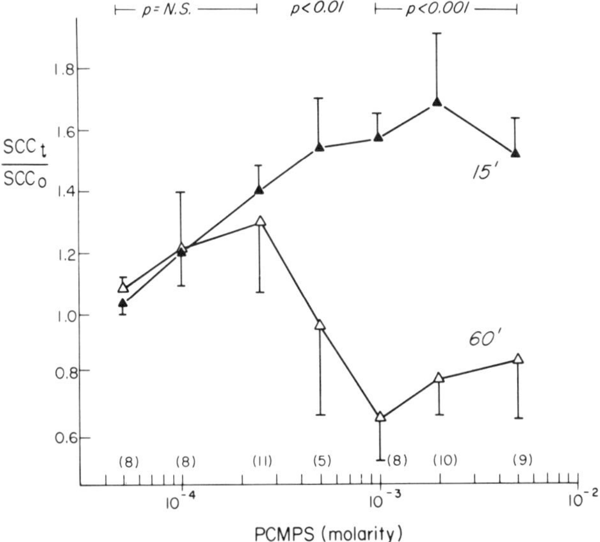

Fig. 1. Concentration dependence of the SCC response to PCMPS. The ratio of the SCC at time t (SCC_t) divided by that at time zero (SCC_0) is plotted on the ordinate as a function of the molar concentration of PCMPS in the luminal medium on the abscissa. Time zero is taken as the time of addition of PCMPS to the media. The SCC_t/SCC_0 ratio at t = 15 min is denoted by -▲- and the ratio at t = 60 min by -△-. The vertical lines are equal to 1 SEM and the number of hemibladders for each time point is in parentheses. The p value of the differences in the mean ratios at 15 and 60 min was calculated by the Student "t" test.

proportional to the log concentration of PCMPS, up to 6×10^{-4} M. No further increase in SCC was obtained at concentrations greater than 6×10^{-4} M; the maximum increase in SCC was ~60%. At 5×10^{-5}, 10^{-4}, and 2.5×10^{-4} M PCMPS, the increase in SCC was sustained for 60 min. At higher concentrations there was a secondary inhibition in SCC which fell to about 70% of the basal value at the end of 60 min. These results suggest that at low concentrations (i.e., $<2.5 \times 10^{-4}$ M), titration of membrane SH groups with PCMPS facilitates Na^+ transport across the apical boundary. The delayed fall in SCC at higher concentrations may be a consequence of nonspecific toxic effects or of distruption of the intracellular junctions, thereby permitting back-diffusion of Na^+ from the intercellular spaces.

The possibility that PCMPS and vasopressin (or cyclic AMP) might enhance the SCC by effects on the same apical pathways was explored by using these reagents in combination. PCMPS was added to a final concentration of 2.5×10^{-4} M, since this concentration yields a response that is more than half-maximal at 15 min and is reasonably well sustained at 60 min (Fig. 1). In substrate-free media, pretreatment with PCMPS for 25 min followed by addition of vasopressin (20 mU/ml) inhibited the vasopressin-dependent increment, at the peak, by about 40% (Fig. 2). An almost identical result was obtained with cyclic AMP and theophylline (an inhibitor of the phosphodiesterase that hydrolyzes cyclic AMP) in substrate-free conditions (Fig. 3). In glucose-enriched media, however, the combined response to PCMPS and vasopressin was simply the sum of the individual effects (Fig. 4). These findings imply that the inhibitory effects of PCMPS on the response to vasopressin depends in some way on the energy supply for Na^+ transport. Additivity of the effects of PCMPS and vasopressin was studied in a further set of experiments in which PCMPS was added 8 min after the vasopressin

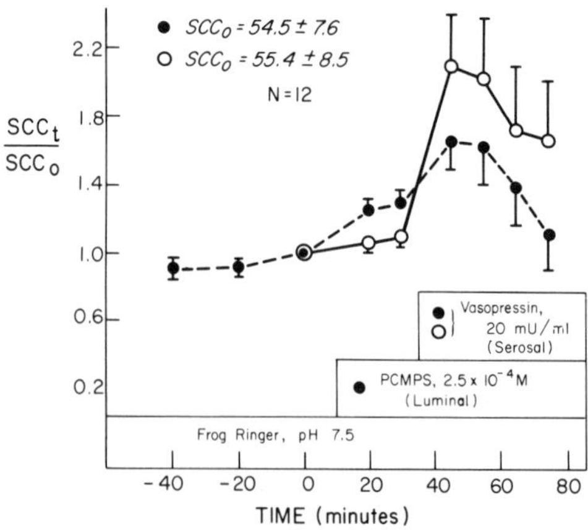

Fig. 2. The SCC response to PCMPS and vasopressin. The SCC_t/SCC_0 ratio is on the ordinate and time on the abscissa. PCMPS was added to the luminal media, at $t = 10$ min, of one of each pair of hemibladders (-●-) and vasopressin at $t = 35$ min to the serosal media of both hemibladders. Mean ± 1 SEM at $t = 0$ (SCC_0) and number of pairs are given. The length of each line is equal to 1 SEM.

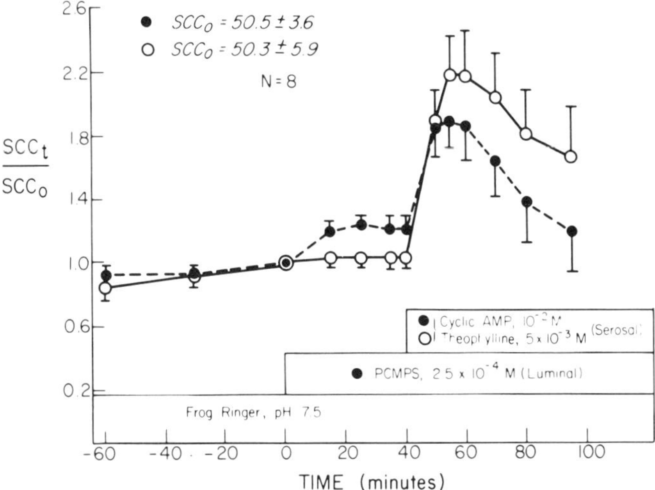

Fig. 3. The SCC response to PCMPS and cyclic AMP plus theophylline. The conventions used in this figure are given in the legends of Figs. 1 and 2.

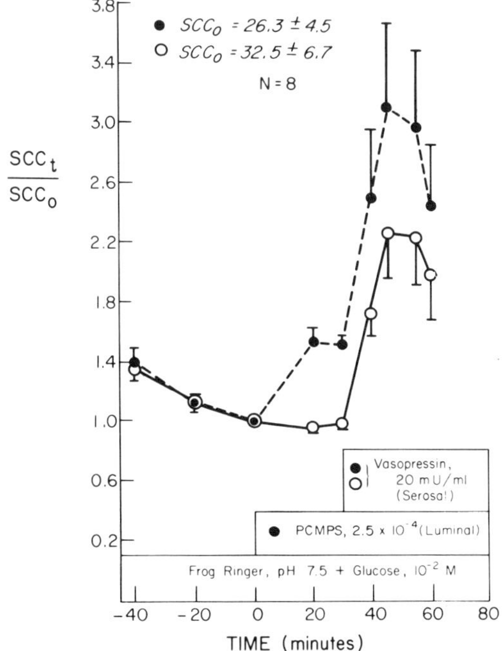

Fig. 4. The SCC response to PCMPS and vasopressin in glucose-enriched media. The conventions used in this figure are given in the legends of Figs. 1 and 2.

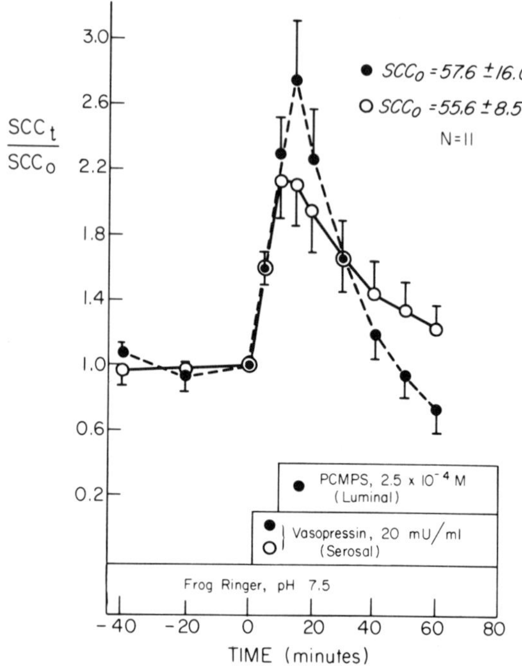

Fig. 5. The SCC response to vasopressin and PCMPS. The conventions used in this figure are given in the legends of Figs. 1 and 2.

(Fig. 5). It appears that the effect of PCMPS and vasopressin in concert can be ascribed to a summation of their separate effects (cf. Figs. 4 and 5). At a concentration of 20 mU/ml, vasopressin elicits a maximal effect on SCC (Petersen and Edelman, 1964). Since PCMPS was present at a concentration sufficient to elicit half-maximal or greater effects, these results imply that these reagents may act via parallel pathways or at different sites in the epithelium.

2. *NTCB*

At concentrations less than 10^{-3} M, NTCB [a new reagent first synthesized by Degani and Patchornik (1971)] had no effect on SCC. Fifteen to twenty minutes after the addition of NTCB, to a final concentration of 10^{-3} M, SCC increased transiently, 25% above the baseline value (Fig. 6). The maximum appeared at 15–20 min; at 1 hr and thereafter, the SCC of the treated hemibladder tended to be less than that of the paired control. The delayed fall in SCC, however, was greater with PCMPS then with NTCB.

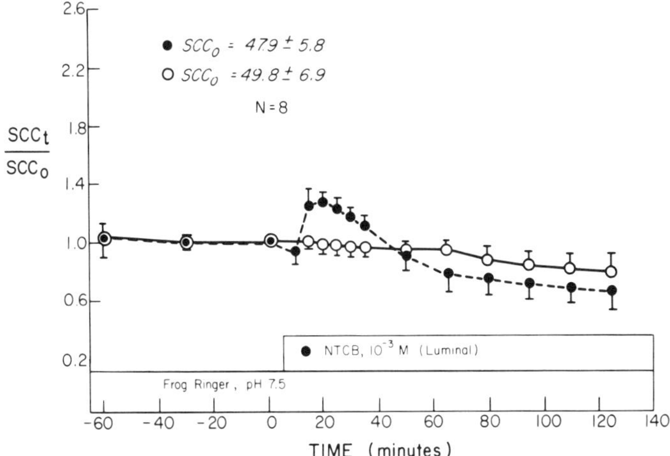

Fig. 6. The SCC response to NTCB. The conventions used in this figure are given in the legends of Figs. 1 and 2.

Concentrations greater than 10^{-3} M were not tested, since NTCB was not soluble in water at high concentrations. In substrate-free media, pretreatment with NTCB (10^{-3} M) for 45 min resulted in marked inhibition of the subsequent response to vasopressin (20 mU/ml), as shown in Fig. 7. Moreover, the inhibitory effect on the response to vasopressin was abolished by

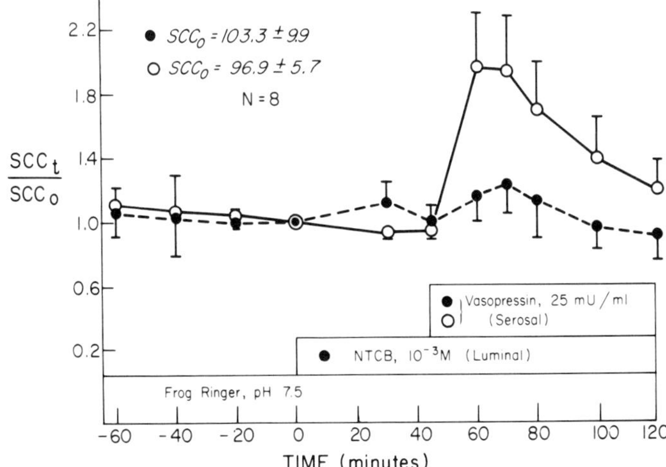

Fig. 7. The SCC response to NTCB and vasopressin. The conventions used in this figure are given in the legends of Figs. 1 and 2.

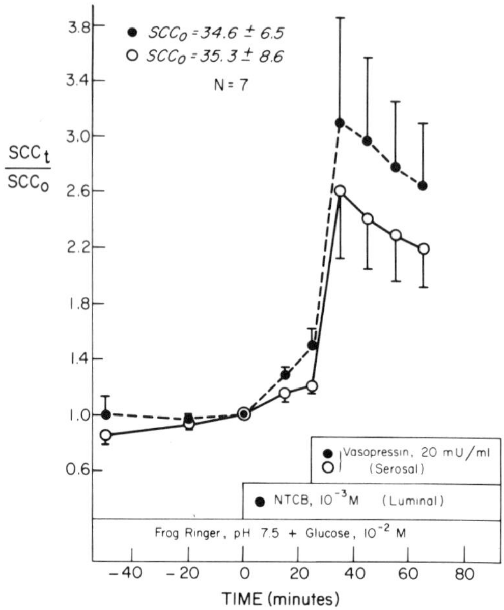

Fig. 8. The SCC response to NTCB and vasopressin in glucose-enriched media. The conventions used in this figure are given in the legends of Figs. 1 and 2.

adding glucose (10^{-2} M) to the media (Fig. 8). Thus the effects of NTCB and PCMPS were quite similar in that both evoked transient increases in SCC with about the same time course and impaired the response to vasopressin only in substrate-free media. Strikingly dissimilar effects, however, were seen when vasopressin was added 10 min before NTCB. As shown in Fig. 9, the combined effects of NTCB and vasopressin were almost twofold greater than that of vasopressin alone. Synergistic effects of about the same magnitude were seen on simultaneous addition of NTCB and cyclic AMP (+ theophylline) (Fig. 10). The latter results imply that enhancement of the response to vasopressin is at the level of the action of cyclic AMP rather than at the level of interaction of vasopressin and the adenyl cyclase system.

3. *DTNB*

In contrast to the results obtained with PCMPS and NTCB, DTNB (final concentration 5×10^{-4} M in the luminal media) had no effect on the baseline SCC, nor on the response to vasopressin (20 mU/ml) (Fig. 11). Even at 10^{-3} M, DTNB had no significant effect on basal Na^+ transport for 165 min. Despite the lack of any pharmacological action, DTNB titrates

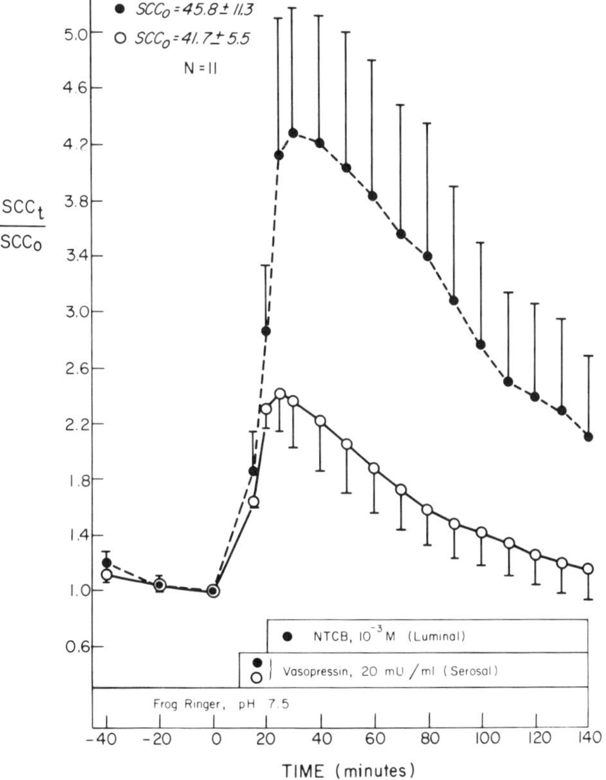

Fig. 9. The SCC response to vasopressin and NTCB. The conventions used in this figure are given in the legends of Figs. 1 and 2.

a measurable fraction of the apical membrane SH groups as judged by the rate of generation of TNB. The tolerance to titration with DTNB implied that, at 5×10^{-4} M, this reagent might be suitable for assessing the role of apical membrane SH groups in processes involving changes in apical Na^+ conductance.

C. Effects of Amiloride on Titratable SH Groups

Addition of amiloride to the luminal media produces a readily reversible, complete inhibition of active Na^+ transport across the toad bladder, apparently by blockade of the apical transit pathways (Bentley, 1968; Ehrlich and Crabbé, 1968). Thus, if SH-containing proteins are involved in the mechanism

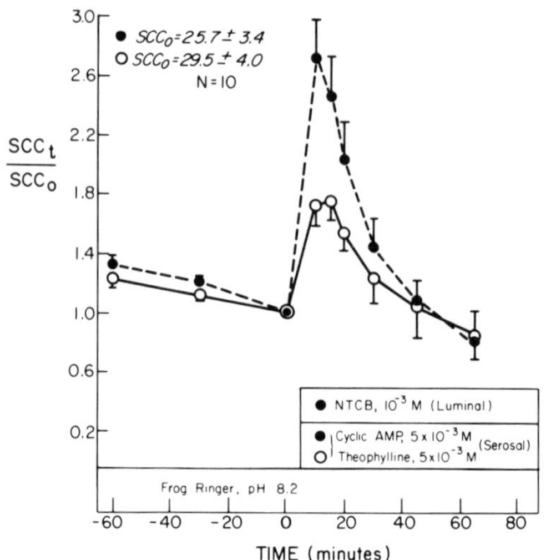

Fig. 10. The SCC response to cyclic AMP plus theophylline and NTCB. The conventions used in this figure are given in the legends of Figs. 1 and 2.

of apical Na^+ conductance, and if this subpopulation of SH groups constitutes a significant fraction of the total number of titratable groups, blockade with amiloride might reduce the number of SH groups available for reaction with DTNB. This prediction was tested as a "model" of the converse effect anticipated with vasopressin.

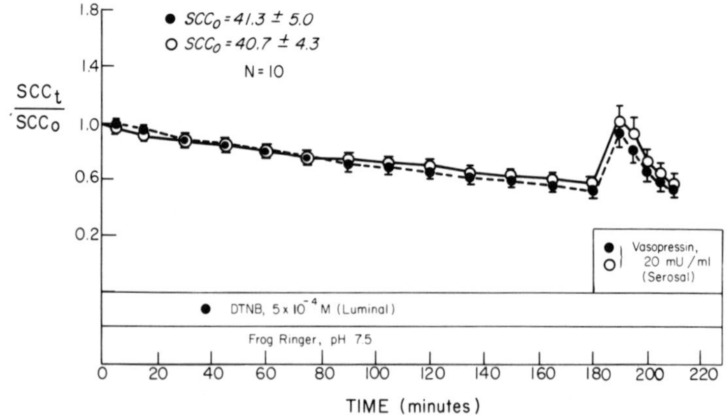

Fig. 11. The SCC response to DTNB and vasopressin. The conventions used in this figure are given in the legends of Figs. 1 and 2.

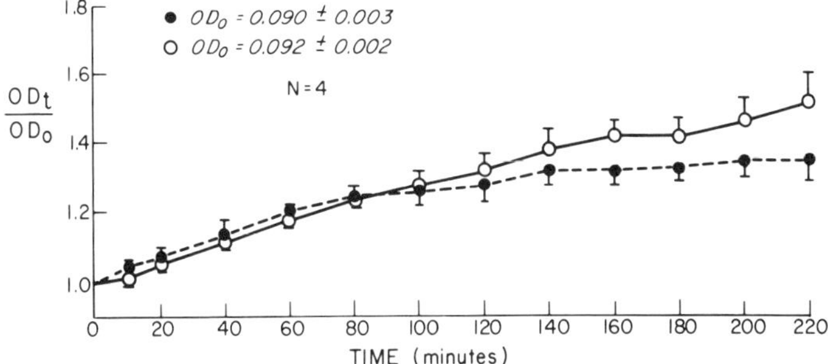

Fig. 12. The effect of amiloride on the conversion of DTNB to TNB. The optical density at 412 nm measured at time t (OD_t) divided by that at time zero (OD_0) is on the ordinate and time in minutes is on the abscissa. The experimental hemibladders (-●-) were exposed to amiloride (10^{-4} M) in the luminal media for 10–15 min, at which time the SCC was essentially nil. DTNB (final concentration 5×10^{-4} M) was then added to the luminal media of all chambers (time zero) and the change in OD was read at 10-min intervals as described in the text. The points are the mean values and the vertical lines are 1 SEM. The differences in the OD ratios are statistically significant at 220 min ($p < 0.05$).

Amiloride (final concentration 10^{-4} M) was added to the luminal media 10–15 min before titration with DTNB (5×10^{-4} M), which was initiated at the time that the SCC was essentially nil. As shown in Fig. 12, the rate of titration of available SH groups, presumably in the apical membrane, was indistinguishable in the treated and control hemibladders for the first 90 min. From 90 to 220 min, DTNB continued to titrate SH groups in the control hemibladders but not in hemibladders treated with amiloride. These results suggest that the SH groups involved in the Na^+ transit pathway may not be immediately available for titration with DTNB.

D. Effects of Vasopressin on Titratable SH Groups

Vasopressin apparently stimulates transepithelial Na^+ transport, at least in part, by facilitating the transit of Na^+ across the apical membrane boundary (Frazier *et al.*, 1962; Civan and Frazier, 1968). This hypothesis implies structural changes which may be detectable by a change in the number of SH groups available for titration. In these experiments, DTNB (5×10^{-4} M) was added to the luminal media at the same time as the addition of vasopressin (final concentration 30 mU/ml) or the diluent to the serosal media. The results shown in Fig. 13 indicate that vasopressin increased significantly the number of SH groups available for titration within 5 min, which was the first titration point, and that this difference persisted for at least 1 hr. In parallel experiments, we tested the possibility that vasopressin

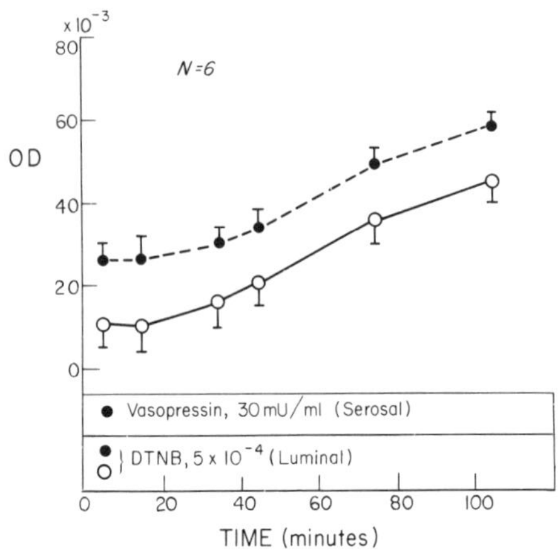

Fig. 13. The effect of vasopressin on the conversion of DTNB to TNB. The OD at 412 nm is on the ordinate and time is on the abscissa. Vasopressin was added to the serosal media of the experimental hemibladders (-●-) and, simultaneously, DTNB was added to the luminal media of all. The points are the mean values and the vertical lines are 1 SEM.

simply rendered the apical membrane leaky to diffusible SH-containing compounds (e.g., glutathione, methionine). Thus aliquots of luminal media were removed periodically from the chambers after addition of vasopressin or diluent to the serosal media and titrated with DTNB (5×10^{-4} M) in cuvettes. No SH groups were detectable in the aliquots of the luminal fluids up to 25 min after addition of either the diluent or vasopressin. These results suggest that the immediate increase in titratable SH groups shown in Fig. 13 was not a result of leakage of soluble SH-containing compounds into the luminal media.

V. DISCUSSION

The transit of Na^+ across the apical plasma membrane of the toad bladder has the characteristics of passive diffusion through selective channels (or via carriers) of limited capacity (Frazier *et al.*, 1962; Leaf and Sharp, 1971). Although the molecular mechanism is not known, evidence has been presented that vasopressin, as well as its putative intermediate, cyclic AMP, increases active Na^+ transport across the toad bladder by facilitating the passage of Na^+ through the apical diffusion barrier (Frazier *et al.*, 1962; Civan and Frazier, 1968). The present study was designed to explore the roles of SH-containing compounds, presumably proteins, of the apical plasma membrane in both basal (hormone-independent) and vasopressin-dependent transepithelial Na^+ transport.

The effect of PCMPS and NTCB on the SCC were quite similar and characterized by an increase (~25–60 %), attained by 10–15 min after addition of the reagents to the luminal media, followed by a progressive decline to below the control levels by 1 hr—especially at the highest concentrations of the reagents. In separate studies, Spooner and Edelman (submitted for publication) found that the increase in SCC induced by PCMB was the same as the increase in net Na^+ flux measured with $[^{22}Na^+]$ and $[^{24}Na^+]$. The exclusion of PCMB, PCMPS, and NTCB from the interior of the epithelium, inferred from measurements of TCA-soluble SH content, implies that these reagents enhance net transepithelial Na^+ transport by increasing apical Na^+ conductance. The stimulatory effect may be a consequence of direct modifications in the proteins that determine apical Na^+ conductance or of indirect modifications secondary to other events, such as release of Ca^{2+} from key surface sites or derangement of an orderly array of plasma membrane proteins. Evaluation of these various possibilities remains for future studies. The delayed fall in SCC evoked by PCMPS and NTCB may result from leakage of key intermediates from the intracellular compartment across an excessively permeable apical membrane or from disruption of the tight intercellular junctions that seal off the passive diffusion pathway between cells (Loewenstein *et al.*, 1965). In contrast to the results obtained with PCMPS and NTCB, DTNB had no effect on either basal SCC or the response to vasopressin, although significant numbers of SH groups were titrated as judged by the appearance of TNB in the luminal media. It would appear, therefore, that titration of selected groups or classes of SH groups influence apical Na^+ conductance while others are indifferent in this respect. This inference of heterogeneity in the functional role played by membrane SH groups in ion penetration in the toad bladder is in accord with similar findings on the roles of various SH groups in the regulation of ion fluxes across red cell membranes (Rega *et al.*, 1967; Rothstein, 1970). Our results also implied heterogeneity in the participation of SH groups in the response to vasopressin or cyclic AMP.

In substrate-free media, pretreatment with PCMPS or NTCB inhibited the rise in SCC elicited by vasopressin. Addition of glucose to the media protected against these inhibitory effects. The protective effect of glucose indicates that the energy state of the epithelium determines the expression of the inhibitory effect. Since glucose has no such effect on the influence of PCMPS or NTCB on the basal SCC, it is possible that separate SH-dependent pathways are involved in basal and hormone-stimulated Na^+ transport. This inference is also supported by the finding that the increases in SCC produced by PCMPS or NTCB and vasopressin or cyclic AMP were additive since the latter were present at saturating concentrations, i.e., a common pathway would limit the response to that elicited by vasopressin or cyclic AMP alone. The basal and hormone-stimulated pathways need not, of course, be dissimilar in kind. Additional evidence implying the existence of parallel apical pathways in basal and hormone-regulated Na^+ transport was provided by the marked synergistic effects of adding vasopressin shortly before NTCB or of adding cyclic AMP and NTCB simultaneously. Potentiation of the response to vasopressin or its intermediate could be a result of making SH groups accessible to titration by NTCB that were inaccessible in the hormone-depleted state. Obviously, one possible mechanism for doing so would be a cyclic-AMP-dependent conformational change in Na^+ "gate" or "carrier" proteins, which is augmented by SH titration. Regardless of the mechanism of augmentation of the response to hormone by NTCB, the inference that the hormone opens inaccessible or slowly accessible SH groups to titration is supported by the finding of a vasopressin-dependent increase in the number of SH groups available for titration with DTNB, within 5 min. The possibility that this increase is a trivial expression of leakage of SH groups from the cell interior was contradicted by the failure to detect soluble SH-containing compounds in the media. This argument, however, can not be given as "proof" since the assays for leakage of soluble SH compounds were not done simultaneously with assays of membrane SH groups—indeed, by their nature, these measurements cannot be carried out simultaneously.

Our results support the view that augmentation of apical Na^+ conductance is an important part of the response to vasopressin (or its intermediate cyclic AMP). Moreover, the specificity of the synergistic effect of NTCB (i.e., in contrast to PCMPS) raises the hope of differential labeling, with NTCB, of the sites involved in hormone-dependent Na^+ transit, as a guide to isolation and purification of these proteins. Labeling with PCMPS could serve to distinguish between hormone-specific and nonspecific membrane sites.

VI. SUMMARY

The role of reactive SH groups (presumably in proteins) of the apical plasma membrane in transepithelial Na^+ transport was studied in the isolated urinary bladder of the toad. On the basis of assays for TCA-soluble SH compounds (e.g., glutathione, methionine), PCMB, PCMPS, NTCB, and DTNB did not penetrate the intracellular compartment from the luminal media either in control or vasopressin-treated bladders. In contrast, PCMB from the serosal side and NEM from the luminal side titrated significant fractions of the TCA-soluble SH compounds. We conclude, therefore, that PCMB, PCMPS, NTCB, and DTNB are suitable reagents for studies on the physiological properties of apical plasma membrane SH groups. Titration of apical membrane SH groups with PCMPS, NTCB, and DTNB revealed heterogeneity in functional responses: PCMPS and NTCB elicited transient, 25–60% increases in SCC. In substrate-free media, pretreatment with these reagents inhibited the increase in SCC produced by vasopressin or cyclic AMP (+ theophylline). In glucose-enriched media, the responses to combinations of vasopressin and PCMPS or NTCB were additive, implying activation via parallel pathways. Simultaneous addition of vasopressin or cyclic AMP (+ theophylline) and NTCB resulted in marked synergism, presumably as a result of unmasking of SH groups by the hormone (or the intermediate). These results suggest that basal Na^+ transport is regulated in part by SH compounds in the apical membrane that are distinct, although not necessarily different in kind, from those involved in the response to vasopressin.

Acknowledgments

We are grateful for the able technical assistance provided by Sara Izen, Viven Ho, Florette Yen, and John Inciardi.

REFERENCES

Banerjee, S. P., and Sen, A. K., 1969, On the mechanism of inhibition of ($Na^+ + K^+$)-ATPase by N-ethylmaleimide and ethacrynic acid, *Fed. Proc.* **28**:589.

Bentley, P. J., 1968, Amiloride: A potent inhibitor of sodium transport across the toad bladder, *J. Physiol.* **195**:317–330.

Civan, M. M., and Frazier, H. S., 1968, The site of the stimulatory action of vasopressin on sodium transport in toad bladder, *J. Gen. Physiol.* **51**:589–605.

Degani, Y., and Patchornik, A., 1971, Selective cyanylation of sulfhydryl groups. II. On the synthesis of 2-nitro-5-thiocyanato-benzoic acid, *J. Org. Chem.* **36**:2727–2728.

Ehrlich, E. N., and Crabbé, J., 1968, The mechanism of action of amipramizide, *Pflügers Arch.* **302**:79–96.

Ellman, G. L., and Lysko, H., 1967, Disulfide and sulfhydryl compounds in TCA extracts of human blood and plasma, *J. Lab. Clin. Med.* **70**:518–527.

Farah, A., Yamodis, N. D., and Pessah, N., 1969, The relation of changes in sodium transport to protein-bound disulfide and sulfhydryl groups in the toad bladder epithelium, *J. Pharm. Expt. Ther.* **170**:132–144.

Frazier, H. S., Dempsey, E. F., and Leaf, A., 1962, Movement of sodium across the mucosal surface of the isolated toad bladder and its modification by vasopressin, *J. Gen. Physiol.* **45**:529–543.

Green, F. A., 1967, Erythrocyte membrane sulfhydryl groups and RH antigen activity, *Immunochemistry* **4**:247–257.

Koefoed-Johnsen, V., and Ussing, H. H., 1958, The nature of the frog skin potential, *Acta Physiol. Scand.* **42**:298–308.

Leaf, A., and Sharp, G. W. G., 1971, A discussion on active transport of salts and water in living tissues, *Roy. Soc. Lond., Phil. Trans. B* **262**:323–332.

Loewenstein, W. R., Socolar, S. J., Higashino, S., Kanno, Y., and Davidson, N., 1965, Intercellular communication: Renal, urinary bladder, sensory, and salivary gland cells, *Science* **149**:295–296.

Lowry, O. H., Rosebrough, N. J., Farr, A. L., and Randall, R. J., 1951, Protein measurement with the Folin phenol reagent, *J. Biol. Chem.* **193**:265–275.

Petersen, M. J., and Edelman, I. S., 1964, Calcium inhibition of the action of vasopressin on the urinary bladder of the toad, *J. Clin. Invest.* **43**:583–594.

Rega, A. F., Rothstein, A., and Weed, R. I., 1967, Erythrocyte membrane sulfhydryl groups and the active transport of cations, *J. Cell. Physiol.* **70**:45–52.

Rothstein, A., 1970, Sulfhydryl groups in membrane structure and function, *in* "Current Topics in Membranes and Transport" (F. Bronner and A. Kleinzeller, eds.), Vol. 1, pp. 135–176, Academic Press, New York.

Skou, J. C., 1965, Enzymatic basis for active transport of Na^+ and K^+ across cell membrane, *Physiol. Rev.* **45**:596–617.

Spooner, P. M., and Edelman, I. S., Studies on the effect of aldosterone on electrical resistance of toad bladder, Submitted for publication.

Ussing, H. H., 1963–1964, Transport of electrolytes and water across epithelia, *in* "Harvey Lectures," Series 59, pp. 1–30.

Ussing, H. H., and Zerahn, K., 1951, Active transport of sodium as the source of electric current in the short-circuited frog skin, *Acta Physiol. Scand.* **23**:110–127.

Van Steveninck, J., Weed, R. I., and Rothstein, A., 1965, Localization of erythrocyte membrane sulfhydryl groups essential for glucose transport, *J. Gen. Physiol.* **48**: 617–632.

Chapter 5

Biogenesis of Chloroplast Membranes in *Chlamydomonas reinhardi*: Chloroplast-Controlled Transfer of Cytoplasmic Proteins to the Developing Chloroplast Membranes as Visualized by Quantitative Radioautography[1]

Israel Goldberg,[2] Ilan Friedberg,[3] and Itzhak Ohad

Department of Biological Chemistry
The Hebrew University
Jerusalem, Israel

I. INTRODUCTION

Transfer of proteins from the cytoplasm to organelles such as mitochondria or chloroplasts has been implied by many studies (Eytan and Ohad, 1970; Hoober, 1970a, b; Sebald *et al.*, 1973; Hoober *et al.*, 1969; Goodenough and Levine, 1970; Smillie and Scott, 1969; Surzycki *et al.*, 1970; Mason and Schatz, 1973; Woodward *et al.*, 1970; Rubin and Tzagoloff, 1973). Although it has been shown that transfer of proteins from microsomes to the outer mitochondrial membrane can occur *in vitro* (Kadenbach, 1967), such direct evidence has not been obtained so far for proteins of the internal membrane system of the chloroplast. One main difficulty in the study of interorganelle transfer of proteins and its control resides in the fact that isolated microsomes, mitochondria, or chloroplasts are not active enough in synthesis of proteins *in vitro* to permit such studies. However, it is possible

[1] This work is part of a series, the previous paper of which appeared in *J. Biol. Chem.* **249**:738–744 (1974).

[2] Present address: Department of Applied Microbiology, Hadassah Medical School, The Hebrew University, Jerusalem.

[3] Present address: Department of Electron Microscopy, Tel Aviv University, Israel.

to take advantage of radioautographic methods, as has been done in order to measure transport and localization of proteins (Jamieson and Palade 1968a, b) or lipids (Ohad *et al.*, 1967; Goldberg and Ohad, 1970b) within intracellular organelles.

It has been shown recently that proteins of cytoplasmic origin are utilized by the cell for the formation of the internal membrane system of the chloroplast in *Chlamydomonas reinhardi* (Eytan and Ohad, 1970; Hoober, 1970a, 1972). The experimental evidence in support of this view was gathered mainly from results obtained with specific protein inhibitors and use of appropriate mutants (Eytan and Ohad, 1970; Hoober 1970a, 1972; Hoober *et al.*, 1969). From these studies, it appears that the synthesis of membrane proteins within the chloroplast is at least partially controlled by the presence of membrane proteins of cytoplasmic origin (Eytan and Ohad, 1970). Recently, Hoober has shown that proteins that can be identified on the basis of their electrophoretic behavior as chloroplast membrane proteins of cytoplasmic origin appear to accumulate in a soluble form when synthesized in the presence of chloramphenicol (Hoober, 1972).

However, due to the fact that these proteins were found in the soluble fraction of a total cell homogenate, one cannot establish whether they have been accumulated in the cytoplasm or have been already transported to the chloroplast but not incorporated into the membranes. There is practically no information on the possibility that systhesis or accumulation of proteins of chloroplast origin exerts any influence on transport of cytoplasmic proteins to their final location in the chloroplast membranes. In the present work we have investigated this possibility, using electron microscopic autoradiography. It is demonstrated that proteins synthesized in the cytoplasm are transferred to the chloroplast membranes, and the transport of certain protein species seems to require concomitant synthesis of chloroplast proteins. It appears to us that this information might contribute to our understanding of the degree of chloroplast autonomy, a subject which is under investigation in many laboratories, as well as to the understanding of the mechanisms controlling the development of the chloroplast membranes.

II. MATERIALS AND METHODS

Cell growth, harvesting, and greening experiments were as described previously (Goldberg and Ohad, 1970a). Two *Chlamydomonas reinhardi* mutants were used, the *y-1* mutant, which was characterized (Ohad *et al.*, 1967) and utilized for greening experiments previously (Ohad *et al.*, 1967; Goldberg and Ohad, 1970a, b), and a spontaneous yellow mutant, obtained in our laboratory from an arginine-requiring *Chlamydomonas* mutant (*Arg-2*) kindly donated by Dr. J. S. Sussenbach and Dr. P. J. Strijkert, Philips Research Lab., N.V. Philips Gloeilampenfabrieken, Eindhoven, Nederland (Sussenbach and Strijkert, 1969). This mutant, which will be referred to hereafter as *Arg-2y*, shows an identical morphology, pattern of chlorophyll dilution during growth in the dark, and chlorophyll and protein synthesis during greening in the light[4] as described for the *y-1* mutant (Ohad *et al.*, 1967; Goldberg and Ohad, 1970a, b). The *Arg-2y* cells were usually grown on the same acetate mineral medium utilized for the *y-1* mutant with addition of arginine (20 μg/ml). For all greening experiments, cells were harvested after 6-7 generations of growth in the dark, when chlorophyll content was about 1 $\mu g/10^7$ cells.

A. Radioactive Labeling and Autoradiography

^{3}H-Arginine (1.85 Ci/mmol, Radiochemical Centre, Amersham, England) was added at a final concentration of 1.1×10^{-2} mM. The high specific radioactivity was necessary to ensure a final incorporation of about 1 dpm/cell during a pulse labeling of 9-20 min. This, in turn, induced the appearance of an average of 30 radioautographic grains/cell section after about two weeks of exposure. Cells were incubated either in the dark or light (700 ft-c) in the presence or absence of inhibitors of protein synthesis. Radioactive arginine was added at the time indicated and the cells were further incubated as described (see the footnotes to the tables).

Pulse-labeled cells were pelleted by centrifugation (2000 g, 3 min) and fixed immediately. For chase experiments, the labeled cells were washed by centrifugation, as above, at 25°C in fresh growth medium containing 50 μg/ml nonradioactive arginine, and were further incubated, in presence or absence of inhibitors of protein synthesis, as described below.

Fixation was in 2% OsO_4 in 0.075 M phosphate buffer, *p*H 7 at 4°C overnight. Dehydration and embedding in Epon were done as described by Luft (1961). The fixation and the alcohol and propylene oxide supernatants from the dehydrated pellets, were pooled for each sample and analyzed for radioactivity, in order to account for loss of radioactivity during the processing (Goldberg and Ohad, 1970b). Thin sections (600 Å) were cut with an Ultrome III LKB microtome using glass or diamond knives (Sugg. Rondikin Co. Honolulu, Hawaii). Cell sections were covered with Ilford L-4 emulsion (Ilford, Essex, England) and developed following the procedure of Caro and van Tubergen (1962) modified as described by Goldberg and Ohad (1970b). Cell sections were stained with uranyl acetate and lead citrate (Reynolds,

[4] I. Goldberg and I. Friedberg, unpublished results.

1963). Electron micrographs were obtained with a Philips EM 300 electron microscope at an electron magnification of ×7000.

Statistical analysis of the radioautographic experiments was carried out as described previously (Goldberg and Ohad, 1970b). The data are expressed as mean values ± confidence interval (SEMxt). All calculations were corrected by the *t*-test using a confidence interval of 95 % and degrees of freedom between 17 and 30. For analysis, 268 sections of individual whole cells were used and a total of 8056 radioautographic grains were counted.

To calculate the relative specific radioactivity of a subcellular compartment, its relative radioactivity (percentage of total grains found over the cell section) was divided by its relative area (percentage of total area of the cell section). The confidence interval was calculated for the average of the sum of the relative specific activities of a given subcellular compartment from all micrographs analyzed. The confidence interval for the ratio of relative specific radioactivities of two different subcellular compartments was calculated for the average of the sum of relative ratios which were found within each separate cell section.

B. Chemical Determinations

Total chlorophyll was measured in 80 % acetone extracts as described by Arnon (1949). Protein was measured by the procedure of Lowry *et al.* (1951), and protein radioactivity was determined by measuring the amount of ^{3}H incorporated from ^{3}H-arginine into a hot-TCA precipitate (Eytan and Ohad, 1970). The precipitates were dissolved in a solution containing 0.1 N NaOH and 0.2 % sodium dodecyl sulfate. Aliquots of 0.1–0.5 ml were mixed with 8 ml of Bray phosphorus solution (Bray, 1960) and counted in a scintillation counter (Packard Tricarb Model 3003, Packard Instrument Co. Inc., Downers, Grove, Illinois). Quenching was measured in each sample by the addition of known internal standards.

All reagents and solvents used in this work were of analytical grade.

III. RESULTS

Addition of chloramphenicol (CAP) to greening cells 3 hr after onset of illumination reduces the rate of synthesis of both chlorophyll (Fig. 1A) and protein (Fig. 1B). The initial change in the rate of synthesis of both components after addition of the inhibitor is difficult to evaluate and seems to be higher than that obtained 3.5 hr later. This might be due, at least

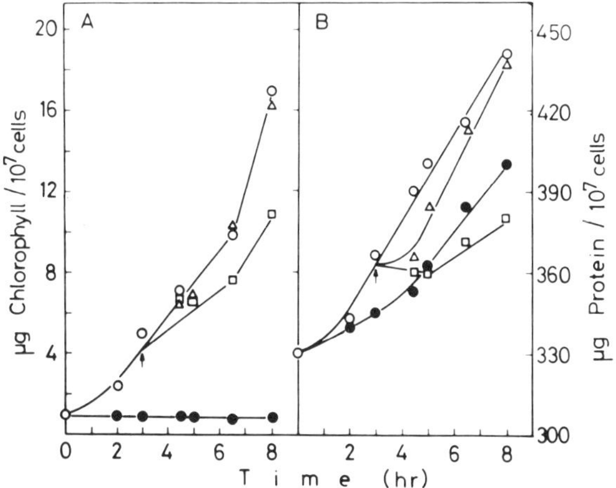

Fig. 1. Effect of addition and removal of chloramphenicol on (A) chlorophyll and (B) protein synthesis during greening of *y*-1 cells. Dark-grown cells were washed and resuspended in fresh growth medium at a final concentration of 8.2×10^6 cells/ml. Cell suspensions were incubated in the dark or in the light, for 3 hr. At that time, chloramphenicol (350 μg/ml) was added to the part of the cell suspensions incubated in the light. After 50 min of further incubation, the cells were washed by centrifugation in chloramphenicol-free medium and incubation continued in the light with (□) or without (△) a second addition of chloramphenicol as above. Samples run in parallel were used for pulse labeling by ^{3}H-arginine and chase (see Table III). Dark circles, cells incubated in the dark; open circles, cells incubated in the light.

partially, to the manipulations of the cultures, including centrifugation washings and transfer to fresh medium. The final reduction in the rate of chlorophyll and protein synthesis was about 50%. After the removal of the inhibitor, the initial rate of greening was resumed (Fig. 1).

The relative distribution of radioactivity in protein as compared with the total radioactivity found in the cells after different treatments was measured. In experiments in which the amount of radioactivity in proteins was determined it was found that after pulse-labeling for 9 min almost 90% of the radioactivity present in the cells was found in proteins. In experiments in which the labeling was carried out for 20 min, the radioactivity in proteins was between 70 and 80% of the total radioactivity incorporated into the cells. Of the total radioactivity present in the cells, less than 7% was lost during the fixation and dehydration. Chloroform–methanol (2:1 v/v) or acetone (80%) extracts of washed cells contained less than 2% of the total radioactivity. Thus one can conclude that most of the radioautographic grains represent radioactivity in proteins. Examination of radioautograms of H^3-arginine-labeled cells (Figs. 2–4) shows that the radioautographic grains

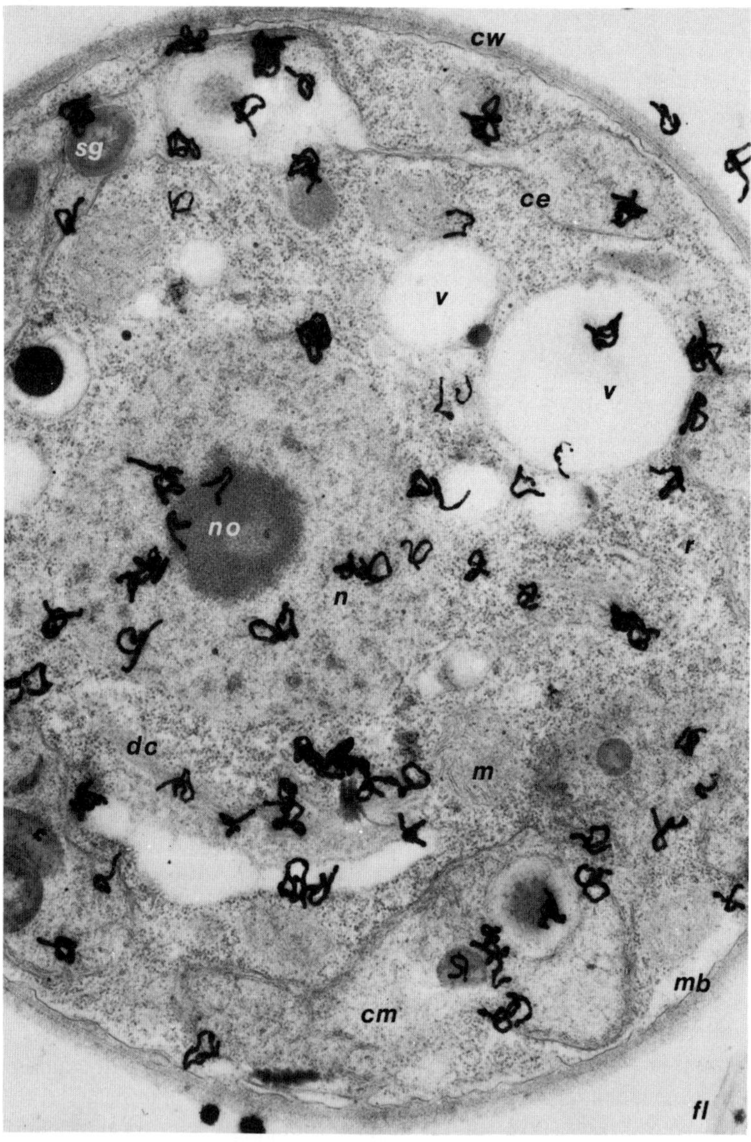

Fig. 2. Sample radioautograph used for analysis of distribution of radioactivity in proteins labeled with ^{3}H-arginine, during greening of *y*-1 cells. Cells incubated for 3.5 hr and labeled for 20 min in the dark. Radioautographic grains are found over the chloroplast, mitochondria (m), nucleus (n), and nucleolus (no), dyctiosome (dc), and ribosome-containing areas (r). ×16,000. Reduced 20% for reproduction. General abbreviations used in Figs. 2–4: ce, chloroplast envelope; cm, chloroplast matrix; dc, dictiosome; es, eyespot; fl, flagella; m, mitochondria; mb, cell membrane; n, nucleus; ne, nuclear envelope; no, nucleolus; pt, paired thylakoids; py, pyrenoid; pyt, pyrenoid tubules; sg, starch granules; v, vacuole; cw, cell wall.

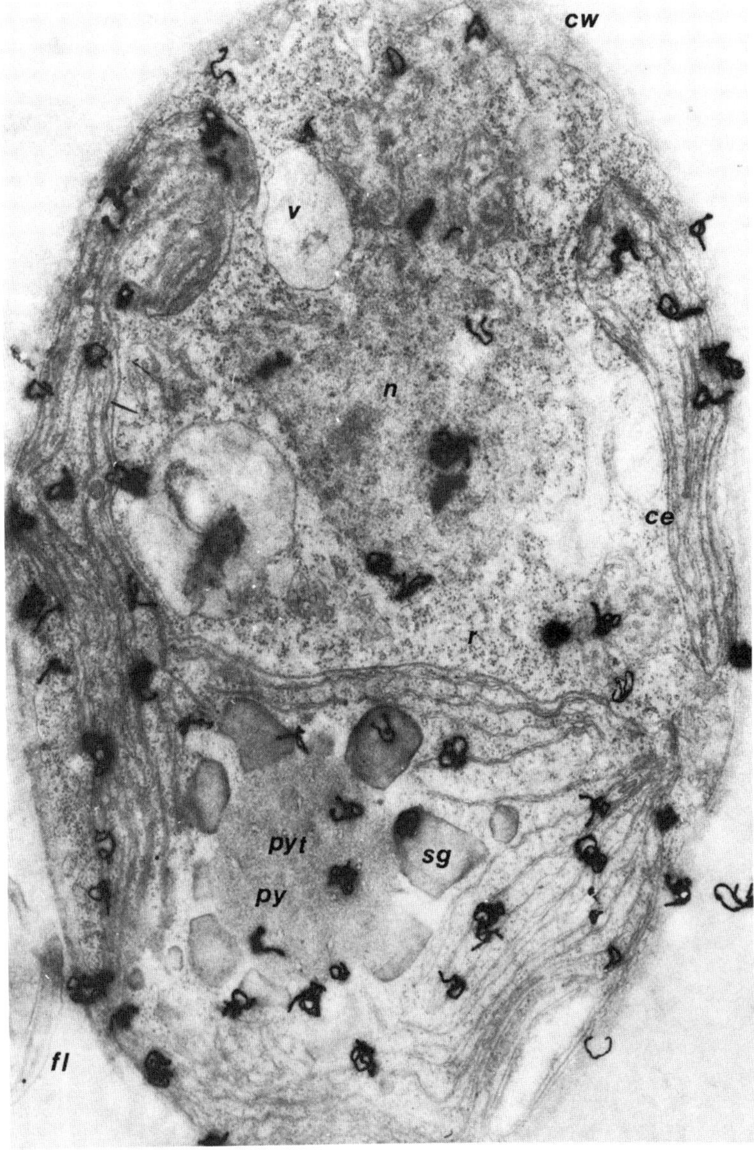

Fig. 3. Cells were incubated for 3.5 hr and labeled for 20 min in the light. The radioautographic grains are preferentially located over the chloroplast. Within the chloroplast most of the grains are found in close apposition to the photosynthetic lamellae. ×20,000. Reduced 20 % for reproduction.

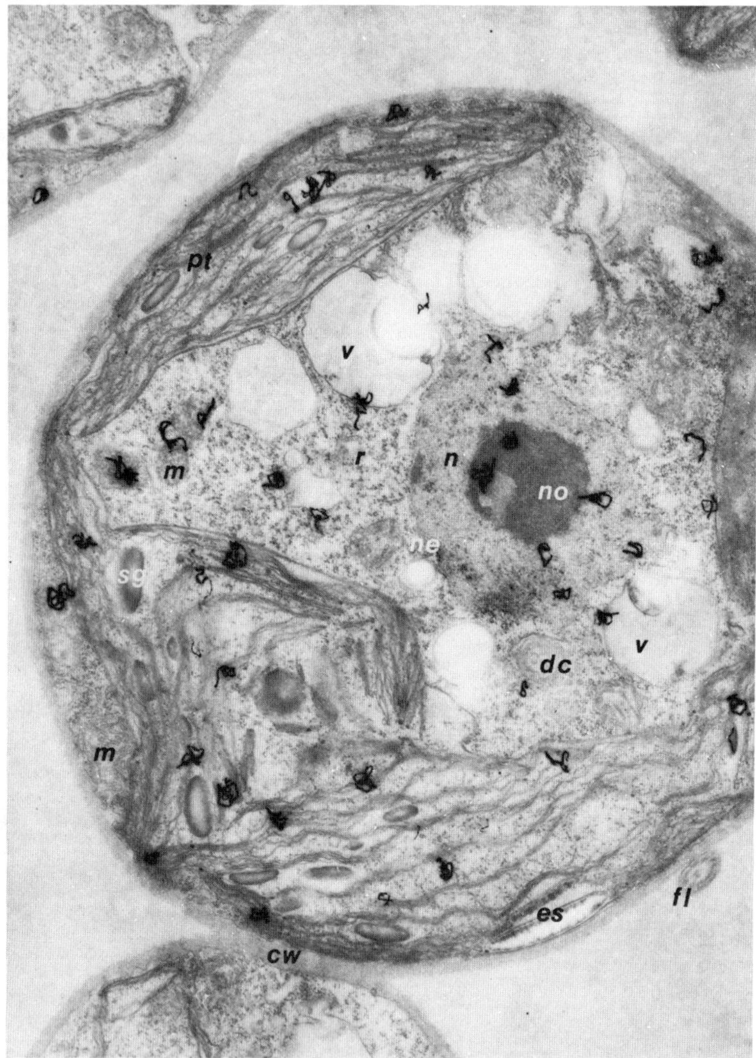

Fig. 4. Same as Fig. 3. Within the chloroplast almost all the radioautographic grains are located on membranes. Notice the relatively high labeling of the nucleus and nucleolus. Grains are also found over mitochondria (m), dyctiosome (dc), and vacuole membranes (v). ×16,000. Reduced 20% for reproduction.

within the chloroplast of light-exposed cells were preferentially located close to or superimposed on the profiles of thylakoids and grana (Figs. 3 and 4). Relatively few grains were observed over the free matrix of the chloroplast. Statistical analysis of the distribution of radioautographic grains within the chloroplast indicates that 82.6% ± 4.5% of the grains were located on or within 0.1 μm of the membrane profiles at the end of 20 min of labeling in the light. After a chase period of 210 min in the light, this distribution of grains remained unchanged (82.9% ± 5.4%).

Within the cytoplasm most of the grains were located on the mitochondria, nucleus, and Golgi membranes. Membranes of the vacuoles were also labeled. Relatively few grains were found over the ribosome-containing area of the cytoplasm matrix. Within the nucleus, both the nucleoplasm and nucleolus were labeled.

The general preservation of the cells was good and no major changes of the morphological features already described (Ohad *et al.*, 1967; Goldberg and Ohad, 1970b) for *C. reinhardi* cells were noticed. Figures 2–4 represent typical examples of radioautograms utilized for statistical analysis of grain distribution.

The incorporation of arginine into proteins in pulse labeling experiments was inhibited as expected when cycloheximide (CHI) or CAP was added to the media or when the greening cells were labeled after transfer to the dark (Eytan and Ohad, 1970; Hoober, 1970a). However, the relative specific radioactivity of the chloroplast in cells incubated in the light or in the light with addition of CHI was higher than that of cells greening in the presence of CAP or labeled after transfer to the dark (Tables I and III). On the other hand, the relative specific radioactivity of the mitochondria was higher in cells incubated in the dark as compared with mitochondria from cells incubated in the light or in the light with addition of CAP or CHI (Table II). In all cases, there was no significant difference in the labeling of the nucleus or cytosol, with the exception that the cytosol was more radioactive in cells treated with CAP in the light (Table II).

The above results demonstrate that the specific radioactivities of the chloroplast and mitochondria are differentially affected by light, dark, CHI, and CAP. The relative labeling of the chloroplast proteins was inhibited by CAP [see also Kadenbach (1967)]. Since it was demonstrated that certain protein species of these organelles are manufactured in the cytoplasm and then have to be transported into the chloroplast, the question arose as to whether interference with the normal synthetic pattern of proteins by CAP would cause disturbance of the transport of cytoplasmic proteins to the chloroplast. To test this possibility, the distribution of radioactivity was analyzed immediately after a pulse-labeling of cells incubated in the light, dark, or light with CAP, and after a chase period of 210 min under different

Table I

Distribution of Radioautographic Grains over Chloroplast and Protoplasm of *Chlamydomonas reinhardi Arg-2y* Cells Pulse-Labeled with ^{3}H-Arginine in Presence of Chloramphenicol (CAP) and Cycloheximide (CHI) during the Greening Process[a]

Treatment	Total radioactivity in proteins, cmp $\times$ $10^{-5}/10^7$ cells	Total grains counted	Percentage of grains in chloroplast of total grains/cell, %	Chloroplast relative specific radioactivity	Ratio of relative specific radio-activity of chloroplast to protoplasm
Light	6.92	571	65.35 ± 7.05	1.44 ± 0.11	2.50 ± 0.42
Light → Dark	3.30	738	54.08 ± 6.56	1.26 ± 0.12	1.48 ± 0.27
Light → CAP	3.92	459	53.93 ± 6.41	1.32 ± 0.15	1.85 ± 0.46
Light → CHI	2.76	569	70.59 ± 8.11	1.48 ± 0.16	3.60 ± 0.60

[a] Dark-grown *Arg-2y* cells were washed and resuspended in fresh growth medium, without arginine, at a final concentration of 9.2×10^6 cells/ml. The cell suspension was incubated in the light for 3.5 hr and then divided into four parts. One part was transferred to the dark while the others were incubated in the light with or without addition of chloramphenicol (600 μg/ml) or cycloheximide (0.4 μg/ml). After 30 min, cells were pulse-labeled with ^{3}H-arginine (1.85 Ci/mmol, 20 μCi/ml) for 9 min. For experimental details see Section II.
Specific activity is defined as percentage of grains of total number of grains on the cell area, divided by percentage of area of the total cell area.

Table II

Distribution of Radioautographic Grains over the Nucleus and Mitochondria of *Chlamydomonas reinhardi Arg-2y* Cells Pulse-Labeled with ^{3}H-Arginine in Presence of Chloramphenicol (CAP) and Cycloheximide (CHI) during the Greening Process[a]

Treatment	Nucleus specific radioactivity	Mitochondria specific radioactivity	Cytosol (not including nucleus and mitochondria) specific radioactivity
Light	1.21 ± 0.31	2.40 ± 0.23	0.56 ± 0.06
Light → Dark	1.38 ± 0.30	4.65 ± 0.90	0.41 ± 0.07
Light → CAP	1.01 ± 0.29	2.74 ± 0.28	0.69 ± 0.06
Light → CHI	0.81 ± 0.23	3.24 ± 0.60	0.56 ± 0.07

[a] See footnote to Table I.

Table III

Distribution of Radioautographic Grains over Chloroplast and Protoplasm of *Chlamydomonas reinhardi y*-1 Cells Pulse-Labeled with ^{3}H-Arginine during the Greening Process[a]

Treatment	Chlorophyll, μg/10^7 cells	Total radioactivity in proteins, cpm $\times$ 10^{-5}/10^7 cells	Total grains counted	Percentage of grains in chloroplast of total grains/cell	Chloroplast relative specific radioactivity	Ratio of relative specific radioactivity of chloroplast to protoplasm
Pulse light	6.55	4.12	1371	61.80 ± 7.05	1.42 ± 0.10	2.23 ± 0.29
210-min Chase light	17.00	3.07	345	75.20 ± 6.40	1.62 ± 0.17	4.25 ± 1.07
Pulse dark	1.33	4.90	423	31.53 ± 4.10	0.92 ± 0.13	0.84 ± 0.32
210-min Chase dark	1.05	4.85	787	45.97 ± 5.31	1.22 ± 0.14	1.54 ± 0.50
Pulse light + CAP	6.60	3.42	743	48.22 ± 4.27	1.21 ± 0.10	1.46 ± 0.25
210-min Chase light + CAP	11.40	3.25	643	46.01 ± 5.02	1.31 ± 0.20	1.85 ± 0.62
210-min Chase light − CAP	16.40	3.50	329	63.72 ± 5.99	1.70 ± 0.22	3.60 ± 1.02

[a] Same experimental conditions as in Fig. 1. Pulse-labeling with ^{3}H-arginine was carried out for 20 min after 3.5 hr from onset of incubation. Specific activity is defined as percentage of grains of total number of grains on the cell area, divided by percentage of area of the total cell area.

Table IV

Distribution of Radioautographic Grains over the Nucleus and Mitochondria of *Chlamydomonas reinhardi y*-1 Cells Pulse-Labeled with ^{3}H-Arginine during the Greening Process

Treatment	Nucleus specific radioactivity	Mitochondria specific radioactivity	Cytosol (not including nucleus and mitochondria) specific radioactivity
Pulse light	1.55 ± 0.17	1.65 ± 0.30	0.56 ± 0.09
210-min Chase light	1.25 ± 0.40	2.22 ± 0.41	0.37 ± 0.08
Pulse dark	2.46 ± 0.96	2.96 ± 0.59	0.38 ± 0.07
210-min Chase dark	1.75 ± 0.54	2.63 ± 0.29	0.39 ± 0.06
Pulse light + CAP	1.24 ± 0.24	2.46 ± 0.31	0.56 ± 0.08
210-min Chase light + CAP	1.50 ± 0.30	2.40 ± 0.26	0.46 ± 0.05
210-min Chase light − CAP	1.90 ± 0.65	2.54 ± 0.40	0.31 ± 0.05

[a] Same experimental conditions as in Table III. Specific activity is defined as percentage of grains of total number of grains on the cell area, divided by percentage of area of the total cell area.

conditions. The results of this experiment show that during the chase in the light there was an increase in the relative radioactivity of the chloroplast (Table III). The relative radioactivity of the chloroplast is lower when labeling is done in presence of CAP and remains practically unchanged if CAP is not removed during the chase (Table III). The relative radioactivity of the mitochondria seems not to have been affected by CAP as compared with the light control (Tables II and IV). The relative radioactivity of the mitochondria was not significantly changed when CAP was present during the chase (Table IV). When CAP was removed during the chase period, an increase occurred only in the relative radioactivity of the chloroplast, which seems to have been due to a reduction of the relative radioactivity of the cytosol (Tables III and IV). A significant change in the relative radioactivity of chloroplast and mitochondria was not detected when the chase period was shorter than 210 min (30 min and 120 min). No significant changes were observed in the relative radioactivity of the nucleus (Table IV), nor in the relative radioactivity of the chloroplast or mitochondria when labeling and chase periods were carried out in the dark (Tables III and IV).

IV. DISCUSSION

The validity of the interpretation of results obtained in this work rests on several assumptions: (1) The marker used for labeling is largely specific for proteins, (2) labeling is due to net synthesis and not to turnover and reutilization of labeled precursors unevenly distributed between the different subcellular compartments (3) localization of radioautographic grains can be accurate enough to distinguish between the different structures under investigation. In the present work, the above criteria were fulfilled to a satisfactory degree. Thus arginine was chosen as a precursor for protein labeling since this amino acid is not metabolized and further utilized for synthesis of carbohydrates, nucleotides, or lipids (Sussenbach and Strijkert, 1969). Indeed, it was found that about 90% of the label was incorporated into proteins in a 9-min pulse and about 80% of the radioactivity of proteins incorporated in this system under the conditions of our experiments is due to net synthesis (Goldberg and Ohad, 1970b). The resolution of radioautography permits localization of the radioautographic grains within $\pm 0.2\ \mu m$ from the radioactive source (Caro and van Tubergen, 1962; Salpeter *et al.*, 1969), and physical translocation of grains due to sample manipulation can be ruled out, as also shown in previous work in which ^3H-acetate was used as a marker for lipids (Goldberg and Ohad, 1970a, b). In addition, most of the radioactivity present in the cells at the end of the pulse was preserved during preparation of samples for radioautography, less than 7% having been lost.

The results indicate that during the greening of dark-grown cells most of the radioactivity incorporated into the cells is localized in the chloroplast. The specific radioactivity of the mitochondria is higher in dark-incubated cells and in general is higher than that of the chloroplast. This would imply that a very active process of protein synthesis takes place in the mitochondria. However, one should notice that the radioactivity of the mitochondria amounts to less than 15% of that of the cell, while that of the chloroplast is between 32 and 75%. In addition, most of the radioactivity of the chloroplast in light-incubated cells is localized on membranes which occupy only a small fraction of the total chloroplast area in greening cells. Since calculation of specific activity is based on the total area of the organelle, this will result in a lowering of the calculated value for the chloroplast while increasing it for the mitochondria, in which the ratio of the surface occupied by membrane to total area is higher than that of the developing chloroplast.

The results shown in this work, which are summarized in Fig. 5, clearly indicate that radioactive proteins are transferred from the cytoplasm to the chloroplast. The transfer seems to be reduced when chloramphenicol is present in the incubation medium, resulting in a relative accumulation of radioactive protein within the cytosol. However, this accumulated radioactivity can be transferred to the chloroplast. This implies that synthesis of proteins by the chloroplast ribosomes is necessary in order to permit transport of proteins from the cytoplasm into the chloroplast. Results not shown here indicate that when cells greened in the presence of chloramphenicol

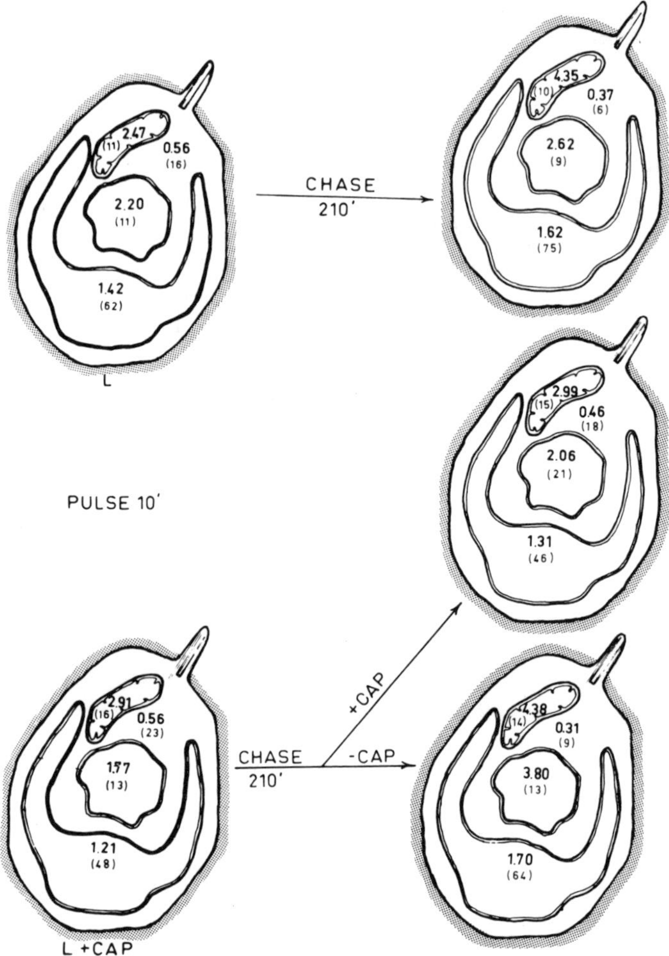

Fig. 5. Schematic representation of the results obtained in this work. The following cell compartments are shown: the bilobate or cup-shaped chloroplast contains within its cup the nucleus, a mitochondrion, and the rest of the cytosol. The numbers indicate specific activities (autoradiographic grain/unit area). The numbers in parenthesis represent percentage of total radioactivity incorporated (grains/cell) found in each compartment.

(600 μg/ml) are labeled for short periods of time (1 min) one can still find a similar distribution of grains between the chloroplast and the cytoplasm as observed at the end of a 9- or 20-min pulse. This would indicate that some proteins of cytoplasmic origin can be synthesized and transferred very rapidly to the chloroplast. These could be proteins of the *L*-protein type

(Eytan and Ohad, 1970; Hoober, 1970a) which were found to be synthesized in the presence of CAP (Eytan and Ohad, 1970; Hoober, 1970a) and localized in the chloroplast membranes even after a short pulse. On the other hand, there seems to be a second class of proteins which are synthesized in the cytoplasm and which are not transported into the chloroplast when CAP is present. Their transfer seems to be much slower since it was detected only after a 210-min chase. The distribution of radioactive grains within the chloroplast indicates that about 82% are located on or in close proximity to membranes. Since this figure was not significantly changed with an addition of 25% in chloroplast radioactivity achieved by transfer of proteins from the cytoplasm, one can assume that a similar portion of the transferred proteins is also located within or close to the chloroplast membranes. Due to uncertainty in the autoradiographic method, one cannot disregard the possibility that a greater amount of the transferred proteins is soluble or not membrane bound. Further experimental data are required in order to establish this possibility. Several soluble proteins located in the chloroplast are known to originate in the cytosol (Smillie and Scott, 1969; Surzycki *et al.*, 1969; Schiff, 1970; Ebbon and Tait, 1969). However, one cannot yet ascribe an identity to the slowly moving membrane proteins detected in this work.

It has previously been shown that membranes formed in the presence of chloramphenicol are photosynthetically inactive and become activated by incubation of the cells after removal of the inhibitor (Eytan and Ohad, 1970; Hoober *et al.*, 1969). This can occur even under conditions in which no further protein synthesis takes place in the cytoplasm. Proteins incorporated into chloroplast membranes under these conditions have been considered to be of chloroplast origin (Eytan and Ohad, 1970; Wallach *et al.*, 1970). The incorporation of these proteins within the membranes is correlated with repair of photosynthetic activity. Since the transfer of slowly moving cytoplasmic proteins into the chloroplast occurs under the same conditions, one cannot exclude the possibility that these proteins might also participate in the reactivation process. The results shown here clearly demonstrate that synthesis of proteins by the chloroplast ribosomes is continuously required for transfer of certain cytoplasmic proteins into the chloroplast and thus demonstrate the existence of mutual control between chloroplast and cytoplasm in the process of synthesis of chloroplast membranes. The type of control exerted in this case does not seem to be due to lack of photosynthetic products caused by the failure of membranes produced in presence of chloramphenicol to photosynthesize since the transfer of *L*-protein from the cytoplasm to the chloroplast occurs under the same conditions. The mechanism of this control and the role of the slow transferred cytoplasmic proteins remains to be established.

VI. SUMMARY

The light-induced formation of the photosynthetic membranes (greening) in *y-1* mutant of *Chlamydomonas reinhardi* requires synthesis of new proteins which become incorporated into the growing membranes. It has been shown previously (Eytan and Ohad, 1970) that proteins synthesized by both chloroplast and cytoplasmic ribosomes concur in the formation of functional photosynthetic membranes, indicating the presence of a mechanism permitting the specific transfer of membrane proteins synthesized in the cytoplasm into the chloroplast. Transfer of such proteins cannot yet be identified by the usual biochemical techniques unless they become part of the growing photosynthetic membranes. However, it is possible to follow their synthesis and translocation between the different cellular compartments by use of quantitative electron microscopic radioautography. In the present work, the radioautographic grain distribution among chloroplast, chloroplast membrane, nucleus, mitochondria, and the remainder of the cytoplasm (cytosol) was carried out following short radioactive pulse-labeling and chase during greening of dark-grown mutants in the presence or absence of protein synthesis inhibitors. The results indicate that transport of some of the proteins of cytoplasmic origin to their final location within the chloroplast is at least partially controlled by concomitant synthesis of proteins by the chloroplast ribosomes.

ACKNOWLEDGMENTS

We would like to thank Dr. G. Eytan and Dr. R. Jennings for critical discussions and suggestions throughout the work and for reading the manuscript, and Mrs. J. Reichler and A. Vilenz for skillful technical assistance.

REFERENCES

Arnon, D. L., 1949, Copper enzymes in isolated chloroplasts. Polyphenoloxidase in *Beta Vulgaris*, *Plant Physiol.* **24**:1–15.

Bray, G. A., 1960, A simple efficient liquid scintillator for counting aqueous solutions in a liquid scintillation counter, *Anal. Biochem.* **1**:279–285.

Caro, L. G., and van Tubergen, R. P., 1962, High-resolution autoradiography. I. Methods, *J. Cell. Biol.* **15**:173–188.

Ebbon, J. G., and Tait, G. H., 1969, Studies on *S-Adenosylmethionine*–magnesium protoporphyrin methyltransferase in *Euglena gracilis* strain Z, *Biochem. J.* **111**: 573–582.

Eytan, G., and Ohad, I., 1970, Biogenesis of chloroplast membranes. III. Cooperation between cytoplasmic and chloroplast ribosomes in the synthesis of photosynthetic lamellar proteins during the greening process in a mutant of *Chlamydomonas reinhardi y-1*, *J. Biol. Chem.* **245**:4297–4307.

Goldberg, I., and Ohad, I., 1970a, Biogenesis of chloroplast membranes. IV. Lipid and pigment changes during synthesis of chloroplast membranes in a mutant of *Chlamydomonas reinhardi y-1*, *J. Cell. Biol.* **44**:563–571.

Goldberg, I., and Ohad, I., 1970b, Biogenesis of chloroplast membranes. V. A radioautographic study of membrane growth in a mutant of *Chlamydomonas reinhardi y-1*, *J. Cell. Biol.* **44**:572–591.

Goodenough, V. W., and Levine, R. P., 1970, Chloroplast structure and function in ac-20, a mutant strain of *Chlamydomonas reinhardi*. III. Chloroplast ribosomes and membrane organization, *J. Cell. Biol.* **44**:547–562.

Hoober, J. K., 1970a, Sites of synthesis of chloroplast membrane polypeptides in *Chlamydomonas reinhardi y-1*, *J. Biol. Chem.* **245**:4327–4334.

Hoober, J. K., 1970b, Formation of chloroplast membranes in *Chlamydomonas reinhardi y-1*, *J. Cell Biol.* **47**:90a.

Hoober, J. L., 1972, A major polypeptide of chloroplast membranes of *Chlamydomonas reinhardi*. Evidence for synthesis in the cytoplasm as a soluble component, *J. Cell Biol.* **52**:84–96.

Hoober, J. K., Siekevitz, P., and Palade, G. E., 1969, Formation of chloroplast membranes in *Chlamydomonas reinhardi y-1*. Effects of inhibitors of protein synthesis, *J. Biol. Chem.* **244**:2621–2631.

Jamieson, J. D., and Palade, G. E., 1968a, Intracellular transport of secretory proteins in the pancreatic exocrine cell. III. Dissociation of intracellular transport from protein synthesis, *J. Cell Biol.* **39**:580–588.

Jamieson, J. D., and Palade, G. E., 1968b, Intracellular transport of secretory proteins in the pancreatic exocrine cell. IV. Metabolic requirements, *J. Cell Biol.* **39**:589–603.

Kadenbach, B., 1967, Synthesis of mitochondrial proteins: demonstration of a transfer of proteins from microsomes into mitochondria, *Biochim. Biophys. Acta* **134**:430–442.

Lowry, O. H., Rosebrough, N. J., Farr, A. L., and Randall, R. J., 1951, Protein measurement with the folin phenol reagent, *J. Biol. Chem.* **193**:265–275.

Luft, J. H., 1961, Improvements in epoxy resin embedding methods, *J. Biophys. Biochem. Cytol.* **9**:409–419.

Mason, T. L., and Schatz, G., 1973, Cytochrome c-oxidase from baker's yeast. II. Site of translation of the protein components, *J. Biol. Chem.* **248**:1355–1360.

Ohad, I., Siekevitz, P., and Palade, G. E., 1967, Biogenesis of chloroplast membranes. I. Plastid dedifferentiation in a dark-grown algal mutant (*Chlamydomonas reinhardi*), *J. Cell Biol.* **35**:521–552.

Ohad, I., Siekevitz, P., and Palade, G. E., 1967, Biogenesis of chloroplast membranes. II. Plastid differentiation during greening of a dark-grown algal mutant (*Chlamydomonas reinhardi*), *J. Cell Biol.* **35**:553–584.

Reynolds, E. S., 1963, The use of lead citrate at high pH as an electron-opaque stain in electron microscopy, *J. Cell Biol.* **17**:208–213.

Rubin, M. S., and Tzagoloff, A., 1973, Assembly of the mitochondrial membrane systems X. Mitochondrial synthesis of three of the subunit proteins of yeast cytochrome oxidase, *J. Biol. Chem.* **248**:4275–4279.

Salpeter, M. M., Bachmann, I., and Salpeter, E. E., 1969, Resolution in electron microscope radioautography, *J. Cell Biol.* **41**:1–20.

Schiff, J., 1970, Developmental interactions among cellular compartments in *Euglena*, *in* "Symposia of the Society for Experimental Biology. Control of Organelle Development" (P. L. Miller, ed.), pp. 277–301, Cambridge University Press.

Sebald, W., Machleidt, W., and Otto, J., 1973, Products of mitochondrial protein synthesis in *Neurospora crassa*. Determination of equimolar amounts of three products in cytochrome oxidase on the basis of amino-acid analysis, *Eur. J. Biochem.* **38**:311–324.

Smillie, R. M., and Scott, S., 1969, Organelle biosynthesis: the chloroplast, *in* "Progress in Molecular and Subcellular Biology" (F. E. Hahn, ed.), Vol. 1, pp. 136–202, Springer-Verlag, Berlin, Heidelberg, New York.

Surzycki, S. J., Goodenough, V. W., Levine, R. P., and Armstrong, J. J., 1970, Nuclear and chloroplast control of chloroplast structure and function in *Chlamydomonas reinhardi*, *in* "Symposia of the Society for Experimental Biology. Control of Organelle Development" (P. L. Miller, ed.), pp. 13–35, Cambridge University Press.

Sussenbach, J. S., and Strijkert, P. J., 1969, Arginine metabolism in *Chalmydomonas reinhardi*. On the regulation of the arginine biosynthesis, *Eur. J. Biochem.* **8**:403–407.

Wallach, D., Bar-Nun, S., and Ohad, I., 1970, Photophostphorylation activity in a *Chlamydomonas reinhardi* mutant *(y-1)* during light induced formation of its chloroplast membranes, *Israel J. Chem.* **8**:161.

Woodward, D. O., Edwards, D. L., and Flavell, R. B., 1970, Nucleocytoplasmic interactions in the control of mitochondrial structure and function in *neurospora*, *in* "Symposia of the Society for Experimental Biology. Control of Organelle Development" (P. L. Miller, ed.), pp. 55–69, Cambridge University Press.

Chapter 6

Nerve Excitability—Toward an Integrating Concept

Eberhard Neumann

Max-Planck-Institut of Biophysical Chemistry
Goettingen, Germany

and

David Nachmansohn

Departments of Neurology and Biochemistry
Columbia University
New York, New York

"*The driving force of quantitative biological study is, however, our mystical conviction that Nature is one.*" (*Aharon Katchalsky, 1969*)

I. INTEGRAL PHYSICOCHEMICAL MODEL FOR NERVE EXCITATION

A. Introduction

It is well known that bioelectricity and nerve excitability are manifested electrically in stationary membrane potentials and the various forms of transient potential changes such as, e.g., the action potential (Hodgkin, 1964; Tasaki, 1968). Although these electrical properties reflect membrane processes, bioelectricity and excitability are intrinsically coupled to the metabolic activity of the excitable cells. Now, inherent to all living cells is a high degree of coupling between energy and material flows. But already on the subcellular level of membranes intensive chemodiffusional flow coupling occurs and apparently time-independent properties reflect balance between active and passive flows of cell components.

The specific function of the excitable membrane requires the maintenance of nonequilibrium states (Katchalsky, 1967). The intrinsic nonequilibrium nature of membrane excitability is most obviously reflected in the asymmetric ion distributions across excitable membranes. These nonequilibrium distributions are metabolically mediated and maintained by "active transport" (e.g., Na^+/K^+ exchange pump). A fundamental requirement for such an active chemodiffusional flow coupling is spatial anisotropy of the coupling medium (Prigogine, 1968), including the membrane structure (Katchalsky, 1967).

Anisotropy is apparently a characteristic property of biological membranes. Furthermore, cellular membranes including nerve membranes may be physicochemically described in terms of a layer structure. For instance, it was found that the internal layer of the axonal membrane contains proteins required for excitability. Proteolytic action on this layer causes irreversible loss of this property (Tasaki and Singer, 1966).

The membrane components of excitable membranes include ionic constituents: fixed charges (some of them may serve to facilitate locally perm-selective ion diffusion), mono- and divalent metal ions, especially Ca^{2+} ions, and in particular ionic side groups of membrane proteins directly involved in the control of electrical properties. Local differences in the distribution of fixed charges may create membrane regions with different permeability characteristics, and (externally induced) ionic currents may amplify either the accumulation or the depletion of ions within the membrane. Structural inhomogeneity of excitable membranes may thus lead to the observed nonlinear dependencies between certain physical parameters, e.g. between current intensity and potential (Cole, 1968). A well-known example of this nonlinearity is the current rectification in resting stationary states of excitable membranes. Correspondent to current rectification, there is a straightforward dependence of the membrane potential on the logarithm of ion activities only in limited concentration ranges (Hodgkin and Keynes, 1955; Tasaki and Takenada, 1964). Thus various chemical and physical membrane parameters show the intrinsic structural and functional anisotropy of excitable membranes. This recognition reveals the approximate nature of any classical equilibrium and nonequilibrium approach to electrochemical membrane parameters on the basis of linearity and involving the assumption of a homogeneous isotropic membrane. Since, however, details of the membrane anisotropy are not known, any exact quantitative nonequilibrium analysis of membrane processes and ion exchange currents across excitable membranes faces great difficulties.

B. Stationary Membrane Potentials

It is only within the framework of a number of partially unrealistic, simplifying assumptions that we can approximately describe the electrochemical behavior of excitable membranes within a limited range of physicochemical state variables (Agin, 1967). Explicitly, we use the formalism of classical irreversible thermodynamics restricted to linearity between flows and driving forces. The application of the Nernst–Planck equation to the ion flows across excitable membranes represents such a linear approximation widely used to calculate stationary membrane potentials.

There is experimental evidence that large contributions to the resting stationary membrane potential $\Delta\psi_r$ are attributable to ion selectivities. Indeed, it appears that excitable membranes have developed dynamic structures which are characterized by permselectivity for certain ion types and ion radii. This property can be described by "Nernst" terms which, however, are strictly valid only for electrochemical equilibria. These Nernst terms can be calculated by application of the Nernst–Planck equation relating the ion flow J_i to the gradient of the electrochemical potential $\nabla\tilde{\mu}_i$ of the ion type i,

$$J_i = U_i C_i \nabla(-\tilde{\mu}_i) \tag{1}$$

where U_i is the ionic mobility and C_i is the molar concentration. The electrochemical potential of ion i of valency Z_i is defined by

$$\tilde{\mu}_i = \mu_i{}^0 + RT \ln a_i + Z_i F\psi \tag{2}$$

where $\mu_i{}^0$ is the standard chemical potential and a_i the thermodynamic activity, which in dilute solution may be approximated by $a_i \simeq C_i$; RT is the (molar) thermal energy, F is the Faraday constant, and ψ is the electrical potential.

In the frame of the linear model the flux component $J_i(x)$ perpendicular to the membrane surfaces at the point x within the membrane is given by

$$J_i(x) = U_i(x)C_i(x)\,(d/dx)\,(-\tilde{\mu}_i) \tag{3}$$

Integration of Eq. (3) within the membrane boundaries ($x = 0$ and $x = d$, with d the membrane "thickness") in a closed form requires restrictive assumptions. Despite the experimentally suggested inhomogeneity of excitable membranes, the following approximations are used: (1) the ionic mobilities are the same throughout the membrane and (2) space charge effects are negligible, so that we may assume approximate microscopic electroneutrality. Thus $\nabla^2\psi \simeq 0$ within the membrane. The Laplace equation $\nabla^2\psi = 0$ is equivalent to the constant-field condition that usually is not an appropriate physical assumption (Agin, 1967; Zelman and Shih, 1972).

For the condition of stationarity, i.e., at zero net membrane current ($I_m = 0$), the uniform model treatment yields the Nernst contributions

$$(\Delta\psi_{\mathrm{N}})_{I_m=0} = \frac{RT}{F}\sum_i \frac{t_i}{Z_i}\ln\frac{a_i^{(o)}}{a_i^{(i)}} \tag{4}$$

where t_i is the transference number representing the fraction of membrane current carried by ion type i. (The fraction t_i can vary between 0 and 1.)

Although stationary states of excitable membranes always reflect balances between active transport processes and passive "leakage" fluxes, we can safely neglect flow contributions associated with active transport as long as the time scale considered does not exceed the millisecond range.

The electrical potential difference of an excitable membrane under "resting" conditions (stationary case of $I_m = 0$) can thus be approximated by

$$\Delta\psi_r = (\Delta\psi_{\mathrm{N}})_{I_m=0} \tag{5}$$

This relationship is useful for a limited number of practical cases.

It should be realized that the measured potential difference (measured, for instance, with calomel electrodes in connection with salt bridges) cannot be used to calculate the exact value for the average electric field $\bar{E}$ across the permeation barrier of thickness d:

$$\bar{E} = -(1/d)\,\Delta\psi' \tag{6}$$

The electrical potential difference $\Delta\psi'$ associated with the permeation barrier results from different sources. Although these sources are mutually coupled, we can formally separate the various contributions. Due to the presence of fixed charges at membrane surfaces (Segal, 1968), there are Donnan potentials $\Delta\psi_{\mathrm{D}}^{(o)}$ and $\Delta\psi_{\mathrm{D}}^{(i)}$ from the inside (i) and outside (o) interfaces between membrane and environment. Furthermore, there are interdiffusion potentials $\Delta\psi_U$ arising from differences in ionic mobilities within ion exchange domains of intramembraneous fixed charges (Tasaki, 1968). We can sum these terms to obtain $\Delta\psi_{\mathrm{D}U} = \Delta\psi_{\mathrm{D}}^{(o)} + \Delta\psi_{\mathrm{D}}^{(i)} + \Delta\psi_U$. Thus

$$\Delta\psi' = \Delta\psi_{\mathrm{N}} + \Delta\psi_{\mathrm{D}U} \tag{7}$$

Compared to $\Delta\psi_{\mathrm{D}U}$, the "Nernst" contributions to $\Delta\psi$ appear relatively large. Inserting Eqs. (4) and (7) in Eq. (6), we obtain for the resting state

$$(\bar{E})_{I_m=0} = -\frac{1}{d}\left(\Delta\psi_{\mathrm{D}U} + \frac{RT}{F}\sum_i \frac{t_i}{Z_i}\ln\frac{a_i^{(o)}}{a_i^{(i)}}\right) \tag{8}$$

Among the various ions known to contribute to $\Delta\psi$ (*in vivo*) are the metal ions Ca^{2+}, Na^+, and K^+, protons, and Cl^- ions. In the resting state the value of the stationary membrane potential $\Delta\psi_r$ is different for different cells and tissues. Distribution and density of fixed charges, ion gradients, and the extent to which they contribute to $\Delta\psi_r$ differ for different excitable membranes.

Variations of the natural ionic environment lead to alterations in the Donnan and interdiffusion terms. Since ions are an integral part of the membrane structure, for instance, as counterions of fixed charges, any change in ion type and salt concentration, but also variations in temperature and pressure, may cause changes in the lipid phase (Traeuble and Eibl, 1974) and in the conformation of membrane proteins. As a consequence of such structural changes membrane "fluidity" and permeability properties may be considerably altered. It is known that, for instance, an increase in the external Ca^{2+} ion concentration increases the electrical resistance of excitable membranes [see, e.g., Cole (1968)].

C. Transient Changes of Membrane Properties

A central problem in bioelectricity is the question: What is the mechanism of the various transient changes of membrane parameters associated with nerve activity? Before discussing a specific physicochemical model for the control of electrical activity it appears necessary to recall a few fundamental electrophysiological and chemical observations on excitable membranes.

1. Transient Changes in $\Delta\psi$

The various transient expressions of nerve activity, such as depolarizations and hyperpolarizations of the resting stationary potential level, reflect changes of dielectric capacitive nature, interdiffusional ion redistributions, and "activations" or "inactivations" of different gradients or changes of the extent to which these gradients contribute to the measured parameters. At constant pressure and temperature, the ion gradient contributions to $\Delta\psi$ can be modeled as variations of the t_i parameters.

[It should be realized that, in general, the actually measured parameters, such as potential differences, ionic conductivity, or membrane resistance, may contain larger contributions from cell membranes (of glia or Schwann cells) between the excitable membrane and the measuring electrodes.]

a. Threshold Behavior. Transient changes in $\Delta\psi$ can be caused by various physical and chemical perturbations originating from adjacent membrane regions, from adjacent excitable cells, or from external stimuli.

Phenomenologically, we differentiate sub- and suprathreshold responses of the excitable membrane to stimulation. Subthreshold changes are, for instance, potential changes that attenuate with time and distance from the site of perturbation; this phenomenon has been called electrotonic spread. If, however, the intensity of depolarizing stimulation exceeds a certain threshold value, a potential change is triggered that does not attenuate but propagates as such along the entire excitable membrane. This suprathreshold response is called "regenerative" and has been called action potential or nerve impulse.

b. Stimulus Characteristics. In order to evoke an action potential, the intensity of the stimulation has to reach the threshold with a certain minimum velocity (Cole, 1968). If the external stimulus, for instance, is a current of gradually increasing intensity I, the minimum slope condition for the action potential can be written $dI/dt \geqslant (dI/dt)^{\min}$. The observation of a minimum slope condition for the generation of action potentials is of crucial importance for any mechanistic approach to excitability.

On the other hand, when rectangular current pulses are applied, there is an (absolute) minimum intensity I_{th}, called rheobase, and a minimum time interval Δt of current application. It is found that the product of suprathreshold intensity and minimum duration is approximately constant. Thus, for $I \geqslant I_{\text{th}}$,

$$I\,\Delta t \simeq \text{const} \tag{9}$$

Equation (9) describes the well-known strength–duration curve. Whereas the strength–duration product does not depend on temperature, the temperature coefficient of the rheobase (dI_{th}/dT related to a temperature increase of 10°C) Q_{10} is generally about 2 [see, e.g., Cole (1968)].

The value of I_{th} is history dependent. The threshold changes as a function of previous stimulus intensity and duration. Hyperpolarizing and depolarizing subthreshold prepulses also change the threshold intensity of the stimulating current. See Section IC1(d).

It appears that, in general, the induction of the action potential requires the *reduction* of the intrinsic membrane potential $\Delta\psi_r$ below a certain threshold value $\Delta\psi_{\text{th}}$. This potential decrease $\Delta(\Delta\psi) = \Delta\psi_r - \Delta\psi_{\text{th}}$ is usually about 15–20 mV and has to occur in the *form of an impulse*, $\int \Delta(\Delta\psi)\,dt$, in which a minimum condition between membrane potential and time must be fulfilled:

$$\frac{d(\Delta\psi)}{dt} \geqslant \left[\frac{d(\Delta\psi)}{dt}\right]^{\min} \tag{10}$$

The condition for the initiation of an action potential can be written

$$\Delta\psi_r - \int \left[\frac{d(\Delta\psi)}{dt}\right]^{\min} dt = \Delta\psi_{\text{th}}$$

or in the more general form

$$\varDelta\psi_r - (1/\varDelta t)\int \varDelta(\varDelta\psi)\,dt \leqslant \varDelta\psi_{\text{th}} \tag{11}$$

The potential change $\varDelta(\varDelta\psi)$ is equivalent to a change in the intrinsic field E across the membrane of thickness d. Thus $\varDelta E = -\varDelta(\varDelta\psi)/d$ and from Ohm's law we have $d(\varDelta E) = -R_m I$, where R_m is the membrane resistance.

Including Eq. 4, we can write

$$\int \varDelta(\varDelta\psi)\,dt = -d\int \varDelta E\,dt = \int R_m I\,dt$$

$$= \frac{RT}{z_i F}\int \varDelta \ln \frac{a_i^{(o)}}{a_i^{(i)}}\,dt \tag{12}$$

With Eq. (12) we see how the intrinsic membrane potential can be changed: by an (external) field (voltage or current) pulse or by an ion pulse involving those ions that determine the membrane potential [cf. Eq. (4)]. Such an ion pulse can be produced if the concentration of K^+ ions on the outside of excitable membranes is sufficiently increased within a sufficiently short time interval (impulse condition). Similarly, a pH change can cause action potentials (Tasaki *et al.*, 1965).

As seen in Eq. (4), the membrane potential is dependent on temperature. Therefore temperature (and also pressure) changes alter the stationary membrane potential in excitable membranes. Moreover, action potentials can be evoked by thermal and mechanical shocks [see, e.g., Julian and Goldman (1962)].

Furthermore, under natural conditions electrical depolarization produces an action potential only if the excitable membrane is kept above a certain (negative) level of polarization; in squid giant axons the minimum membrane potential is about −30 to −40 mV. This directionality of membrane polarization and the requirement of field reduction for regenerative excitation are a further manifestation of the intrinsic anisotropy of excitable membranes.

c. Propagation of Local Activity. In many excitable cells $\varDelta\psi_r$ is about −60 to −90 mV, where the potential of the cell interior is negative. Assuming an average membrane thickness $d = 100$ Å, we calculate from Eq. (6) an average field intensity of approximately 60–90 kV/cm [see however, Carnay and Tasaki (1971)]. The field vector in the resting stationary state is directed from the outside to the inside across the membrane of the excitable cell. (It is this electric field which partially compensates the chemical potential gradient of the K^+ ions.) Any perturbation which is able to change locally the intrinsic membrane potential, i.e., the membrane field, will also affect adjacent parts of the membrane or even adjacent cells. If the field change remains below threshold or does not fulfill the minimum slope condition, there will be only subthreshold attenuation of this field change. A nerve impulse in its rising phase may, however, reduce the membrane field in adjacent parts to such an extent and within the required minimum time interval that suprathreshold responses are triggered. Recall that electric fields represent long-range forces (decaying with distance)!

d. Refractory Phases. Another important observation suggestive for the nature of bioelectricity is the refractory phenomenon. If a nerve fiber is stimulated a second time shortly after a first impulse was induced, there is either an impulse that propagates much more slowly or only a subthreshold change. Immediately after stimulation a fiber is *absolutely refractory*; no further impulse can be evoked. This period is followed by a *relatively refractory phase* during which the threshold potential is more positive and only a slowly propagated impulse can be induced. The various forms of history-dependent behavior are also called accommodation or adaptation of the fiber to preceding manipulations.

2. Time Constants of $\Delta\psi$ Changes

If rectangular current pulses of fixed duration but of variable amplitude and amplitude directions are locally applied to an excitable membrane, a series of local responses is obtained. Figure 1 shows schematically the time course of these local responses upon hyper- and depolarizing stimulations: subthreshold potential changes and action potential (Katz, 1966).

It appears that the threshold level reflects a kind of instability. When the stimulus is switched off, fluctuations either result in the action potential (path $c \rightarrow c'$) or in a return to the resting level (path $c \rightarrow c''$). If, for phase c,

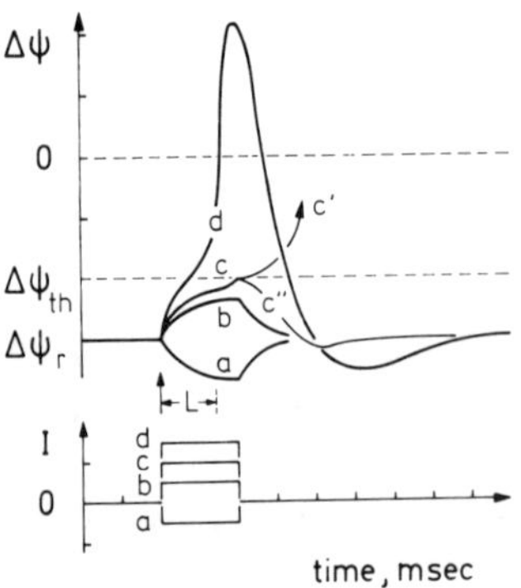

Fig. 1. Sub- and suprathreshold responses of the membrane potential $\Delta\psi$ to stimulating rectangular current pulses of fixed duration but variable intensity I. (a) Inward current pulse causing hyperpolarization, (b) subthreshold outward current ($I < I_{th}$) causing depolarization, (c) threshold outward current (I_{th}) causing either an action potential (path c′) or return to resting stationary potential $\Delta\psi_r$ (path c″), (d) suprathreshold outward pulse ($I > I_{th}$) causing action potential. L, latency phase of the action potential; $\Delta\psi_{th}$, threshold potential (Neumann, 1974).

a stimulus of the same intensity were to be longer lasting, an action potential would develop immediately from the hump visible in the late phase of stimulation.

In view of a rather continuous transition from a subthreshold response to the action potential, the notion of a threshold in the context of an all-or-none law for the action potential becomes less rigorous.

It has been found that certain sections of the time course of sub- and suprathreshold responses can be described with linear differential equations of first order. Furthermore, the time constants τ_m associated with these sections are very similar for subthreshold potential changes as well as for the initial rising phase (also called latency; see Fig. 1) of the action potential.

The actual values for τ_m are different for different excitable membranes and depend on temperature. In squid giant axons the value of τ_m at about 20°C is about 1 msec; the corresponding value for lobster giant axons is about 3 msec. The temperature coefficient ($d\tau_m^{-1}/dT$ related to a temperature increase of 10°C) Q_{10} is about 2 (Cole, 1968).

As outlined by Cole, a time constant of the order of milliseconds can hardly be modeled by simple electrodiffusion: An ion redistribution time of 1 msec requires the assumption of an extremely low ionic mobility of 10^{-8} cm sec^{-1}/V cm^{-1}. However, even in "sticky" ion exchange membranes ionic mobilities are not below 10^{-6} cm sec^{-1}/V cm^{-1}. Furthermore, the temperature coefficients of simple electrodiffusion are only about 1.2–1.5.

Thus the magnitude and temperature coefficient of τ_m suggest that electrodiffusion is not rate-limiting for a large part of the ion redistributions following perturbations of the membrane field.

This conclusion is supported by various other observations. The time course of the action potential can be formally associated with a series of time constants, all of which have temperature coefficients of about 2–3 [see, e.g., Cole (1968)]. The various phases of the action potential are prolonged with decreasing temperature. In the framework of the classical Hodgkin–Huxley phenomenology prolonged action potentials should correspond to larger ion movements. However, it has been recently found that in contrast to this prediction, the amount of ions actually transported during excitation decrease with decreasing temperature (Landowne, 1973).

It thus appears that ion movements (caused by perturbations of stationary membrane states) are largely rate-limited by membrane processes. Such processes may comprise phase changes of lipid domains or conformational changes of membrane proteins. Changes of these types usually are cooperative and temperature dependencies are particularly pronounced within the cooperative transition ranges.

Prolonged potential changes and reduced ion transport at decreased temperatures can be readily modeled if at lower temperatures configurational rearrangements of membrane components involve smaller fractions of the membrane than at higher temperatures.

There is a formal resemblance between the time constant τ of a phase change or a chemical equilibrium and the *RC* term of an electrical circuit with resistance *R* and capacitance *C* (Oster *et al.*, 1973). If a membrane process is associated with the thermodynamic affinity *A* and a rate $J_r = d\xi/dt$, where ξ is the fractional advancement of this process, we can formally define a reaction resistance $R_r = (\partial A/\partial J_r)$ and an average reaction capacitance $\bar{C}_r = -(\partial A/\partial \xi)^{-1}$. The time constant is then given by $\tau = R_r\bar{C}_r$.

The time course of electrical parameters can thus be formally modeled in terms of reaction time constants for membrane processes of the type discussed above. For the chemical part of the observed threshold change in the membrane potential, $\Delta\psi = I_m R_m\ [1 - \exp(-t/\tau_m)]$, caused by the rectangular current pulse of the intensity I_m across the membrane resistance R_m, we have $\tau_m = R_r C_r$.

As already mentioned, the time constants for subthreshold changes and for the first rising phase of the action potential are almost the same. Furthermore, the so-called "Na^+-ion activation–inactivation curves (in voltage clamp experiments) for both sub- and suprathreshold conditions are similar in shape, although much different in amplitude" (Plonsey, 1969). These data strongly suggest that sub- and suprathreshold responses are essentially based on the same control mechanism; it is then most likely the molecular organization of the control system that accounts for the various types of response.

3. *Proteins Involved in Excitation*

Evidence is accumulating that proteins and protein reactions are involved in transient changes of electrical parameters during excitation. As briefly mentioned, the action of proteases finally leads to inexcitability. Many membrane parameters, such as ionic permeabilities, are *p*H dependent. In many examples this dependence is associated with a *pK* value of about 5.5, suggestive for the participation of carboxylate groups of proteins.

Sulfhydryl reagents and oxidizing agents interfere with the excitation mechanism, e.g., prolonging the duration and finally blocking the action potential (Huneeus-Cox *et al.*, 1966). The membrane of squid giant axons stain for SH groups provided the fibers had been stimulated (Robertson, 1970). These findings strongly suggest the participation of protein-specific redox reactions in the excitation process.

Of particular interest are the effects of ultraviolet radiation on nodes of Ranvier. The spectral radiation sensitivity of the rheobase and of the fast, transient inward current (normally due to Na^+ ions) in voltage clamp is very similar to the ultraviolet absorbance spectrum for proteins (v. Muralt and Stämpfli, 1953; Fox, 1972). The results of Fox also indicate that the fast, transient inward component of the action current is associated with only a small fraction of the node membrane.

The conclusions on locally limited excitation sites are supported by the results obtained with certain nerve poisons. Extremely low concentrations of tetrodotoxin reduce and finally abolish the fast inward component of voltage clamp currents [for review see Evans (1972)]. The block action of this toxin is *p*H dependent and is associated with a *pK* value of about 5.3.

It thus appears that the ionic gateways responsible for the transient inward current involve protein organizations comprising only a small membrane fraction.

4. *Impedance and Heat Changes*

a. Impedance Change. The impedance change accompanying the action potential is one of the basic observations in electrophysiological studies on excitability (Cole, 1968). It is generally assumed that the impedance change reflects changes in ionic conductivities (ionic permeabilities). Figure 2 shows schematically that the conductance first rises steeply and in a second phase decays gradually toward the stationary level. In contrast to the pronounced resistance change, the membrane capacity apparently does not change during excitation (Cole, 1970). This result, too, suggests that only a small fraction of the membrane is involved in the rather drastic permeability changes during excitation.

b. Heat Exchange Cycle. The action potential is accompanied by relatively large heat changes (Abbot *et al.*, 1958). These heat changes can be thermodynamically modeled in terms of a cyclic variation of membrane states. The complex chain of molecular events during a spike can be simplified

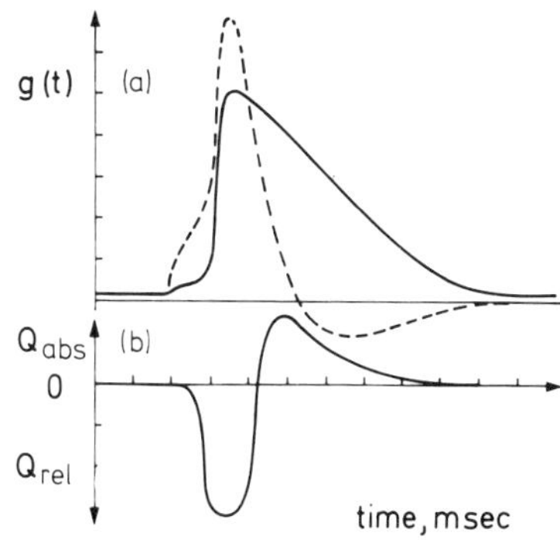

Fig. 2. Schematic representation of (a) conductivity change $g(t)$ accompanying the action potential (dashed line) and (b) heat exchange cycle. Q_{rel} is the heat released during the rising phase and Q_{abs} is the heat absorbed during the falling phase of the action potential (Neumann, 1974).

by the sequence of state changes $A \rightarrow B \rightarrow A$, where A represents the resting stationary state and B the transiently excited state of higher ionic permeability.

The heat changes occur under practically isothermal–isobaric conditions. If we now associate the Gibbs free energy change ΔG with the (overall) excitation process $A \rightleftharpoons B$, we can write

$$\Delta G = \Delta H - T\,\Delta S \tag{13}$$

where ΔH is the reaction enthalpy (as heat exchangeable with the environment) and ΔS the reaction entropy.

More recently, it has been confirmed that the rising phase of the action potential is accompanied by heat release Q_{rel}, while during the falling phase the heat Q_{abs} is reabsorbed (Howarth *et al.*, 1968); see Fig. 2. In our simple A–B model, the first phase is associated with

$$\Delta G_{A\to B} = \Delta H_{A\to B} - T\,\Delta S_{A\to B}$$

and the second one with

$$\Delta G_{B\to A} = \Delta H_{B\to A} - T\,\Delta S_{B\to A}$$

In general, for a cyclic process (where the original state is restored), $\oint dG = 0$, or

$$\Delta G_{A\to B} + \Delta G_{B\to A} = 0$$

In the hypothetical case of ideality (reversibility), $\Delta H = Q$ and

$$Q_{\mathrm{rel}} + Q_{\mathrm{abs}} = 0.$$

Since no *natural* process occurs ideally (i.e., completely reversibly) there are always irreversible contributions. This means a part of $\Delta G_{A\to B}$ and of $\Delta G_{B\to A}$ will dissipate into heat. We can formally split ΔG into a reversible (exchangeable) contribution ΔG^{rev} and an irreversible contribution ΔG^{irr} (Neumann, 1973). Thus we have

$$\Delta G_{A\to B} = \Delta G^{\mathrm{rev}}_{A\to B} + \Delta G^{\mathrm{irr}}_{A\to B}$$

$$\Delta G_{B\to A} = \Delta G^{\mathrm{rev}}_{B\to A} + \Delta G^{\mathrm{irr}}_{B\to A}$$

By definition $\Delta G^{\mathrm{irr}} = -T\,\Delta S^{\mathrm{irr}} \leqslant 0$, since the change in the inner entropy ΔS^{irr} is always larger than or equal to zero (see, e.g., Prigogine, 1968). The measured heats are then given by

$$Q_{\mathrm{rel}} = \Delta H^{\mathrm{rev}}_{A\to B} + \Delta G^{\mathrm{irr}}_{A\to B}$$

$$Q_{\mathrm{abs}} = \Delta H^{\mathrm{rev}}_{B\to A} + \Delta G^{\mathrm{irr}}_{B\to A}$$

Since in a cycle $\Delta H^{\text{rev}}_{A\to B} + \Delta H^{\text{rev}}_{B\to A} = 0$, we find for the "difference" between heat released and heat absorbed

$$\Delta Q = Q_{\text{rel}} + Q_{\text{abs}} = (\Delta G^{\text{irr}}_{A\to B} + \Delta G^{\text{irr}}_{B\to A}) = -T\,\Delta S^{\text{irr}} \leqslant 0$$

Due to irreversible contributions we have for the absolute values $|Q_{\text{rel}}| > |Q_{\text{abs}}|$ (Q_{rel} counting negative!). It is found experimentally that $|Q_{\text{abs}}| \simeq 0.9\,|Q_{\text{rel}}|$. As stressed by Guggenheim (1949), $\Delta G^{\text{irr}} < 0$ or $\Delta S^{\text{irr}} > 0$, *only if phase changes and/or chemical reactions are involved.* Then for our case we can write

$$\Delta Q = \Delta G^{\text{irr}} = -\sum_j A_j \xi_j < 0$$

where A is the affinity and ξ is the extent of membrane processes j involved (Neumann, 1973).

Since the action potential most likely involves only a small fraction of the excitable membrane, the measured heat changes Q_{rel} and Q_{abs} appear to be very large. Since on the other hand the mutual transition $A \to B \to A$ "readily" occur, the value of ΔG ($=\Delta G_{A\to B} = -\Delta G_{B\to A}$) cannot be very large. In order to compensate a large ΔH (here $\sim Q$), there must be a large value for ΔS; see Eq. (13). This means that the *entropy change associated with the membrane permeability change during excitation is also very large.*

It is, in principle, not possible to deduce from heat changes the nature of the processes involved. However, large configurational changes—equivalent to a large overall ΔS—in biological systems frequently arise from conformational changes of macromolecules or macromolecular organizations such as membranes or from chemical reactions. In certain polyelectrolytic systems such changes even involve metastable states and irreversible transitions of domain structures (Neumann, 1973).

The large absolute values of Q and the irreversible contribution ΔQ suggest structural changes and/or chemical reactions to be associated with the action potential. Furthermore, there are observations indicating the occurrence of metastable states and nonequilibrium transitions in excitable membranes at least for certain perfusion conditions (Tasaki, 1968).

5. *Summary*

1. Due to various similarities, sub- and suprathreshold responses appear to reflect a common basic mechanism (more complex than simple electrodiffusion).

2. Proteins and, in some examples, redox reactions are involved in excitation.

3. Large changes in heat and membrane resistance accompanying the action potential point to configurational rearrangements of membrane components: phase changes, conformational transitions, or chemical reactions.

4. Evidence is accumulating suggesting that excitability comprises only a small membrane fraction.

5. Many features of subthreshold responses can be analyzed as membrane relaxations to small perturbations of stationary states (tractable with linear differential equations of first order). Suprathreshold responses are basically nonlinear and can be seen as relaxations to (locally) large perturbations of the excitable membrane.

D. The Cholinergic System and Excitability

Among the earliest known macromolecules associated with excitable membranes are some proteins of the cholinergic system. The cholinergic apparatus comprises acetylcholine (AcCh), the synthesis enzyme choline-*O*-acetyltransferase (Ch-T), acetylcholine-esterase (AcCh-E), acetylcholine-receptor (AcCh-R), and a storage site (S) for AcCh. Since details of this system are discussed in Section II, here only a few aspects essential for our integral model of excitability are discussed.

1. Localization of the AcCh System

Chemical analysis has revealed that the concentration of the cholinergic system is very different for various excitable cells. For instance, squid giant axons contain much less AcCh-E (and Ch-T) than axons of lobster walking legs [see, e.g., Brzin *et al.* (1965)]; motor nerves are generally richer in cholinergic enzymes than are sensory fibers (Gruber and Zenker, 1973).

The search for cholinergic enzymes in nerves was greatly stimulated by the observation that AcCh is released from isolated axons of various excitable cells, provided inhibitors of AcCh-E such as physostigmine (eserine) are present. This release is appreciably increased upon electrical stimulation (Calabro, 1933; Lissak, 1939) or when axons are exposed to higher external K^+ ion concentrations (Dettbarn and Rosenberg, 1966). Since larger increases of external K^+ concentration depolarize excitable membranes, reduction of the membrane potential appears to be a prerequisite for AcCh liberation from the storage site.

Evidence for the presence of axonal AcCh storage sites and receptors is still mainly indirect. However, direct evidence is accumulating for extrajunctional AcCh receptors (Porter *et al.*, 1973); binding studies with α-bungarotoxin (a nerve venom with a high affinity to AcCh-R) indicate the presence of receptor-like proteins in axonal membrane fragments (Denburg *et al.*, 1972).

There are still questions as to the presence and localization of the cholinergic system. However, the differences in the interpretations of chemico-analytical data and the results of histochemical light and electron microscope investigations are gradually being resolved; there appears to be a progressive confirmation of the early chemical evidence for a ubiquitous cholinergic system. For instance, AcCh-E reaction products can be made visible in the excitable membranes of more and more nerves formerly called noncholinergic [see, e.g., Koelle (1971)]. Catalysis products of AcCh-E are demonstrated in pre- and postsynaptic parts of excitable membranes [see, e.g., Koelle (1971) and Lewis and Shute (1966)]. In a recent study, stain for the α-bungarotoxin–receptor complex was found to be visible also in presynaptic parts of axonal membranes [see Fig. 1 in Porter *et al.* (1973)]. These findings suggest the presence of the cholinergic system in both junctional membranes and thus render morphological support for the results of previous studies on the pre- and postsynaptic actions of AcCh and inhibitors and activators of the cholinergic system (Masland and Wigton, 1940; Riker *et al.*, 1959).

2. *The Barrier Problem*

Histochemical, biophysical, biochemical, and, particularly, pharmacological studies on nerve tissue face the great difficulty of an enormous morphological and chemical complexity. It is now recognized that due to various structural features not all types of nerve tissue are equally suited for certain investigations.

The excitable membranes are generally not easily accessible. The great majority of nerve membranes are covered with protective tissue layers of myelin, of Schwann or glia cells. These protective layers insulating the excitable membrane frequently comprise structural and chemical barriers that impede the access of test compounds to the excitable membrane. In particular, the lipid-rich myelin sheats are impervious to many quarternary ammonium compounds, such as AcCh and curare.

In some examples, such as the frog neuromuscular junction, externally applied AcCh or the receptor inhibitor curare have relatively easy access to the synaptic gap, whereas the excitable membranes of the motor nerve and the muscle fiber appear to be largely inaccessible. On the other hand, the cholinergic system of neuromuscular junctions of lobsters are protected against external action of these compounds, whereas the axons of the walking legs of lobster react to AcCh and curare [see, e.g., Dettbarn (1967)].

Penetration barriers also comprise absorption of test compounds within the protective layers. Furthermore, chemical barriers in the form of hydrolytic enzymes frequently cause decomposition of test compounds before they can reach the nerve membrane. For instance, phosphoryl phosphatases in the Schwann cell layer of squid giant axons cause hydrolysis of organophosphates such as the AcCh-esterase inhibitor diisopropylfluorophosphate (DFP), and impulse conduction is blocked only at very high concentrations of DFP (Hoskin *et al.*, 1966).

A very serious source of error in concentration estimates and in interpretations of pharmacological data resides in procedures that involve homogenization of lipid-rich nerve tissue [see, e.g., Nachmansohn (1969)]. For instance, homogenization liberates traces of inhibitors (previously applied) which despite intensive washing still adhered to the tissue. Even when the excitable membrane was not reached by the inhibitor, membrane components react with the inhibitor during homogenization (Hoskin *et al.*, 1969). Thus block of enzyme activity observed after homogenization is not necessarily an indicator for block during electrical activity [see, e.g., Katz (1966), p. 90].

As demonstrated in radiotracer studies, failure to interfere with bioelectricity is often concomitant with the failure of test compounds to reach the excitable membrane. Compounds like AcCh or *d*-tubocurarine (curare) act on squid giant axons only after (enzymatic) reduction of structural barriers (Rosenberg and Hoskin, 1963). Diffusion barriers even after partial reduction are often the reason for longer incubation times and higher concentration of test compounds as compared to less protected membrane sites. In this context it should be mentioned that the enzyme choline-*O*-acetyltransferase, sometimes considered to be a more specific indicator of the cholinergic system, is frequently difficult to identify in tissue and is *in vitro* extremely unstable (Nachmansohn, 1963).

In the light of barrier and homogenization problems it appears obvious that any statements on the absence of the cholinergic system or on the failure of blocking compounds to interfere with excitability are only useful if they are based on evidence that the test compound had actually reached the nerve membrane.

3. *Electrogenic Aspects of the AcCh System*

Particularly suggestive for the bioelectric function of the cholinergic system in axoms are electrical changes resulting from eserine and curare application to nodes of Ranvier, where permeability barriers are less pronounced (Dettbarn, 1960a, b). Similar to the responses of certain neuromuscular junctions, eserine prolongs potential changes also at nodes; curare

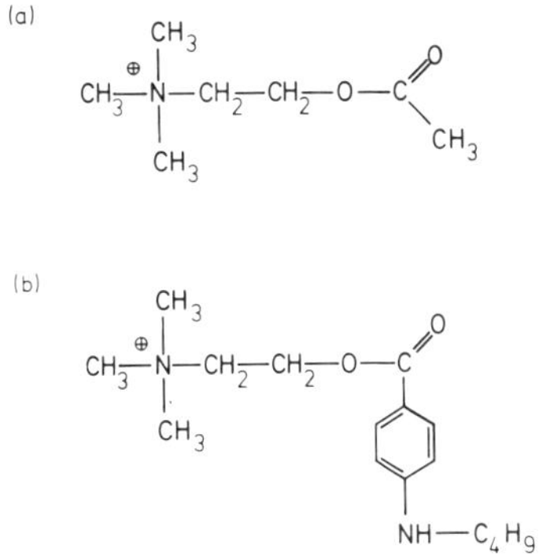

(a) $CH_3-\overset{\oplus}{N}(CH_3)_2-CH_2-CH_2-O-C(=O)-CH_3$

(b) $CH_3-\overset{\oplus}{N}(CH_3)_2-CH_2-CH_2-O-C(=O)-C_6H_4-NH-C_4H_9$

Fig. 3. Chemical structure of (a) the acetylcholine ion and (b) the tetracaine ion. Note that the structural difference is restricted to the acid residue: (a) CH_3 and (b) the aminobenzoic acid residue rendering tetracaine lipid-permeable.

first reduces the amplitude of the nodal action potential and then also decreases the intensity of subthreshold potential changes in a similar way as known for frog junctions.

In all nerves the generation of action potentials is readily blocked by (easily permeating) local anesthetics such as procaine or tetracaine. Due to structural and certain functional resemblance to AcCh, which is particularly pronounced for tetracaine (see Fig. 3), these compounds may be considered as analogs of AcCh. In a recent study it is convincingly demonstrated how (by chemical substitution at the ester group) AcCh is successively transformed from a receptor activator (reaching junctional parts only) to the receptor inhibitor tetracaine, reaching readily junctional and axonal parts of excitable membranes (Bartels, 1965). Local anesthetics are also readily absorbed in lipid bilayer domains of biomembranes [see for review Seeman (1972)].

Not only does external application of AcCh without esterase inhibitors face diffusion barriers, but esterase activity increases the local proton concentration (Dettbarn, 1967); the resulting changes in pH may contribute to changes in membrane potential.

Only very few nerve preparations appear to be suited to demonstrate a direct electrogenic action of externally applied AcCh. Diffusion barriers and differences in local concentrations of the cholinergic system may be the reason that the impulse condition for the generation of action potentials cannot be everywhere fulfilled [see Section I C1(b)]. Some neuroblastoma cells produce subthreshold potential changes and action potentials upon electrical stimulation as well as upon AcCh application (Harris and Dennis, 1970; Nelson *et al.*, 1971; Hamprecht, 1974).

There are thus, without any doubt, many pharmacological and chemical similarities between synaptic and axonal parts of excitable membranes *as far as the cholinergic system is concerned.* On the other hand, there are various differences. But it seems that these differences can be accounted for by structural and chemical factors.

With regard to this problem, two extreme positions of interpretation may be distinguished. On the one side, more emphasis is put on the differences between axonal and junctional membranes. An extreme view considers the responses of axons to AcCh and structural analogs as pharmacological curiosities (Ritchie, 1963), and, in general, the cholinergic nature of excitable membrane is not recognized and acknowledged. However, the presence of the cholinergic system in axons and the various similarities to synaptic behavior suggest the same basic mechanism for the cholinergic system in axonal and synaptic parts of excitable membranes.

Since there is as yet *no direct experimental evidence that AcCh crosses the synaptic gap*, the action of AcCh can be alternatively assumed to be restricted to the interior of the excitable membrane of junctions and axons. This assumption is based on the fact that no trace of AcCh is detectable outside the nerve unless AcCh-esterase inhibitors are present. According to this alternative hypothesis, *intramembranous* AcCh combines with the receptor and causes permeability changes mediated by conformational changes of the AcCh-receptor.

This is the basic postulate of the chemical theory of bioelectricity (see Section II), attributing the primary events of all forms of excitability in biological organisms to the cholinergic system: in axonal conduction, for subthreshold changes (electronic spread) in axons and pre- and postsynaptic parts of excitable membranes.

In the framework of the chemical model the various types of responses—excitatory and inhibitory synaptic properties—are associated with structural and chemical modifications of the *same basic mechanism involving the cholinergic system*. Participation of neuroeffectors like the catecholamines or GABA and other additional reactions within the synapse possibly give rise to the various forms of depolarizing and hyperpolarizing potential changes in postsynaptic parts of excitable membranes. The question of coupling between pre- and postsynaptic events during signal transmission cannot be answered for the time being. It is, however, suggestive to incorporate in transmission models the transient increase of the K^+ ion concentration in the synaptic gap after a presynaptic impulse (see also Section 3).

4. *Control Function of AcCh*

Recall that transient potential changes such as the action potential result from permeability changes caused by proper stimulation. However, a (normally proper) stimulus does not cause an action potential if (among others) certain inhibitory analogs of the cholinergic agent such as, e.g., tetracaine are present. Tetracaine also reduces the amplitudes of subthreshold potential changes; in the presence of local anesthetics, e.g., procaine, mechanical compression does not evoke action potentials (Julian and Goldman, 1962). It thus appears that the (electrical and mechanical) stimulus does *not directly* effect sub- and suprathreshold permeability changes, suggesting preceding events involving the cholinergic system.

If the AcCh-esterase is inhibited or the amount of this enzyme is reduced by protease action, subthreshold potential changes and (postsynaptic) current flows as well as the action potential are prolonged [see, e.g., Takeuchi and Takeuchi (1972) and Dettbarn (1960a, b)]. Thus AcCh-E activity appears to play an essential role in terminating the transient permeability changes. The extremely high turnover number of this enzyme (about 1.4×10^4 AcCh molecules per sec, i.e., a turnover time of 70 μsec) is compatible with a rapid removal of AcCh (Nachmansohn, 1959).

In summary, the various studies using activators and inhibitors of the cholinergic system indicate that both initiation and termination of the permeability changes during nerve activity are (active) processes associated with AcCh. It seems, however, possible to decouple the cholinergic control system from the ionic permeation sites or gateways. Reduction of the external Ca^{2+} ion concentration appears to cause such a decoupling; the result is an increase in potential fluctuations or even random, i.e., uncontrolled, firing of action potentials [see, e.g., Cole (1968)].

E. The Integral Model

In the previous sections some basic electrophysiological observations and biochemical data were discussed that any adequate model for bioelectricity has to integrate. It is stressed that among the features excitability models have to reproduce are the threshold behavior, the various similarities of sub- and suprathreshold responses, stimulus characteristics, and the various forms of conditioning and history-dependent behavior.

An attempt at an integral model of excitability has been recently developed (Neumann *et al.*, 1973) on the basis of the classical chemical model elaborated by Nachmansohn (see Section II).

In the present account we explore some previously introduced concepts for the control of electrical membrane properties by the cholinergic system. Among these fundamental concepts are the notion of a basic excitation unit, the assumption of an AcCh storage site particularly sensitive to the electric field of the excitable membrane, and the idea of a continuous sequential translocation of AcCh through the cholinergic proteins (AcCh cycle). Finally, we proceed toward a formulation of various excitation parameters in terms of nonequilibrium thermodynamics.

1. Key Processes

In order to account for the various interdependencies between electrical and chemical parameters, it is necessary to distinguish among a minimum number of single reactions associated with excitation.

A possible formulation of some of these processes in terms of chemical reactions was given previously (Neumann *et al.*, 1973). The reaction scheme is briefly summarized.

(1) Supply of AcCh to the membrane storage site S following synthesis (formally from the hydrolysis products choline and acetate).

$$\mathrm{S} + \mathrm{AcCh} = \mathrm{S}_1(\mathrm{AcCh}) \tag{14}$$

For the uptake reaction two assumptions are made: (i) The degree of AcCh association with the binding configuration S_1 increases with increasing membrane potential (cell interior negative), and (ii) the uptake rate is limited by the conformational transition from state S to S_1, and is slow as compared to the following translocation steps.

Vesicular storage of AcCh as indicated by Whittaker and co-workers [see, e.g., Whittaker (1973)] is considered as additional storage for membrane sites of high AcCh turnover, for instance, at synapses.

(2) Release of AcCh from the storage form S_1(AcCh), for instance, by depolarizing stimulation:

$$\mathrm{S}_1(\mathrm{AcCh}) = \mathrm{S}_2 + \mathrm{AcCh} \tag{15}$$

Whereas S_1(AcCh) is stabilized at large (negative) membrane fields, S_2 is more stable at small intensities of the membrane field. The field-dependent conformational changes of S are assumed to gate the path of AcCh to the AcCh-receptors.

The assumptions for the dynamic behavior of the storage translocation sequence

$$\mathrm{S} + \mathrm{AcCh} \underset{k_{-1}}{\overset{k_1}{\rightleftharpoons}} \mathrm{S}_1(\mathrm{AcCh}) \underset{k_{-2}}{\overset{k_2}{\rightleftharpoons}} \mathrm{S}_2 + \mathrm{AcCh} \tag{16}$$

can be summarized as follows: The rate constant k_2 for the release step is larger than the rate constant k_1 for the uptake, and also $k_2 \gg k_{-2}$. (See also Section IE5.)

(3) Translocation of released AcCh to the AcCh-receptor R and association with the Ca^{2+}-binding conformation $R_1(Ca^{2+})$. This association is assumed to induce a conformational change to R_2 that, in turn, releases Ca^{2+} ions:

$$\mathrm{R}_1(\mathrm{Ca}^{2+}) + \mathrm{AcCh} = \mathrm{R}_2(\mathrm{AcCh}) + \mathrm{Ca}^{2+} \tag{17}$$

(4) Release of Ca^{2+}-ions is assumed to change structure and organization of gateway components G. The structural change from a closed configuration G_1 to an open state G_2 increases the permeability for passive ion fluxes.

(5) AcCh hydrolysis. Translocation of AcCh from R_2(AcCh) to the AcCh-esterase E involving a conformational transition from E_1 to E_2:

$$R_2(\text{AcCh}) + E_1 = E_2(\text{Ch}^+, \text{Ac}^-, \text{H}^+) + R_2 \tag{18}$$

The hydrolysis reaction causes the termination of the permeability change by reuptake of Ca^{2+},

$$R_2 + \text{Ca}^{2+} = R_1(\text{Ca}^{2+}) \tag{19}$$

concomitant with the relaxation of the gateway to the closed configuration G_1.

Thus the reactions (18) and (19) "close" a reaction cycle which is formally "opened" with reactions (14) and (16).

Since under physiological conditions (i.e., without esterase inhibitor) no trace of AcCh is detectable outside the excitable membrane (axonal and synaptic parts), the sequence of events modeled in the above reaction scheme is suggested to occur in a specifically organized structure of the cholinergic proteins, a structure that is intimately associated with the excitable membrane.

2. *Basic Excitation Unit*

Before proceeding toward a model for the organization of the cholinergic system, it is instructive to consider the following well-known electrophysiological observations.

In a great variety of excitable cells the threshold potential change to trigger the action potential is about 20 mV. This voltage change corresponds to an energy input per charge or charged group within the membrane field of only about one kT unit of thermal energy (k is Boltzmann's constant; T is the absolute temperature) at body temperature. If only one charge or charged group would be involved, thermal motion should be able to initiate the impulse. Since random "firing" is very seldom, we have to conclude that several ions and ionic groups have to "cooperate" in a concerted way in order to cause a suprathreshold permeability change.

Furthermore, there are various electrophysiological data which suggest at least two types of gateways for ion permeation in excitable membranes [for summary see Cole (1968)]: a rapidly operating ion passage normally gating passive flow of Na^+ ions (into the cell interior) and permeation sites that normally limit passive K^+ ion flow.

There are various indications, such as the direction of potential change and of current flow, suggesting that the rising phase of the action potential has predominantly contributions from the "rapid gateway"; the falling phase of the overall permeability change involves larger contributions of the K-ion gateways [see also Neher and Lux (1973)]. There is certainly coupling between the two gateway types: electrically through field changes and possibly also through Ca^{2+} ions transiently liberated from the "rapid gateways." As explicitly indicated in Eqs. (17) and (19), Ca^{2+} ion movement precedes and follows the gateway transitions. Recent electrophysiological studies on neuroblastoma cells confirm the essential role of Ca^{2+} ions in subthreshold potential changes and in the gating phase of the action potential (Spector *et al.*, 1973).

At the present stage of our model development we associate the direct cholinergic control of permeability changes only to the rapidly operating gateway G. As seen in Fig. 2, the rising phase of the conductane change caused by the permeability increase is rather steep. This observation also supports a cooperative model for the mechanism of the action potential.

The experimentally indicated functional cooperativity together with the (experimentally suggested) locally limited excitation sites suggests a structural anchorage in a cooperatively stabilized membrane domain.

In order to account for the various boundary conditions discussed above, we have introduced the notion of a *basic excitation unit* (BEU). Such a unit is suggested to consist of a gateway G that is surrounded by the cholinergic control system. The control elements are interlocked complexes of storage S, receptor R, and esterase E, and are called SRE assemblies. These assemblies may be organized in different ways and, for various membrane types, the BEU's may comprise different numbers of SRE assemblies. As an example, the BEU schematically represented in Fig. 4 contains six SRE units controlling the permeation site G.

The core of the BEU is a region of dynamically coupled membrane components with fixed charges and counterions, such as Ca^{2+} ions. Figure 4 shows that the receptors of the SRE assemblies form a ringlike array. We assume that this structure is cooperatively stabilized and, through Ca^{2+} ions, intimately associated with the gateway components. In this way the Ca^{2+} dependent conformational dynamics of the receptors is coupled to the transition behavior of the gateway.

The receptor ring of a BEU is surrounded by the "ring" of the storage sites and (spatially separated) by the "ring" of the AcCh esterases. The interfaces between the different rings define local reaction spaces through which AcCh is exchanged and translocated.

The BEU's are assumed to be distributed over the entire excitable membrane; the BEU density may vary for different membrane parts. The high density of cholinergic proteins found in some examples may be due to clustering of BEU's.

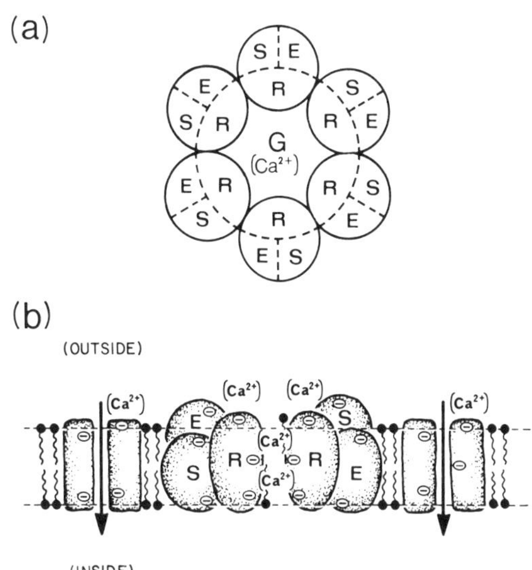

Fig. 4. Scheme of the AcCh-controlled gateway G. (a) Basic excitation unit (BEU) containing, in this example, six SRE assemblies, viewed perpendicular to the membrane surface. S, AcCh-storage site; R, AcCh-receptor protein; E, AcCh-esterase. (b) Cross section through a BEU flanked by two units which model ion passages for K^+ ions; the arrows represent the local electric field vectors due to partial permselectivity to K^+ ions in the resting stationary state. The circled minus signs symbolize negatively charged groups of membrane components (Neumann, 1974).

Different membrane types may not only vary in the number of SRE assemblies per BEU but also in the type and organization of the gateway components, thus assuring permselectivities for various ion types, particularly in synaptic parts of excitable membranes. It is only the cholinergic control system, the SRE element, which is assumed to be same for all types of rapidly controlled gateways for passive ion flows.

a. Action Potential. In the framework of the integral model, the induction of an action potential is based on cooperativity between several SRE assemblies per BEU. In order to initiate an action potential, a certain critical number of receptors $\overline{m}^c$ per BEU has on average to be activated within a certain critical time interval Δt^c, (impulse condition). During this time interval at least, say, four out of six SRE assemblies have to process AcCh in a concerted manner. Under physiological conditions only a small fraction of BEU's is required to generate and propagate the nerve impulse.

b. Subthreshold Responses. Subthreshold changes of the membrane are seen to involve only a few single SRE assemblies of a BEU. On average not more than one or two SRE elements per BEU are assumed to contribute to the measured responses (within time intervals of the order of Δt^c).

The (small) permeability change caused by Ca^{2+} release from the receptor thus results from only a small part of the interface between receptor and gateway components of a BEU: The ion exchanges $AcCh^+/Ca^{2+}$ and Na^+ are locally limited.

In the framework of this model, spatially and temporally attenuating electrical activity, such as subthreshold axonal, postsynaptic, and dendritic potentials, is the sum of spatially and temporally additive contributions resulting from the local subthreshold activity of many BEU's.

Although the permeability changes accompanying local activity are very small (as compared to those causing the action potentials), the summation over many contributions may result in large overall conductivity changes. Such changes may even occur (to a perhaps smaller extent) when the core of the gateway is blocked. It is suggested that compounds like tetrodotoxin and saxitoxin interact with the gateway core only, thus essentially not impeding subthreshold changes at the interface between receptor ring and gateway.

Influx of Ca^{2+} ions, particularly through pre- and postsynaptic membranes, may affect various intracellular processes leading, e.g., to release of hormones, catecholamines, etc.

3. *Translocation Flux of AcCh*

As discussed before, the excitable membrane, as a part of a living cell, is a nonequilibrium system characterized by complex chemodiffusional flow coupling. Although modern theoretical biology tends to regard living organisms only as quasistationary, with oscillations around a steady average, our integral model for the subthreshold behavior of excitable membranes is restricted to stationarity. We assume that the "living" excitable membrane (even under resting conditions) is in a state of continuous subthreshold activity (maintained either aerobically or anaerobically). However, the nonequilibrium formalism developed later in this section can also be extended to cover nonlinear behavior such as oscillations in membrane parameters.

In the framework of the integral model, continuous subthreshold activity is also reflected in a continuous sequential translocation of AcCh through the cholinergic system. The SRE elements comprise reaction spaces with continuous input by synthesis (Ch-T) and output by the virtually irreversible hydrolysis of AcCh. Input and output of the control system are thus controlled by enzyme catalysis.

a. Reaction Scheme. Since AcCh is a cation, translocation can most readily occur along negatively fixed charges, and may involve concomitant anion transport or cation exchange. The reaction scheme formulated in Section IE1 therefore gives only a rough picture. Storage, receptor, and esterase represent macromolecular subunit complexes with probably several binding sites and the exact stoichiometry of the AcCh reactions is not known. As far as local electroneutrality is concerned, it appears in any case more realistic to assume that for one Ca^{2+} ion released (or bound) there are two other monovalent ions bound (or released), e.g., AcCh ion and Na^+ ion.

The conformationally mediated translocation of the AcCh ion A^+ can then be reformulated by the following sequence:

(1) Storage reaction:

$$S_1(A^+) + C^+ = S_2(C^+) + A^+ \tag{20}$$

(C^+ symbolizes a cation; $2C^+$ may be replaced by Ca^{2+}.)

(2) Receptor reaction:

$$A^+ + R_1(Ca^{2+}) = R_2(A^+) + Ca^{2+} \tag{21}$$

(3) Hydrolysis reaction:

$$R_2(A^+) + E_1 = R_2 + E_2(A^+) \rightarrow (Ch^+, Ac^-, H^+) \tag{22}$$

As already mentioned, the nucleation of the gateway transition (causing the action potential) requires the association of a critical number $\bar{n}^c$ of A^+ with the cooperative number of receptors $\bar{m}^c$ in the Ca^{2+} binding form $R_1(Ca^{2+})$ within a critical time interval Δt^c. This time interval is determined by the lifetime of a single receptor–acetylcholine association.

Using formally $\bar{n}^c$ and $\bar{m}^c$ as stoichiometric coefficients, we can write the concerted reaction inducing gateway transition as

$$\bar{n}^c A^+ + \bar{m}^c R_1(Ca^{2+}) = \bar{m}^c R_2(\bar{n}^c A^+) + \bar{m}^c Ca^{2+} \tag{23}$$

Storage and receptor reactions, Eqs. (20) and (21), represent gating processes preceding gateway opening ("Na activation") and causing the latency phase of the action potential. The hydrolysis process causes closure of the cholinergically controlled gateway ("Na inactivation"). In the course of these processes the electric field across the membrane changes, affecting all charged, dipolar, and polarizable components within the field. These field changes particularly influence the storage site and the membrane components controlling the K^+ permeation regions (Adam, 1970). Figure 5 shows a scheme modeling the "resting" stationary state and a transient phase of the excited membrane.

The complexity of the nonlinear flow coupling underlying suprathreshold potential changes may be tractable in terms of the recently developed network

a Resting stationary state

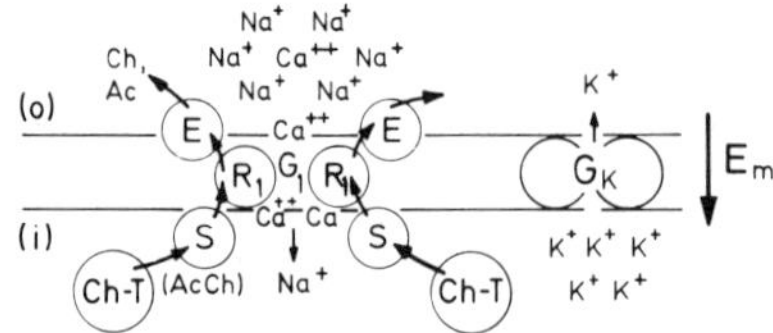

b Excited state

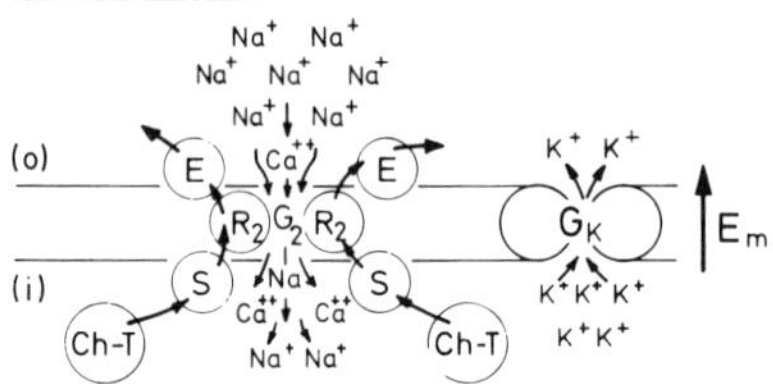

Fig. 5. Schematic representation of a membrane section (a) in the "resting" stationary state and (b) in a transient phase of excitation. In (a), the majority of the acetylcholine receptors are in the Ca^{2+} ion-binding conformation R_1 ; the cholinergically controlled, rapidly operating gateway is in the closed state G_1 and the permeability for Na^+ (and Ca^{2+}) ions is very small as compared to the permeability for K^+ ions through the slow gateway G_K . The electric field vector E_m pointing from the outside boundary (*o*) to the inside boundary (*i*) of the membrane is largely due to the K^+-ion gradient. In (b), most of the receptors are in the acetylcholine-binding conformation R_2 and the rapid gateway is in its open configuration G_2 (Na-activation phase). The change in the electric field (directed outward during the peak phase of the action potential) accompanying the transient Na^+ (and Ca^{2+}) influx causes a transient (slower) increase in the permeability of G_K , thus inducing a (delayed) transient efflux of K^+ ions. Hydrolysis of acetylcholine (AcCh) leads to relaxation of R_2 and G_2 to R_1 and G_1 , restoring the resting stationary state. Translocation of AcCh, occasionally in the resting stationary state and in a cooperatively increased manner after suprathreshold stimulation, through a storage site S of relatively large capacity, receptor, and AcCh-esterase is indicated by the curved arrows. The hydrolysis products choline (Ch) and acetate (Ac) are transported through the membrane, where intracellular choline-*O*-acetyltransferase (Ch-T) may resynthetise AcCh (with increased rate in the refractory phase) (Neumann, 1974).

thermodynamics covering inhomogeneity of the reaction space and nonlinearity (Oster *et al.*, 1973). An attempt at such an approach, which formally includes conformational metastability and hysteretic flow characteristics (Katchalsky and Spangler, 1968; see also Blumenthal *et al.*, 1970) is in preparation (Rawlings and Neumann, 1975).

b. Reaction Fluxes. For the nonequilibrium description of the translocation dynamics we can associate reaction fluxes with the translocation sequence, Eqs. (20)–(22), respectively.

(1) The release flux is defined by

$$J(\mathrm{S}) = d[\bar{n}_r]/dt \tag{24}$$

where $\bar{n}_r$ is the average number of A^+ released into the reaction space between storage and receptor ring.

(2) The receptor flux including association of A^+ and conformation change of R is given by

$$J(\mathrm{R}) = d[\bar{n}]/dt \tag{25}$$

where $\bar{n}$ is the average number of A^+ associated with R.

(3) The esterase (or decomposition) flux is defined by

$$J(\mathrm{E}) = d[\bar{n}_e]/dt \tag{26}$$

where $\bar{n}_e$ is the average number of A^+ processed through AcCh-esterase.

Stationary states of the cholinergic activity are characterized by constant overall flow of AcCh; neither accumulations nor depletions of locally processed AcCh occur outside the limit of fluctuations. Thus, for stationary states,

$$J(\mathrm{S}) = J(\mathrm{R}) = J(\mathrm{E}) = \text{const} \tag{27}$$

Statistically occurring small changes in membrane properties, such as the so-called miniature-end-platepotentials, are interpreted as reflecting amplified fluctuations in the subthreshold activity of the cholinergic system.

Oscillatory excitation behavior observed under certain conditions [see, e.g., Cole (1968)] can be modeled by periodic accumulation and depletion of AcCh in the reaction spaces of the BEU's.

4. *Field Dependence of AcCh Storage*

In the simplest case, a change of the membrane potential affects the chain of translocation events already at the beginning, i.e., at the storage site. Indeed, the observation of AcCh release by electrical stimulation or in response to K^+-ion-induced depolarization support the assumption of a field-dependent storage site for AcCh.

Denoting by $\bar{n}_b$ the amount of AcCh bound on average to S_1, we can define a distribution constant for the stationary state of the storage translocation by $K = \bar{n}_b/\bar{n}_r$. This constant (similar to an equilibrium constant) is a function of temperature T, pressure p, ionic strength I, and electric field E. A field dependence of K requires that the storage translocation reaction involve ionic, dipolar, or polarizable groups.

The isothermal–isobaric field dependence of K at constant ionic strength can be expressed by the familiar relation

$$\left(\frac{\partial \ln K}{\partial E}\right)_{p,T,I} = \frac{\Delta M}{RT} \tag{28}$$

where ΔM is the reaction moment; ΔM is (proportional to) the difference in the permanent (or induced) dipole moments of reaction products and reactants. If a polarization process is associated with a finite value of ΔM, K should be proportional to E^2 (for relatively small field intensities up to 100 kV/cm). Furthermore, a small perturbation of the field causes major changes in K only on the level of higher fields [see, e.g., Eigen (1967)]. It is therefore of interest to recall that, under physiological conditions, excitable membranes generate action potentials only above a certain (negative) potential difference.

The suggestion of a field-induced conformational change in a storage protein to release AcCh derives from recent studies on field effects in macromolecular complexes and biomembranes. It has been found that electrical impulses in the intensity similar to the depolarization voltage changes for the induction of action potentials are able to cause structural changes in biopolyelectrolytes (Neumann and Katchalsky, 1972; Revzin and Neumann, 1974) and permeability changes in vesicular membranes (Neumann and Rosenheck, 1972). In order to explain the results, a polarization mechanism has been proposed that is based on the displacement of the counterion atmosphere of polyelectrolytes or of oligoelectrolytic domains in membrane organizations.

If the conformational dynamics of the storage site indeed involves a polarization mechanism, we may represent the dependence of bound AcCh, $\bar{n}_b$, on the electric field of the membrane as shown in Fig. 6. Increasing the membrane potential increases the amount of bound AcCh and thus also the

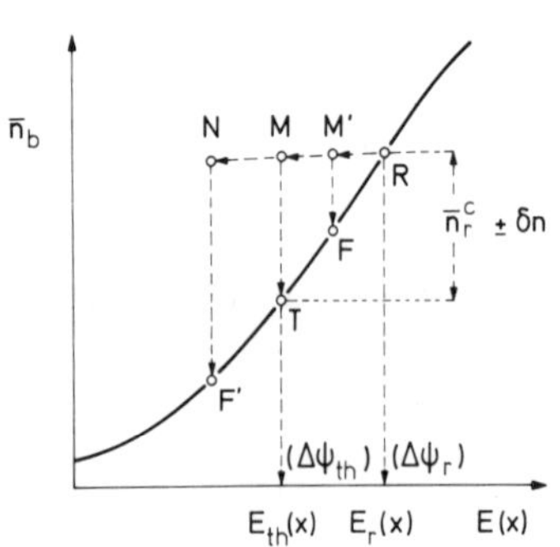

Fig. 6. Model representation of the field-dependent stationary states for AcCh storage. The mean number $\bar{n}_b$ of AcCh ions bound to the storage site at the membrane site x of the release reaction is given as a function of the electric field $E(x)$ (at constant pressure, temperature, and ionic strength). The intervals $M'F$, MT, and NF' correspond to the maximum number of AcCh ions released n_r for three different depolarization steps: (a) a subthreshold change from the resting state R to F, (b) a threshold step R to T, releasing the threshold or critical number n_r^c ($\pm\delta n$, fluctuation), and (c) a suprathreshold step R to F' with $n_r > n_r^c$ (Neumann, 1974).

number of AcCh ions that, after fast reduction of the membrane potential, are additionally (exceeding the stationary level) translocatable to the receptor.

5. Relaxations of AcCh Translocation Fluxes

It is recalled that the receptor reaction [cf. Eqs. (21) and (23)] plays a key role in coupling the control function of AcCh with the permeability change of the gateway. Uptake of AcCh from the storage ring, conformational transition, and Ca^{2+} release comprise a sequence of three single events. It is therefore assumed that the processing of AcCh through the receptor is slower than the preceding step of AcCh release from the storage. The receptor reaction is thus considered to be rate limiting. Therefore any (fast) change in the membrane field will lead to either a transient accumulation or a depletion of AcCh in the reaction space between the S ring and the R ring of a BEU.

In the case of a fast depolarization there is first a transient increase in the storage flux $J(\mathrm{S})$, causing transient accumulation of AcCh in the S–R reaction space. The accumulation rate $J(\mathrm{A}^+)$ is defined by

$$J(\mathrm{A}^+) = d[\bar{n}_a]/dt = J(\mathrm{S}) - J(\mathrm{R}) \tag{29}$$

where the receptor flux is considered as rate limiting also for the (fast) esterase flux $J(\mathrm{E})$. The number $\bar{n}_a$ of transiently accumulated AcCh is calculated by integration of Eq. (29). Recalling the definitions of the single fluxes, Eqs. (24) and (25), $\bar{n}_a = \bar{n}_r - \bar{n}$, and after adjustment of the flux system to stationarity, we obtain $\bar{n}_a = 0$.

Since flux intensities increase with increasing driving forces [see, e.g., Katchalsky (1967)], $J(\mathrm{S})$ will increase with increasing perturbation intensity, thus causing an increase in the rate of all following processes.

It is recalled that in the framework of our integral model the time course of changes in electrical membrane parameters, such as the membrane potential, is controlled by the cholinergic system and the gateway dynamics.

a. Subthreshold Relaxations. Since subthreshold perturbations do not induce the gateway transitions, Eq. (23), the time constant τ_m for the chemical part of the subthreshold relaxations of the membrane potential changes (see Section IC2) is thus equal to the time constant τ_R of the rate-limiting receptor flux. For squid giant axons $\tau_\mathrm{R} = \tau_m \simeq 1$ msec at 20°C.

The relaxation of $J(\mathrm{R})$ to a lasting subthreshold perturbation (e.g., current stimulation) is given by

$$\frac{dJ(\mathrm{R})}{dt} = -\frac{1}{\tau_\mathrm{R}}[J(\mathrm{R}) - J'(\mathrm{R})] \tag{30}$$

where $J'(\mathrm{R})$ is the stationary value of the new flux. Equivalent to Eq. (30), we have for $\bar{n}$

$$\frac{d[\bar{n}]}{dt} = -\frac{1}{\tau_{\mathrm{R}}}([\bar{n}] - [\bar{n}]') \tag{31}$$

describing an exponential "annealing" to a new level of AcCh, $\bar{n}'$, processed through the receptor ring.

It is evident from the reaction scheme (24)–(26) that the time constant τ_{R} is the relaxation time of a coupled reaction system. In order to demonstrate the dependence of τ_{R} on various system parameters, such as the local concentrations of the reaction partners, we can calculate τ_{R} using a few simplifying assumptions.

We recall that the release step is fast as compared to the receptor reaction. Furthermore, for subthreshold perturbations the changes in the local concentration of the metal ions are certainly small in comparison to the concentration changes of the cholinergic reaction partners (buffer condition). We denote by s and r the single binding sites of S and R, and use the simplified reaction scheme

$$s_1(\mathrm{A}^+) + \mathrm{C}^+ \underset{k'_{21}}{\overset{k'_{12}}{\rightleftharpoons}} s_2(\mathrm{C}^+) + \mathrm{A}^+ \tag{32}$$

$$\mathrm{A}^+ + r_1(\mathrm{Ca}^{2+}) \underset{k'_{32}}{\overset{k'_{23}}{\rightleftharpoons}} r_2(\mathrm{A}^+) + \mathrm{Ca}^{2+} \tag{33}$$

Due to the buffer condition, we can approximate $k''_{12} \simeq k'_{12}[\mathrm{C}^+]$ and $k''_{32} = k'_{32}[\mathrm{Ca}^{2+}]$. With $s_1(\mathrm{A}^+) = \bar{n}_b$, $r_2(\mathrm{A}^+) = \bar{n}$, and $[\mathrm{A}^+] = [\bar{n}_r] \simeq [s_2(\mathrm{C}^+)]$, the two reaction fluxes are

$$j(s) = -d[\bar{n}_b]/dt = k''_{12}[\bar{n}_b] - k'_{21}[\bar{n}_r]^2 \tag{34a}$$

$$j(r) = d[\bar{n}]/dt = k'_{23}[\bar{n}_r][r_1(\mathrm{Ca}^{2+})] - k''_{32}[\bar{n}] \tag{34b}$$

Applying normal mode analysis (see Eigen and DeMaeyer, 1963), we obtain the relaxation times $\tau(s)$ and $\tau(r)$:

$$\frac{1}{\tau(s)} = 4k'_{21}[\bar{n}_r]' + k'_{12}[\mathrm{C}^+]$$

$$\frac{1}{\tau(r)} = k'_{23}\left\{[\bar{n}_r]' + \frac{[r_1(\mathrm{Ca}^{2+})]'(k'_{12} + k'_{23}[\bar{n}_r]')}{k'_{12} + 4k'_{21}[\bar{n}_r]'}\right\} + k'_{32}[\mathrm{Ca}^{2+}][\bar{n}]'$$

It is noted that the rate coefficients k' contain conformational contributions; the concentration terms are primed and represent the stationary values of the new flux conditions.

In case of strongly concerted interactions, we can approximate $\tau(s) = \tau_S$ associated with $J(S)$, and $\tau(r) = \tau_R$ associated with $J(R)$.

b. Parameters of Suprathreshold Changes. Recall that the induction of an action potential is associated with three critical parameters: $\bar{n}^c$, $\bar{m}^c$, and $\Delta t^c = \tau_R$.

For a perturbation the intensity of which increases gradually with time, the condition $\bar{n} \geqslant \bar{n}^c$ corresponding to $\bar{n}_r \geqslant \bar{n}_r^c$ (within a BEU) can only be realized if the minimum slope condition leading to

$$dJ(S)/dt \geqslant dJ(S)^{\min}/dt$$

and to the equivalent expression for $J(R)$

$$dJ(R)/dt \geqslant dJ(R)^{\min}/dt$$

is fulfilled [see Section IC1(b), Eq. (10)].

For rectangular (step) perturbations the threshold conditions are

$$J(S) = [\bar{n}_r]/\tau_R \geqslant [\bar{n}_r^c]/\tau_R$$
$$J(R) = [\bar{n}]/\tau_R \geqslant [\bar{n}^c]/\tau_R$$

Since $J(S)$ increases with the intensity of the (step) perturbation, the time intervals Δt ($<\tau_R$) in which $\bar{n}^c$ AcCh ions start to associate with the receptor become smaller with larger stimulus intensities. (See also Fig. 6.)

We can write this "strength–duration" relationship for suprathreshold perturbations in the form

$$J(S)\, \Delta t = [\bar{n}_r^c] \qquad \text{and} \qquad J(R)\, \Delta t = [\bar{n}^c] \tag{35}$$

[compare Eq. (9)]. The time intervals Δt of the receptor activation correspond to the observed latency phases.)

The expressions corresponding to the rheobase (see Section IC1) are

$$J(S)_{\text{th}} = [\bar{n}_r^c]/\tau_R \qquad \text{and} \qquad J(R)_{\text{th}} = [\bar{n}^c]/\tau_R \tag{36}$$

These equations clearly demonstrate the impulse condition [release of $\bar{n}_r \geqslant \bar{n}_r^c$ within Δt^c; cf. Eq. (11)] for the induction of action potentials.

Since $\bar{n}_r^c$ and $\bar{n}^c$ are numbers describing functional cooperativity, the strength–duration products in (35) do not depend on temperature. The fluctuations $\pm\delta n$ for $\bar{n}_r$, however, increase with increasing temperature (and may finally lead to thermal triggering of action potentials). The flux equivalents of the rheobase, Eq. (36), are reaction rates which in general are temperature dependent (having a Q_{10} coefficient of about 2).

There are further aspects of electrophysiological observations which the integral model at the present stage of development can (at least qualitatively) reproduce.

If the membrane potential is slowly reduced, subthreshold flux relaxation of the ratio $\bar{n}_b/\bar{n}_r$ may keep $\bar{n}_r$ always smaller than $\bar{n}_r^c$. Thus, corresponding to experience, slow depolarization does not evoke action potentials (or does so only occasionally).

In order to match the condition $\bar{n}_r > \bar{n}_r^c$ starting from the resting potential, the depolarization has in any case to go beyond the threshold potential, where $\bar{n}_b(M) - \bar{n}_b(T) = \bar{n}_r^c$. (See Fig. 6.)

For stationary membrane potentials $\Delta\psi < \Delta\psi_{\text{th}}$, the maximum number of AcCh ions that can be released by fast depolarization within Δt^c is less than $\bar{n}_r^c$. Thus, corresponding to experience, below a certain membrane potential, near $<\Delta\psi_{\text{th}}$, no nerve impulse can be generated.

c. Refractory Phenomena. After the gateway transition to the open state the receptors of a BEU have to return to the Ca^{2+} binding conformation $R_1(Ca^{2+})$ before a second impulse can be evoked [see Eq. (23)]. Even if $\bar{n}^c$ AcCh ions were already available, the time interval for the transition of $\bar{m}^c$ receptors is finite and causes the observed absolutely refractory phase.

Hyperpolarizing prepulses shift the stationary concentration of $\bar{n}_b$ to higher values (see Fig. 6). Due to an increased "filling degree," the storage site appears to be more sensitive to potential changes (leading, among other factors, to the so-called "off responses"). On the other hand, depolarizing prepulses and preceding action potentials temporarily decrease the actual value of $\bar{n}_b$, thus requiring increased stimulus intensities for the induction of action potentials.

The assumptions for the kinetic properties of the storage site mentioned in the discussion of Eq. (16) are motivated by the accommodation behavior of excitable membranes. The observation of a relatively refractory phase suggests that the uptake of AcCh into the storage form $S_1(A^+)$ is slow compared to the release reaction. Therefore after several impulses there is partial "exhaustion" of the storage site. If, during the (slow) refilling phase there is a new stimulation, $\bar{n}_b$ may still be lower than the stationary level. Therefore the membrane has to be depolarized to a larger extent in order to fulfill the action potential condition $\bar{n}_r \geqslant \bar{n}_r^c$.

6. The AcCh Control Cycle

The cyclic nature of a cholinergic permeability control in excitable membranes by the processing of AcCh through storage, receptor, esterase, and synthetase was already indicated in a reaction scheme developed 20 years ago (Nachmansohn, 1955, Fig. 11). The complexity of mutual coupling between the various cycles directly or indirectly involved in the permeability control of the cholinergic gateway is schematically represented in Fig. 7. In this representation it can be readily seen that manipulation such as external application of AcCh and its inhibitory or activating analogs may interfere at *several sites* of the AcCh cycle. In particular, the analysis of pharmacological and chemical experiments has to face this complexity.

In the previous sections it has been shown that basic parameters of electrophysiological phenomenology can be modeled in the framework of a nonequilibrium treatment of the cholinergic reaction system. The various assumptions and their motivation by experimental observations have been discussed and the cholinergic reaction cycle has been formulated in terms of a chemical reaction scheme.

In conclusion, the integral model at the present level of development

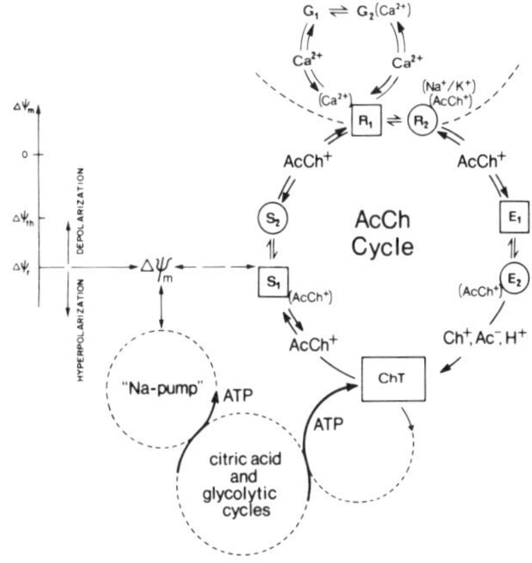

Fig. 7. AcCh cycle for the cyclic chemical control of stationary membrane potentials $\Delta\psi$ and transient potential changes. The binding capacity of the storage site for AcCh is assumed to depend on the membrane potential $\Delta\psi_m$ and is thereby coupled to the "Na^+/K^+ exchange pump" (and the citric acid and glycolytic cycles). The control cycle for the gateway G_1 (Ca^{2+} binding and closed) and G_2 (open) comprises the SRE assemblies (see Fig. 4) and the choline-*O*-acetyltransferase (Ch-T); Ch-T couples the AcCh synthesis cycle to the translocation pathway of AcCh through the SRE assemblies. The continuous subthreshold flux of AcCh through such a subunit is maintained by the virtually irreversible hydrolysis of AcCh to choline (Ch^+), acetate (Ac^-), and protons (H^+) and by steady supply flux of AcCh to the storage from the synthesis cycle. In the resting stationary state, the membrane potential $\Delta\psi_r$ reflects dynamic balance between active transport (and AcCh synthesis) and passive fluxes of AcCh (through the control cycles surrounding the gateway) and of the various ions asymmetrically distributed across the membrane. Fluctuations in membrane potential (and exchange currents) are presumably amplified by fluctuations in the local AcCh concentrations maintained at a stationary level during the continuous translocation of AcCh through the cycle.

appears to cover all the essential pharmacological, electrophysiological, and biochemical data on excitable membranes. The model is expressed in terms of specific reactions, subject to further experimental investigations involving the reaction behavior of isolated membrane components as well as of membrane fragments containing these components, together with the role of structure and organization.

II. BIOCHEMICAL FOUNDATION OF THE INTEGRAL MODEL

A. Cell Membranes

Before discussing biochemical aspects of excitable membranes, it appears necessary to recall a few basic notions of cell membranes in general. It has been recognized that membranes are the site of most vital cell functions, such as those involved in energy supply, active transport, neural function, vision, excitation–contraction coupling, and photosynthesis. Moreover, recent investigations have revealed the fundamental importance of structure and organization in the chemical reactions in living cells. Whereas *classical* biochemistry devoted most of its effort to the isolation and characterization of cell constituents and their behavior in solution, modern biophysical chemistry has recognized that essential characteristics of reactions occurring in solution cannot be simply extrapolated to those of the same chemical processes occurring in the cell. It has become increasingly apparent that rates and extents of reactions are profoundly influenced by structural factors, such as microenvironment (e.g., charged groups surrounding the active site), cooperativity, allosteric effects, allotopy, regulatory factors, and protein–protein and protein–lipid interactions [see, e.g., Loewenstein (1966), Racker (1970), Manson (1971), Rothfield (1971)]. Therefore modern biochemistry aims at analyzing the reaction behavior within cellular structures, in addition to that in solution and on the molecular level.

There has been a major conceptual change in the field of membrane research in the last few years. At the turn of the century, cell membranes were considered mainly as passive barriers in which lipids were essential in preventing easy passage of cell constituents and metabolites. Today it is well established that cell membranes contain many proteins, including enzymes. In some membranes 30–50 different proteins have been isolated and many of them identified. Biomembranes have been recognized as extremely dynamic structures, the site of a great topical variety of chemical processes; Aharon Katchalsky once referred to biomembranes as being powerful biochemical factories.

In many biological membranes greater than two-thirds of the mass is proteins, and about one-third is phospholipids. In a given membrane type, the phospholipids and their chemical properties show a great diversity. This diversity increases when different types of membranes are compared. In addition, there are oligosaccharides and a great variety of small molecules, including different metal ions. However, the remarkable specificity, the diversity, and the efficiency of membrane functions seem to be more readily accounted for when we attribute the dominant role to the membrane proteins rather than to the lipid phase.

In spite of the large amount of information on the chemical composition of membranes, we are still very far from a real knowledge of the molecular organization of the various constituents in the intact membrane. Indeed, recently more than ten different membrane models have been proposed. Without discussing their merits, we would like to mention two models as an illustration of the developments during the last decade: the model of Robertson (1960) of the "unit membrane," and the model of Sjoestrand and Barajas (1970) representing the inner membrane of mitochondria (Fig. 8). The latter model incorporates the many enzymes, showing their subunits and coenzymes, well established to be present in this membrane; it also accounts for the proper relationship between the amount of proteins and that of phospholipids. This model is rather attractive because it integrates a great number of the known membrane constituents.

Recognizing the importance of the proteins and the necessity of preserving them in their native conformation, Sjoestrand and Barajas have introduced new techniques for the preparation and fixation of the specimen for electron microscopy; they avoided the usual standard techniques, which almost certainly denature the proteins. Using the new techniques, they find that the membranes appear to have a thickness of 150–200 Å as compared to the 80–100 Å obtained with the standard procedures. A striking feature of their electron micrographs is the indication of many globular formations within the membrane (Sjoestrand and Barajas, 1968).

Excitable membranes. Excitable membranes have the special ability of transiently changing their permeability in a controlled way with respect to those ions that are the carriers of the electrical exchange currents accompanying membrane potential changes, such as the nerve impulses. As was emphasized by Hill (1960), there appears to be no alternative to the assumption that "*the early production and absorption of heat after a stimulus are largely due to chemical reactions associated with, and following, the permeability cycle.*" Furthermore, various biochemical and electrophysiological observations [see, for instance, Tasaki (1968)] are incompatible with the assumption of simple electrodiffusion processes as an explanation for the generation of the action potential [see also Cole (1965)]. It should be realized that the

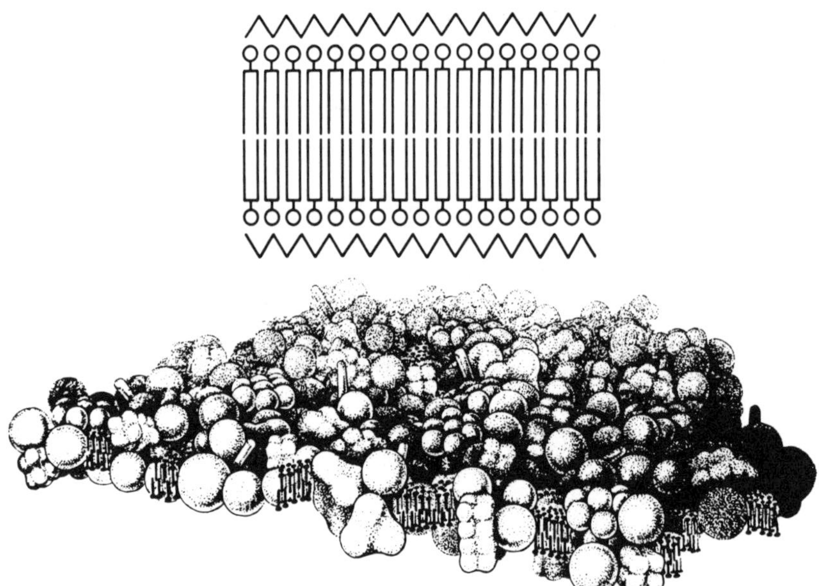

Fig. 8. Top: Model of Robertson (1960) of the "unit membrane," proposed to be 80 Å thick and formed by a bimolecular layer of phospholipids surrounded on the inside and outside by proteins attached to the phospholipids by Coulombic forces. Bottom: Model of Sjoestrand and Barajas (1970) of the inner mitochondrial membrane. The model tries to integrate the enzymes and coenzymes with their subunits biochemically established to be located in this membrane. The phospholipids of this membrane are known to form only a relatively small fraction of this membrane, as indicated in the figure. Thickness about 150–200 Å.

increased ion fluxes during electrical activity appear, indeed, to result from complex intramembrane processes. It is obvious that an elucidation of the mechanism requires the analysis of these membrane processes.

Recent theories describe mechanisms for the permeability changes during nerve activity in terms of cooperative structural changes of macromolecular membrane components (e.g., Tasaki, 1968; Adam, 1970; Blumenthal *et al.*, 1970). These views are very attractive as far as the change in the ion permeability as such, i.e., the gateway behavior, is concerned. However, these macromolecular approaches, too, *ignore* the results of the numerous biochemical studies on bioelectricity and are therefore *selective descriptions covering only a part of the whole phenomenon.* The many data indicating the mutual dependence of electrical parameters and biochemical processes, such as the effects of specific inhibitors of an AcCh processing system in excitable membranes, must be covered and integrated in any model of nerve excitability.

In brief, the various biochemical studies suggest that the permeability changes of excitable membranes are *in vivo* controlled by specific membrane reactions involving the AcCh system.

B. Chemical Hypothesis of Excitability

1. Proteins Processing AcCh

A chemical hypothesis of nerve excitation has been elaborated during the last three decades. The early data have been summarized in a monograph (Nachmansohn, 1959) and later developments in a series of review articles (Nachmansohn, 1963; 1966a, b; 1968; 1970; 1971a, b; 1973). The chemical approach to the study of the mechanisms underlying bioelectricity is essentially based on the analysis of the properties and functions of the proteins processing acetylcholine (AcCh). Instrumental in the experimental investigations has been, since 1937, the use of the electric organs of electric fish, a tissue specialized to an extraordinary degree for the generation of bioelectricity, a specialization almost without a parallel in the animal kingdom. The studies have led to the isolation and characterization of three proteins directly associated with the function of AcCh: AcCh-esterase, AcCh-receptor, and choline-*O*-acetyltransferase.

a. AcCh-esterase. The enzyme AcCh-esterase was first isolated from electric tissue of *Torpedo* in 1938 (Nachmansohn and Lederer, 1939); the protein was purified in the early 1940's and this preparation permitted the analysis of the molecular groups in the active site of the enzyme and led to the explanation of the reaction mechanisms with a variety of compounds and drugs, such as, e.g., the organophosphates. The enzyme was crystallized in 1967 (Leuzinger *et al.*, 1968).

b. AcCh-receptor. To account for a number of experimental observations, an axonal receptor protein capable of changing its conformation upon the reaction with AcCh was first postulated in 1953 (Fig. 9) (Nachmansohn, 1955). Evidence for the protein nature of the receptor was adduced from biochemical and electrophysiological studies of a monocellular electroplax preparation developed by Schoffeniels (Schoffeniels and Nachmansohn, 1957; Schoffeniels, 1957) and refined in the following years by Higman and Bartels (1961, 1962; Higman *et al.*, 1963, 1964). The use of intracellular electrodes and of a particular switching device have permitted precise recording of various electrical parameters, such as simultaneous measurements of the potentials across the excitable and nonexcitable membranes

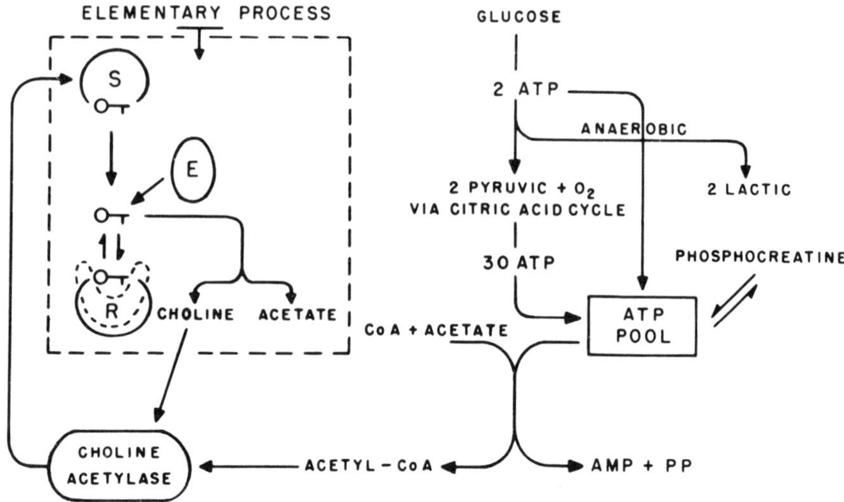

Fig. 9. Sequence of energy transformations associated with conduction, and integration of the acetylcholine system into the metabolic pathways of the nerve cell. The elementary process of conduction may be tentatively pictured as follows: (1) In resting condition acetylcholine (○T) is bound, presumably to a storage protein (S). The membrane is polarized. (2) AcCh is released by current flow (possibly hydrogen ion movements) or any other excitatory agent. The free ester combines with the receptor (R), presumably a protein. (3) The receptor changes its configuration (broken line). This process increases the Na ion permeability and permits its rapid influx. This is the trigger action by which the potential primary source of e.m.f., the ionic concentration gradient, becomes effective and by which the action current is generated. (4) The ester–receptor complex is in dynamic equilibrium with the free ester and the receptor; the free ester is open to attack by acetylcholinesterase (E). (5) The hydrolysis of the ester permits the receptor to return to its original shape. The permeability decreases, and the membrane is again in its original polarized condition.

as well as across the whole cell. Procedures were worked out for determining, with a high degree of reliability, apparent dissociation constants for specific ligands reacting with the receptor protein (Higman *et al.*, 1963). The quantitative analysis of the apparent dissociation constants suggested some details of the nature of the reaction sites in the receptor and revealed both similarities and differences of the active site of the receptor as compared to that of the AcCh-esterase. As an illustration a few examples are given.

Webb (1965) studied reactions of the electroplax with derivates of *N, N'*-bis(diethylaminopropyl)quinone (benzoquinonium) and of *N, N'*-bis(diethylaminoethyl)oxamide-bisbenzylhalide (ambenoniums). The apparent dissociation constants were compared with those obtained with AcCh-esterase (prepared from electric tissue) in solution. This comparison demonstrated that one component of the electroplax reacting with the test compounds was clearly different from the AcCh-esterase. Additionally, these experiments strongly support the postulated protein nature of the receptor.

Studies with a series of aryltrimethylammonium and of *n*-alkyltrimethyl ammonium ions performed by Podleski (1966, 1969) also revealed the enormous differences between the apparent dissociation constants for the receptor and the enzyme. Of particular interest was the use of 1-methyl-acetoxyquinolinium iodide, which belongs to a group of quinolinium derivatives studied by Prince (1966) on AcCh-esterase in solution. These investigations brought a conclusive distinction between the receptor binding site and that of the esterase (Podleski and Nachmansohn, 1966; Podleski, 1967).

Moreover, it was found that compounds reacting with SH groups also affect electrical activity in the electroplax of *Electrophorus* (Karlin and Bartels, 1966). Since sulfhydryl groups are usually associated with proteins, these observations confirm that proteins are associated with bioelectricity (Nachmansohn, 1971a, pp. 55–63). The actual separation of the two proteins was achieved by Changeux and his associates, who isolated the receptor by using α-toxin of *Naja naja* coupled to Sepharose. This group succeeded in separating receptor and enzyme proteins (Meunier *et al.*, 1971). In recent years the receptor has been isolated and purified in many laboratories (e.g., Olsen *et al.*, 1972; Schmidt and Raftery, 1973; Karlin and Cowburn, 1973; Biesecker, 1973; Eldefrawi and Eldefrawi, 1973; Chang, 1974).

c. Choline-O-acetyltransferase. The enzymatic formation of AcCh in a soluble system was discovered in 1942 (Nachmansohn and Machado, 1943). The AcCh synthesis requires the energy of ATP hydrolysis. The enzymatic synthesis of AcCh *in vitro* was the *first* experimental demonstration of an ATP-dependent acetylation. The synthesizing enzyme was first referred to as choline acetylase and is now called choline-*O*-acetyltransferase. This enzyme has been purified to a relatively high degree by standard procedures from squid head ganglia. Recently Husain and Mautner (1973) described a very efficient procedure, using affinity chromatography, which leads to a preparation with a high degree of purity.

d. Storage site. The storage site for AcCh in the excitable membrane is most likely a protein. Only protein would account for the specificity of binding; a relatively high binding constant is suggested by the great difficulty of removing AcCh from the membrane. The protein nature of the storage site was postulated in 1953 (Nachmansohn, 1955).

2. *Distribution of AcCh-esterase and Choline-O-acetyltransferase*

a. AcCh-esterase. Chemical analysis has shown that AcCh-esterase, an esterase well characterized by a number of specific features such as, e.g., a bell-shaped activity–substrate concentration curve, is present in all types of nerve and muscle fibers without exception [for summaries see Nachmansohn (1959, 1963)]. Moreover, indirect biochemical evidence suggested that AcCh-esterase is localized at or near the surface of axons [see, e.g., Nachmansohn and Meyerhof (1941)]; at that time membranes were not yet visualized by electron microscopy. Numerous reports appeared, particularly in the 1950's, which claimed that AcCh-esterase is absent in many excitable cells [see, e.g., Koelle (1963)]. These statements were based on staining techniques with the use of the light microscope. In the specimen slices used in these techniques, the thickness of nerve covering tissue layers may prevent the access of the added substrate (acetylthiocholine) to the enzyme. The absence of staining outside the junction was for more than a decade the only evidence for the assumption that in muscle fibers the enzyme is located exclusively at the motor end plate. In some tissues, in which the enzyme concentration was extremely high when chemical methods were used, the enzyme appeared to be hardly visible or even absent when staining techniques were applied. Even in certain electric organs no AcCh-esterase was detected. All these reports were in sharp contrast to the results obtained with chemical methods demonstrating the presence of AcCh-esterase in all types of excitable fibers from the lowest to the highest forms of life. It is particularly remarkable that not a single exception has been found. It appears plausible to attribute the enhanced staining seen in nerve parts containing junctions to the marked increase in the surface area due to invaginations in the membranes of many synaptic junctions.

b. Choline-O-acetyltransferase. The presence of choline-*O*-acetyltransferase has been demonstrated in a wide variety of excitable tissues, nerve, and muscle fibers. [For summaries see Nachmansohn (1959, 1963).] Again no exception has been found. In general, the concentration of this enzyme is one to two orders of magnitude lower than that of AcCh-esterase. In the framework of the chemical hypothesis of excitability, this is not surprising; the latter appears to be directly associated with electrical activity and may require not only a fast reaction but also a larger margin of excess than an enzyme acting in the recovery period, which may be of long duration. In some fibers choline-*O*-acetyltransferase concentration appears to be low. It was found that in some sensory fibers only about 30 μg of AcCh is formed per gram of fresh tissue per hour. The low concentration in this fiber was considered by Hebb (1957) as a contradiction to the theory that acetylcholine is associated with conduction. Expressing activity of the enzyme in terms of gram of fresh tissue does not, in general, give a satisfactory indication of the actual concentration of AcCh formed at or near the membranes. Membranes may form only a small fraction, say 10^{-4}, of the total mass of the tissue.

Therefore the statement that the concentration is too low to be compatible with the assumption of the role of AcCh in conduction does not appear to be justified [see Nachmansohn (1963)]; in other sensory fibers choline-*O*-acetyltransferase is present in high concentrations (Davis and Nachmansohn, 1964). Of particular importance is the recognition that the enzyme is very unstable. Although the AcCh-esterase is relatively stable, it is nevertheless extremely difficult to obtain quantitative data for the AcCh-esterase concentration in a tissue (see below); the problems of a quantitative evaluation of an unstable enzyme are even greater.

3. *Localization of AcCh-esterase*

When later, in the 1960's, staining techniques were applied in combination with electron microscopy the situation changed drastically. The enzyme was found to be closely associated with the excitable membranes in the conducting parts as well as in the two junctional membranes, those of the nerve terminal and of the postsynaptic membrane. The enzyme was at first found only in the membranes of unmyelinated fibers and seemed to be absent in myelinated nerves. This was again in contrast to the results of chemical analysis. Using the magnetic diver technique, the enzyme was found in myelinated axons isolated from the frog sciatic nerve fiber both at the nodes of Ranvier and in the intermediate section (Brzin and Dettbarn, 1967). Brzin (1966) prepared single axons from this fiber with slices of about 500 Å thickness for examination by electron microscopy. He incubated the slices in the detergent Triton X-100 prior to the addition of the reagents used for the staining. Now the plasma membrane located between the myelin and the axoplasm was regularly and fully stained (Fig. 10). A particularly elegant demonstration of AcCh-esterase in conducting as well as in the *pre-* and *postsynaptic* membranes, using new refined staining techniques, has been recently achieved by Koelle and his associates (Fig. 11).

As mentioned before, AcCh-esterase is chemically detectable in all excitable cells tested so far. However, as in the case of the choline-*O*-acetyltransferase, the AcCh-esterase concentration differs greatly in various types of fibers and in different species. In some cases, such as electric tissue or squid head ganglia, the enzyme concentrations are two to three orders of magnitude higher than in most other tissues, as discussed in the summaries quoted. It is noted that here, too, the chemical determinations were made with homogenized suspensions of the tissue. In the last decade evidence has accumulated that homogenized suspensions do not permit quantitative evaluations of the enzyme present in a tissue (Nachmansohn, 1971a, pp. 37–46). Even in the small particles of a suspension there seems to be enough lipid present to prevent saturation of the enzyme with the substrate. After

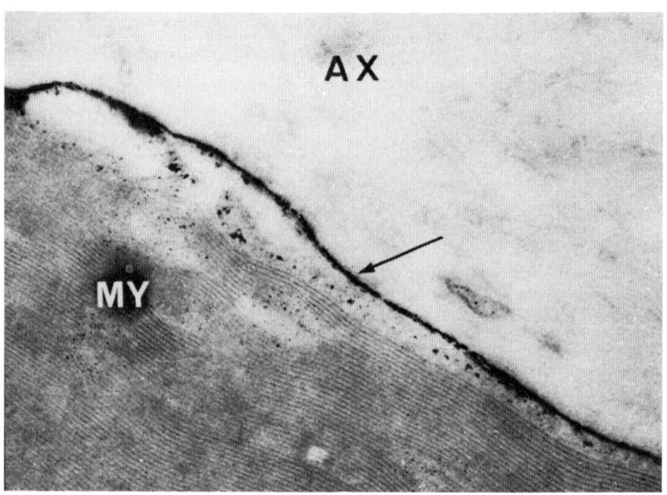

Fig. 10. Large myelinated (MY) ventral root axon (AX) taken from a frog sciatic nerve. The slice was treated with Triton X-100 before the incubation for testing acetylcholinesterase activity with the standard procedure for histochemical staining of the enzyme (adding acetylthiocholine and copper sulfate). The hydrolytic product, thiocholine, forms a precipitate with cooper sulfate. The dense end product is present in the axolemmal (plasma) membrane (arrow). (Brzin, 1966.) Reduced 20% for reproduction.

additional treatment of the suspension with high salt concentration (1–2 M NaCl) or with detergents known to remove a part of the lipids, the activity of the enzyme is found to be much higher than before. This result is not surprising because all nerve tissues and their membranes are rich in lipids, which are generally impervious to AcCh. Thus the figures previously reported for the AcCh-esterase content of nerve tissue are almost certainly lower than are the actual enzyme concentrations of the excitable membranes.

In an electron micrograph of an isolated excitable membrane of an electroplax of *Electrophorus* (Changeux *et al.*, 1969), stained with standard methods for demonstrating the presence of AcCh-esterase, the distribution appears to be uniform all along the membrane (Fig. 12). It should be stressed that about 99% of this membrane is formed by the conducting parts and only a very small fraction are synaptic parts. This apparent uniformity is even more clearly visible in an electron micrograph of the electroplax membrane prepared by Drs. N. Tomas, R. Davis, and G. B. Koelle with the new gold sulfide method (personal communication). The AcCh-esterase concentration in this particular membrane is very high; about 10^{11} molecules of enzyme are present in the excitable membrane of a single electroplax. Thus the AcCh-esterase forms several percent of the mass of the membrane.

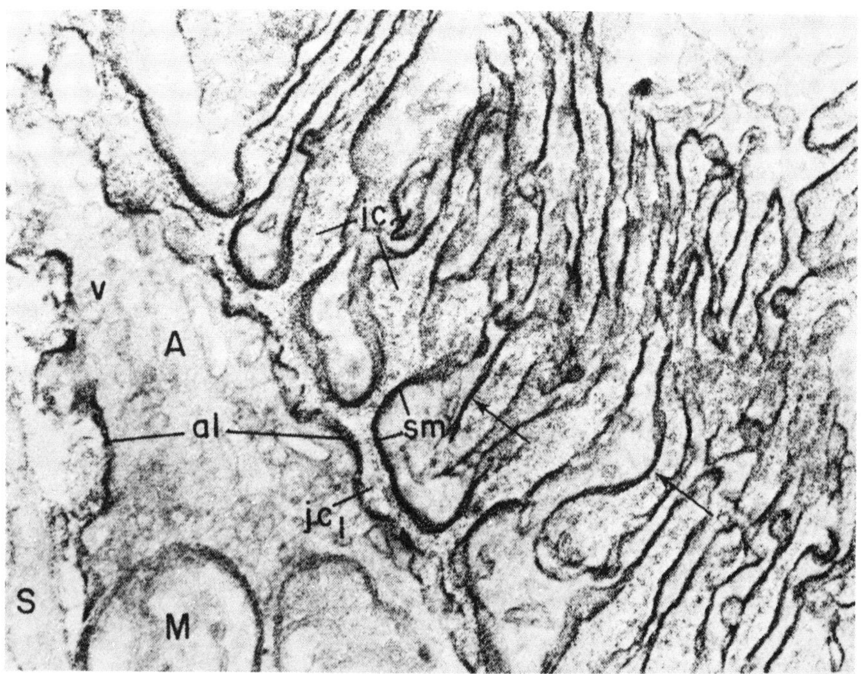

Fig. 11. Electron microscopic histochemical localization of acetylcholinesterase at the motor end plate of mouse intercostal muscle. A high-magnification ($\times 63{,}000$) view of the junctional complex, showing the axonal terminal (A) containing mitochondria (M) and numerous synaptic vesicles (v), the junctional cleft (jc), and junctional folds of the sarcolemma (sm). The electrondense granules, 40–50 Å in diameter, represent gold sulfide, the reaction product of the gold-thiolacetic acid method for the detection of acetylcholinesterase and nonspecific cholinesterase. The axolemma (al) exhibits marked enzymic activity both on the surface facing the primary junctional cleft (jc_1) and at the surface facing the teloglial Schwann cell sheath (S) (the axonal terminal is somewhat separated from the Schwann cell in this micrograph). Where the plane of section is perpendicular to the sarcolemma (arrows) the particles form a dense line about 120–140 Å thick. (Koelle, 1971.) Reduced 20% for reproduction.

The number of receptor molecules in the same membrane appears to be of the same order of magnitude (Karlin *et al.*, 1970).

6. *Role Proposed for the Function of AcCh*

The chemical hypothesis of nerve excitability which has emerged from biochemical and electrophysiological studies is briefly summarized. It is postulated that AcCh is present *within* the excitable membrane bound to a

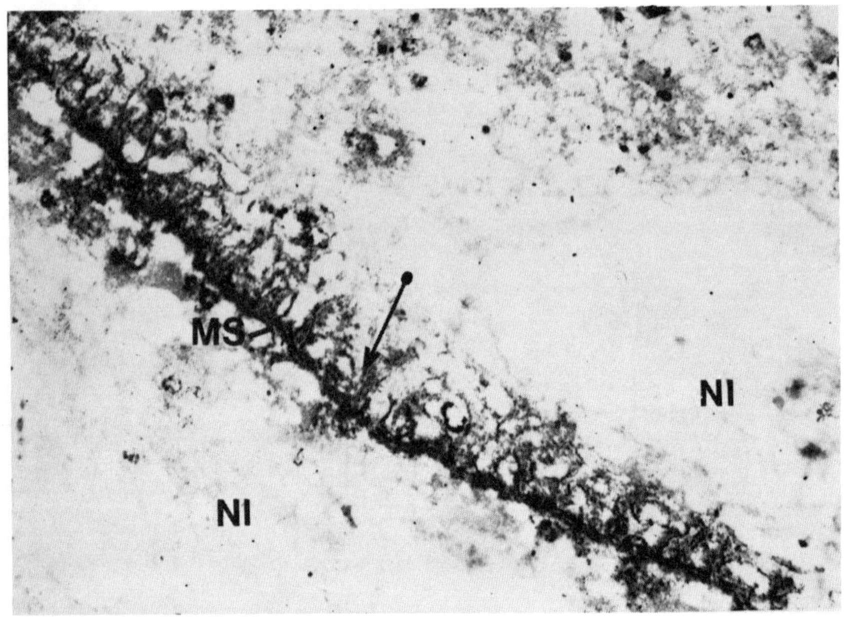

Fig. 12. Electron micrograph of an isolated fragment of excitable membrane from the electroplax of *Electrophorus*, tested for AcCh-esterase activity by standard procedures. The picture shows the uniformity of distribution of AcCh-esterase at the innervated membrane surface (MS). No staining was found in the noninnervated (NI) membrane (×8860). (From Changeux *et al.*, 1969.) Reduced 25% for reproduction.

storage protein. On excitation, AcCh is released *within* the membrane and initiates a series of reactions leading to a large amplification of the normal subthreshold activity. When released from the storage site, AcCh is translocated to the receptor protein (R), inducing a conformational change. This conformational change is assumed to release Ca^{2+} ions, which are bound to the receptor in resting condition. The released Ca^{2+} ions act on the elements in the gateway, producing the changes that permit the flow of ions. These changes may also proceed in several steps. Per molecule of AcCh released, many thousands of ions, probably about 15,000–30,000, move across the membrane in each direction. The conformational change of R is assumed to translocate AcCh to the enzyme AcCh-esterase (E). AcCh is hydrolyzed, in microseconds, and the receptor returns to its original conformation and again binds Ca^{2+} ions; thus the barrier for ion movement is reestablished. But in the meantime the ion flux and the changes of electric field have stimulated adjacent sections of the membrane.

It is assumed that the three proteins, the storage protein (S), the receptor protein (R), and the enzyme protein (E), are interlocked in a complex which is located close to the gateway through which the ions are supposed to move during activity (Nachmansohn, 1973).

In this hypothesis the AcCh system is the essential specific control mechanism covering a series of processes leading to ion permeability changes.

5. *Interdependence between Enzyme, Receptor, and Electrical Activity*

It should be emphasized that the classical chemical theory is *not* based either on the presence or on the high concentration of the enzyme or on the evidence for the receptor in all excitable membranes. The hypothesis derives from the functional relationships observed between the activities of two of the proteins, receptor and enzyme, and the electrical activity. When, for instance, specific competitive inhibitors of either the enzyme or of the receptor are applied to the excitable membrane under appropriate conditions, strong effects on the different electrical parameters have been demonstrated. The direct relationship between the two proteins and axonal conduction has been established with sensory and motor, with so-called "cholinergic" and "adrenergic" fibers, with those of vertebrates and invertebrates, and with muscle fibers (outside synaptic junctions).

The chemical studies have offered pertinent information about the specific proteins involved in the mechanism preceding and controlling the permeability changes in the gateway. They also offer evidence for the cyclic nature of the elementary processes involved in this particular function, an important feature characteristic of cellular mechanisms in general (see Fig. 7). The ability of the enzyme of inactivating the substrate within microseconds is a prerequisite for the theory; this reaction is essential for any postulate of a chemical termination mechanism for the electrical activity, in view of the high propagation speed of the processes.

C. Macromolecular Conformation and Ca^{2+} Ions

It may be useful to recall a few physicochemical aspects of macromolecular conformational changes in connection with Ca^{2+} ions. Structural changes of proteins and macromolecular organizations such as membranes are often cooperative in nature. One of the consequences of cooperativity is the possibility of far-reaching conformational changes by small local changes of environmental conditions. Moreover, conformational changes induced by binding of a ligand at one site may change the reactivity of other, possibly even far remote sites of a macromolecular system (allosteric effects).

The Ca^{2+} ions are particularly effective in inducing large conformational changes, as, for instance, in muscular contractions, and are particularly efficient in systems that contain regions of a relatively high negative surface charge. In such polyelectrolytic regions the osmotic coefficient for Ca^{2+} is of the order of 0.01, i.e., about 99 % of Ca^{2+} counterions are bound (Katchalsky, 1964). The high binding capacity is one of the reasons for the assumption that Ca^{2+} ions play an essential role in maintaining structural and functional integrity of protein and lipoprotein organization. For almost a century Ca^{2+} ions have been assumed to play an essential role in nerve excitability (Brink, 1954). More recently, Tasaki (1968) has emphasized that Ca^{2+} ions are absolutely necessary for nerve excitability.

III. SYNAPTIC TRANSMISSION

A. Problem of the Role of AcCh at Junctions

The well-known hypothesis of neurohumoral transmission attributes to AcCh a function only at synaptic junctions. It is postulated that AcCh is released from the nerve terminal and acts, after crossing the nonconducting gap, as a mediator of the impulse between nerve and nerve or nerve and muscle. This view explicitly postulates fundamentally different mechanisms: chemical transmission of AcCh across junctions, and purely electrical propagation of nerve impulses along fibers. It is this assumption of such a basic difference which appeared unacceptable to many neurobiologists. On the basis of electrical signs, conducting and synaptic parts of excitable membranes share many similar properties; therefore these facts were considered to be hardly compatible with the view of an entirely different mechanism. Not the facts, but only the interpretations were questioned and rejected [see, e.g., Erlanger (1939) and Fulton (1938)]. Since the role of AcCh as a neurohumoral transmitter across junctions is still a widely accepted view, it appears imperative to analyze, at least briefly, the main data on which the original hypothesis was based, to review some more recent data contradicting the original hypothesis and requiring drastic modifications in it, and finally to discuss how in the light of the information presently available one may explain and reconcile the early observations with the integral model of nerve excitability presented here. Rather than a detailed discussion of the peculiar properties of synaptic transmission in general, only those aspects will be stressed that refer to the specific function of AcCh at junctions. This question cannot be omitted or ignored without making it extremely difficult for the reader to understand and evaluate how the two seemingly contrasting views as to the function of AcCh can be reconciled and form part of the integral model. An alternative interpretation of the function of AcCh at the junction will be discussed. It is compatible with the integral model, attributing to AcCh the same role in conducting and synaptic parts of excitable membranes.

B. Early Observations That Suggested a Special Function of AcCh at Synaptic Junctions

1. External Application of AcCh

One of the most fundamental facts supporting a transmitter role of AcCh appeared to be the powerful action of externally applied AcCh on junctions, in contrast to the total failure of AcCh to affect conduction of fibers even in high concentrations. As in the famous experiments of Claude Bernard on curare, frog sciatic nerve fibers were used to test the effect of AcCh on conduction. These fibers are composed of several thousand axons in which the conducting membranes are surrounded by a myelin sheath 30,000–50,000 Å thick; in addition, the entire fiber is surrounded by a sheath of connective tissue. Thus the excitable membrane is well protected against many compounds and particularly against lipid-insoluble quaternary ammonium derivatives such as AcCh and curare. Only at the nodes of Ranvier is myelin absent, although even there one finds a base membrane and some tissue covering the excitable membrane (the plasma membrane) (Fig. 13). When a single axon of this sciatic nerve fiber is isolated, it is possible to measure electrical activity at a single Ranvier node. When curare is applied to such a single axon preparation, it blocks, rapidly and reversibly, the electrical response (Dettbarn, 1960a, b; Dettbarn and Davis, 1963). Thus the action of curare on the conducting membrane parts is similar to that on the synaptic parts; this result is in direct contrast to the classical views maintained since Claude Bernard. The concentrations of curare required for the action on axons are usually higher than at the junction, and vary between 10^{-5} and 10^{-3} M. This is not surprising. As mentioned before, the excitable membrane even at the node of Ranvier is covered by a base membrane and in addition by some other tissue (see Fig. 13); the thickness of the protective layer shows considerable variations in different axons of the same fiber, as shown in electron micrographs (Robertson, 1960). Therefore charged molecules such as curare may have difficulties to varying degrees in reaching the membrane even at the node. This assumption is supported by the observation that the quaternary neostigmine requires ten times higer concentrations to produce the same effects as the tertiary physostigmine,

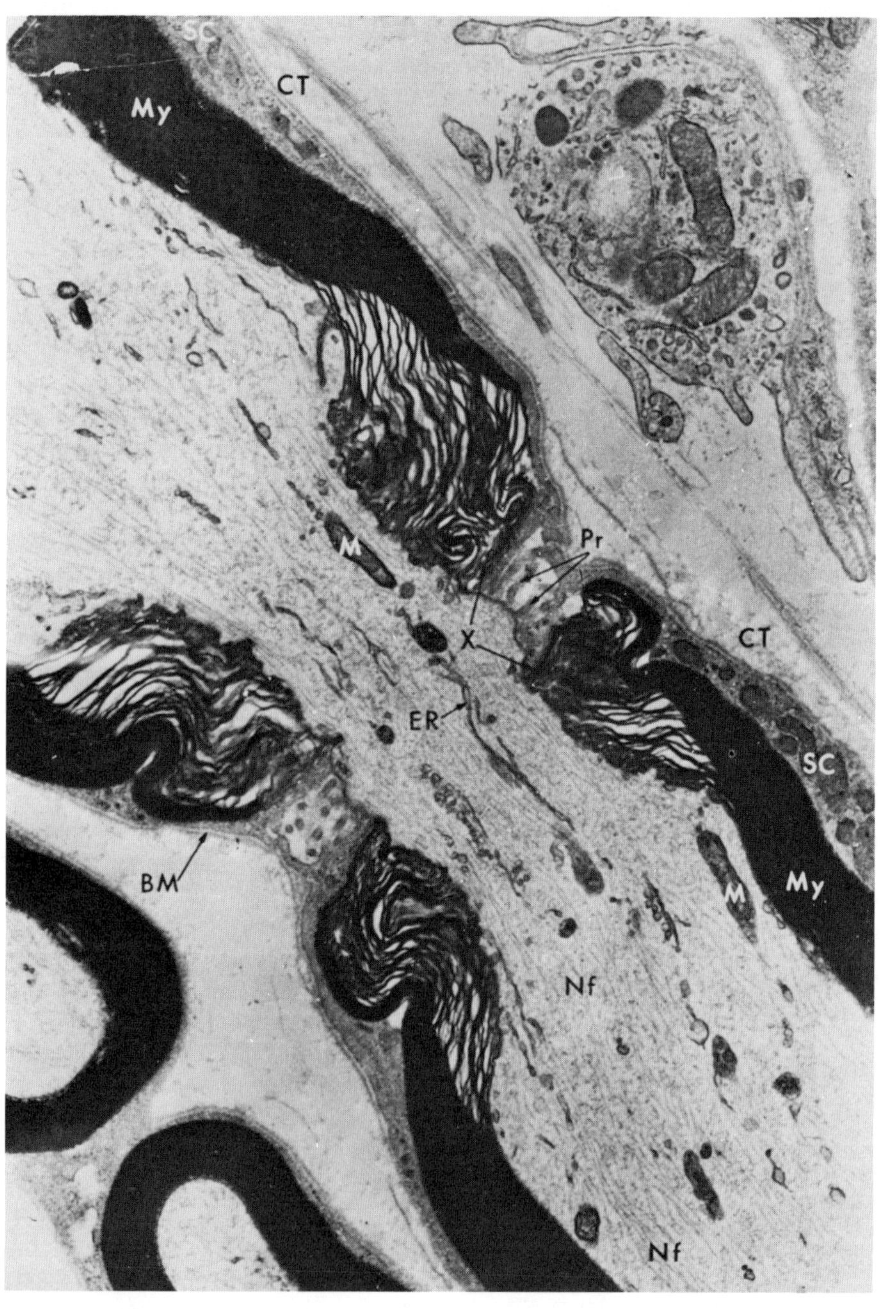
SC
CT
My
M
Pr
X
CT
ER
SC
BM
M
My
Nf
Nf

although both compounds are virtually equally potent inhibitors of AcCh-esterase in solution, with a K_I of about 10^{-7} M. Moreover, physostigmine acts on the electrical activity of the Ranvier node in a way similar to that observed at the junction: Applied in concentrations of 10^{-5}–10^{-6} M, it increases the spike height within seconds; then the spike width increases, soon followed by a decrease of the spike amplitude; eventually conduction fails (Dettbarn, 1960a).

Even the so-called unmyelinated axons are surrounded by Schwann cells rich in lipids; for instance, in the isolated squid giant axon of *Loligo*, the Schwann cell is about 4000 Å thick. Neither AcCh nor curare is able to penetrate into the interior of the axons, as has been repeatedly demonstrated with radioactive and stable isotopes (Bullock *et al.*, 1946; Rothenberg *et al.*, 1948; Hoskin and Rosenberg, 1964). They obviously do not reach the conducting membrane and cannot affect its activity. However, after exposure of the squid giant axon to a few micrograms of phospholipase A for a few minutes, both AcCh and *d*-tubocurarine do affect electrical activity; experiments with radioactively labeled compounds have shown that under these conditions they are found in the axoplasm (Hoskin and Rosenberg, 1964; Rosenberg, 1966). Electron micrographs of axons exposed to phospholipase A showed a slight disintegration of the outer layer of the Schwann cell; in agreement with electrophysiological evidence, the excitable membrane was *not* affected (Martin and Rosenberg, 1968).

In contrast, tertiary analogs of AcCh that are lipid soluble reach the conducting membrane without any pretreatment; they are also found inside the axoplasm. Specific inhibitors of either enzyme or the receptor affect reversibly the electrical activity of the squid giant axon. Table I illustrates the difference between the action of tertiary and quaternary ammonium derivatives on axonal and synaptic parts of the membrane, without or with pretreatment with phospholipase A. It may be mentioned that with local anesthetics that are lipid-soluble structural analogs of AcCh, such as, for instance, tetracaine, a typical competitive action with carbamylcholine and *d*-tubocurarine has been demonstrated at the junction of the mono-

---

Fig. 13. Electron micrograph showing a node of Ranvier of a single fiber from the sciatic nerve of mouse. The sheath of myelin forms a compact tube (My) over most of the internodal area. In the region of the node, fingerlike processes (Pr) of neighboring Schwann cells (SC) interdigitate and cover the nodal area. A basement membrane (BM) and connective tissue fibers (CT) of the endoneurium complete the wrappings of the fiber. At the node the membrane of the axon is free of myelin and is exposed to the interstitial fluids which diffuse through the basement membrane and between the Schwann cell processes. Axoplasm is rich in neurofilaments (Nf) and contains slender elements of the endoplasmic reticulum (ER) and small numbers of mitochondria (M). (Porter and Bonneville, 1964.) Reduced 30% for reproduction.

Table I

Permeability Barriers Protecting Conducting (Axonal) Parts of Excitable Membranes Against the Action of Quaternary (but Not Tertiary) Analogs of Ammonium Derivatives[a]

Compound	Axon (squid)		Synapse (Electroplax)
	Untreated	After exposure to phosph. A	
Tertiary compounds			
Atropine	2×10^{-3}	3×10^{-4}	3×10^{-4}
Physostigmine	7×10^{-3}	1×10^{-3}	0.7×10^{-3}
Procaine[b]	3×10^{-3}	—	1×10^{-3}
Dibucaine[b]	3×10^{-5}	—	3×10^{-5}
Quaternary compounds			
Acetylcholine	$(\geqslant 10^{-1})$	2×10^{-4}	3×10^{-6}
d-Tubocurarine	$(\geqslant 10^{-2})$	3×10^{-5}	3×10^{-6}
Benzoylcholine	2×10^{-2}	—	1×10^{-3}

[a] Exposure of axons to phospholipase A (a few micrograms for a few minutes) strongly reduces the protection by the barriers. Apparently no such barriers exist at synaptic parts of many excitable membranes. The squid giant axon has been used to test the barrier-reducing effect of phospholipase A in conducting (axonal) parts. This axon is unmyelinated; however, it is covered, as are all axons, by a layer of Schwann cells, which in this particular axon is 4000 Å thick. Schwann cells are rich in phospholipids. Reliable values for testing the effects of compounds on the synaptic parts can be obtained from the monocellular electroplax preparation. The molar concentrations of a few representative tertiary compounds are given at which impulse conduction is blocked (without depolarization) in the axons, before and after exposure to phospholipase A. Many quaternary ammonium derivatives, such as AcCh and *d*-tubocurarine (in the concentrations given in parentheses) did not affect conduction at all. However, after exposure of the axon to phospholipase A the concentrations of AcCh and *d*-tubocurarine are close to the concentration values required to block impulse transmission across the synaptic junction. [Data taken from Rosenberg (1967).]

[b] Local anesthetic (analog of acetylcholine).

cellular electroplax preparation; at the conducting parts of this preparation no such competition can be demonstrated, due to the structural barriers preventing carbamylcholine and *d*-tubocurarine from reaching the conducting membrane. But the concentrations of tetracaine blocking electrical activity of conducting and synaptic parts are almost the same (Podleski and Bartels, 1963).

There are a few axons where the conducting parts of the membrane are poorly protected and react directly to AcCh without any previous pretreatment: for instance, the lobster walking leg (Dettbarn and Davis, 1963), and the rabbit vagus after its connective tissue sheath has been removed (Armett and Ritchie, 1960). Recently a hybrid type of neuroblastoma fiber has been found in which an action potential can be produced either by electrical stimulation or by electrophoretic application of AcCh (Hamprecht, 1974).

2. The Artificially Evoked Appearance of AcCh outside Nerve Fibers

Under physiological conditions there is no release of AcCh from the nerve terminals. The appearance in the perfusion fluid of junctions has only been demonstrated under unphysiological conditions. The observations of Loewi (1921) on the "Vagusstoff" described in the textbooks as the classical foundation of the neurohumoral transmitter role of AcCh are only reproducible in severely deteriorated heart preparations, as was shown by Ascher (1925). AcCh is found outside nerve fibers in many other preparations, including, for instance, the ganglion nodosum which has no synapses, when they are severely damaged (Lorente de Nó, 1938). It is known that Otto Loewi tried for more than ten years in the United States to repeat his experiments originally performed in Europe and finally abandoned his efforts. He attributed his failure to a species difference (*Rana pipiens* in the United States).[1] But even if we disregard Loewi's experiments, Dale and his associates emphasized that they were unable to find any trace of AcCh in the junctional perfusion fluid *unless eserine*, a potent inhibitor of AcCh-esterase, was present. Obviously, in the presence of an inhibitor of the rapid and powerful removal mechanism of released AcCh, AcCh will escape from the membrane and appear in the surrounding gap. It should be recognized that the experiments of Dale and his associates are in direct contradiction to Loewi's interpretation of an actual release of the Vagusstoff (AcCh) in the absence of an inhibitor (Loewi, 1921). Nevertheless, the release of AcCh from damaged nerve is still today considered as one of the evidences for a neurohumoral transmitter role.

In view of the insulating barriers surrounding all conducting membrane parts, AcCh appears, even in the presence of eserine, only in the perfusion fluid of junctions. However, preparations in which AcCh may reach the conducting membrane because of incomplete protection (DeLorenzo *et al.*, 1968), such as the axons of the walking leg of lobster, AcCh is released even in resting condition when they are kept in saline solution containing eserine (Dettbarn and Rosenberg, 1966). This is in full agreement with the observations of Lorente de Nó mentioned earlier, in which the appearance of AcCh was demonstrated on several sites outside synaptic junctions. Moreover, it was shown, in the 1930's, that AcCh is released from the cut surfaces of

[1] Bruecke and his associates in Vienna unsuccessfully tried to reproduce Loewi's experiments using *Rana esculenta* (personal communication to D. N. in 1958).

axons into the surrounding fluid when they are stimulated (Calabro, 1933; Bergami, 1936a, b; Bergami *et al.*, 1936) because at the cut axon surface there is no barrier. These findings were repeatedly confirmed and even extended to several types of nerves (e.g., Brecht and Corsten, 1941).

A further experiment may be mentioned which seemed to suggest that AcCh is released on stimulation exclusively from the nerve terminals. Dale and his associates reported that they were unable to find AcCh in the perfusion fluid after the motor nerve had been cut (Dale *et al.*, 1936). However, McIntyre and his associates have demonstrated in very extensive and careful investigations that AcCh appears in the perfusion fluid of muscle that has been stimulated directly, even after complete degeneration of the motor nerve endings; thus there is a release of AcCh from the post- as well as the presynaptic membrane (McIntyre, 1959). In these studies an explanation was found for the negative results of the earlier findings: Occasional failure of AcCh appearance in the perfusion fluid was found to be due to the obliteration of small blood vessels following complete denervation; therefore, the appearance of AcCh in the perfusion fluid of these muscles was sometimes prevented.

In summary, the interpretation of the original observations on which the role of AcCh as a neurohumoral transmitter were based has become untenable in the light of the facts presented.

C. Evidence Supporting a Similar Role of AcCh Cycle in Pre- and Postsynaptic Junctional Membranes

In contrast to the difficulties facing the assumption that AcCh is a neurohumoral transmitter across junctions, the evidence for a similar role of the compound in both conducting and synaptic parts of the membrane has found increasingly strong experimental support. In the 1930's knowledge of the structural aspects of synaptic junctions was extremely limited; the only important aspect which was widely accepted was the existence of a nonconducting gap between nerve terminal and effector cells, although the extent and form of this gap were quite obscure. This situation changed only in the 1960's, mainly due to the rise of electron microscopy. No information existed about the biochemical aspects of cell membranes in general, obviously including excitable membranes in conducting as well as in synaptic parts. The hypothesis of neurohumoral transmission simply *assumed* that AcCh is released from the nerve ending and after crossing the gap is hydrolyzed by an esterase located in the effector cell.

Since that time information about structural aspects and the chemical composition of membranes in general and of excitable membranes in particular, both in conducting and synaptic parts, has increased tremendously.

Barrnett (1962) was the first to demonstrate, by electron microscopy combined with histochemical staining techniques, that AcCh-esterase is located in both junctional membranes, i.e., the nerve terminal and the postsynaptic membrane. This result was consistent with the conclusion based on indirect biochemical data [for a summary see Nachmansohn (1959)]. In the last decade the methods of both electron microscopy and of staining techniques have remarkably improved. A particularly elegant picture of the neuromuscular junction and the localization of AcCh-esterase in pre- and postsynaptic membranes is shown in Fig. 11, taken from Koelle (1971). Besides demonstrating the presence of AcCh-esterase in both junctional membranes, this figure illustrates the complexity of shape and structure of a junction. (The presence of the enzyme in synaptic and conducting membrane parts of the electroplax of *Electrophorus* has been discussed before.)

The presence of an AcCh-receptor and its function in the nerve terminals were first indicated by the results of Masland and Wigton (1940): They found that AcCh, neostigmine, and curate act in a similar way on pre- and postsynaptic membranes of the neuromuscular junction and that AcCh produces antidromic impulses. Since these experiments did not fit the then current notions of neurohumoral transmission, they were completely ignored for more than a decade. But in the 1950's these observations were confirmed and greatly extended by many investigators. Today it is well recognized that nerve terminals are just as sensitive as the postsynaptic membrane to AcCh and its structural analogs [for details see e.g., Riker *et al.* (1959) and Werner and Kuperman (1963)]. Antidromic impulses in motor nerve produced by AcCh and analogs injected into the neuromuscular junctions were recorded at the ventral roots. Thus both AcCh-esterase and -receptor are not only present at both junctional membranes, but they function in the same way in both membranes.

K^+ ion efflux from nerve endings. Cowan (1934) demonstrated a strong efflux of K^+ ions from stimulated axons. Similarly, Feldberg and Vartiainen (1934) found an efflux at synaptic junctions following stimulation. At that time Eccles (1935) considered these findings as evidence that K^+ ions are the transmitters carrying the impulse from the nerve terminal to the postsynaptic membrane. The question of the mechanism by which the K^+ ions are released from the nerve terminal was and could not be raised since no information was available permitting even speculations, neither as to the efflux from axons nor to that from the nerve terminal. It may be useful to mention in this connection that Nastuk (1954) found that there was no synaptic transmission in absence of Na^+ ions.

It was difficult to understand that during many years no current flow from nerve terminals could be detected. This failure was considered as a strong evidence for the assumption that, in constrast to axons, transmission of impulses across junctions requires chemical transmitters. But in 1963, action potentials along nerve terminals were demonstrated by Hubbard and Schmidt (1963); these findings were confirmed by Katz and Miledi (1965). Thus another objection that was used as support for two fundamentally different roles of AcCh in "conduction" and "transmission" has been removed.

D. Alternative Interpretation of the Function of AcCh at Junctions

1. Similarity of the AcCh Function in Conducting and Synaptic Membrane Parts

In view of the remarkable advances in the information about cell membranes in general and the information about the presence and function of the proteins processing AcCh in excitable membranes both in their conducting and synaptic parts, a reevaluation of the role of AcCh at synaptic junctions appears necessary and unavoidable. The observations on which the original hypothesis of AcCh as neurohumoral transmitter were based have found an entirely different explanation and have made the early interpretation untenable (see Section III B). On the other hand, the data in the preceding section strongly support a similar role of AcCh in synaptic as well as in conducting membrane parts.

Thus the alternative interpretation, postulated for many years, appears more satisfactory than the original hypothesis: AcCh released on the arrival of the nerve impulse *within* the nerve terminal from a storage protein acts on the receptor and controls the ion movements across the membranes in a way similar to that proposed for the conducting membrane. The signal given by the release of AcCh is greatly amplified: For 1000 molecules of AcCh released in a nerve terminal, many millions of Na^+ ions will enter the axon and an equivalent amount of K^+ ions will flow into the nonconducting gap. The amount of K^+ ions in a nonconducting gap of about 200–400 Å may easily reach the concentration required to produce the change of potential across the postsynaptic membrane (Neumann *et al.*, 1973). The K^+ concentration in the synaptic cleft will become high enough to induce the conformational change of the storage protein in the postsynaptic membrane, leading to a release of AcCh and thus initiating the same sequence of events as in the conducting and terminal membranes. The BEU's proposed for the axonal membrane are assumed to be present in the nerve terminal as well as in the postsynaptic membrane (Fig. 14).

In summary, the view presented postulates that the AcCh cycle, the basic elementary process controlling the permeability changes required for

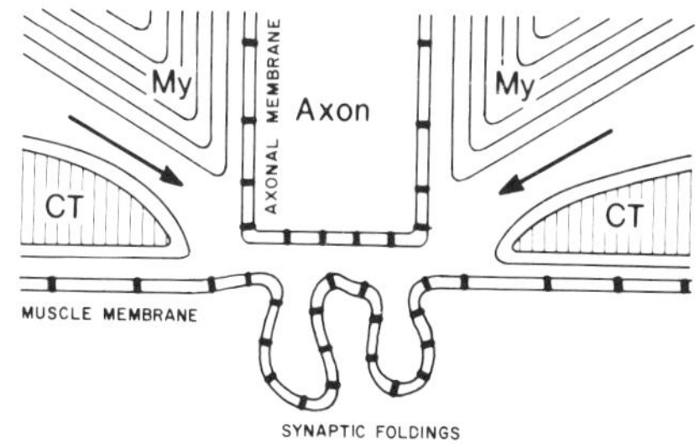

Fig. 14. Scheme of a cross section through a synaptic junction between a neuron and a muscle cell. The bars across the excitable membranes represent cross sections of the suggested basic excitation units (BEU); the density of BEU's is assumed to be higher in the synaptic region than in the axonal parts. My, membrane layers of myelin protecting the axonal membrane; CT, protective layers (e.g., connective tissue) of the muscle membrane. The arrows indicate the sites of relatively easy access for external application of chemicals to the excitable membranes of nerve and muscle.

ion fluxes, is the same in the entire excitable membrane. This concept applies, on the basis of experimental evidence, to excitable membranes throughout the animal kingdom and apparently even to membranes of plants; it is a universal characteristic of excitability. On the other hand, there is no evidence that AcCh at junctions becomes a neurohumoral transmitter across the synaptic cleft in contrast to its suggested function in conducting membranes controlling ion permeability.

2. *Differences between Conduction and Transmission*

However, it must be emphasized that the view of the universal role of AcCh in nerve excitability in no way contradicts the existence of marked and important differences between conduction along axons and transmission across junctions. There are striking differences of structure, shape, organization, and environment. Many of the differences of electrical parameters such as, e.g., the duration of the action potential, may be attributed to such structural factors. Inspecting the complex shape of the neuromuscular junction shown in Fig. 11, we may readily imagine that geometric factors alone must influence the time interval in which the different parts of the postsynaptic membrane are affected, e.g., by presynaptically released K^+ ions. Due to the lack of insulating barriers at most junctions the pharmacological action of externally applied AcCh and curare is usually restricted to junctions. This apparent limitation of action may apply to most drugs.

Finally, it must be recalled that we do not know the molecular organization of cell membranes. This factor, however, may well play an important role in the differences between conduction and transmission. In the framework of our model, the BEU's may not be evenly distributed; the BEU density may be greater at junctions than at axons, but at present no unequivocal quantitative data on esterase and receptor densities are available. In addition, neuroeffectors, such as catecholamines, developed at a later stage of evolution may influence the elementary processes taking place at junctions. These amines may act, for instance, as modulators or regulators of the cholinergic control or on the processes in the gateways, directly or indirectly. The control of the permeability changes in axonal and synaptic parts of the membrane by the AcCh cycle has been occasionally referred to as the "unified concept of transmission and conduction." This is an oversimplification and may cause serious misinterpretation. The notion of a unified role concerns exclusively the specific intramembranous control action of AcCh in junctional and axonal membrane parts. This concept is the main difference between the integral model and the hypothesis of neurohumoral transmission. Attributing to ATP, myosin, and actin a central role in the molecular events during muscular contraction does not exclude marked differences between many parameters of the contractile processes of rabbit striated and smooth muscle, or the muscles of holothuria and worms; a great variety of additional factors are involved which will strongly influence and modify the parameters observed, even if the underlying elementary process is the same.

3. *Excitatory and Inhibitory Effects of Nerve Stimulation*

Nerve stimulation may have excitatory or inhibitory effects. This difference is widely attributed to excitatory and inhibitory transmitters. AcCh usually initiates a series of reactions leading to increased membrane permeability, causing depolarization and excitation. But in some preparations, such as, for instance, in the heart, vagus stimulation or artificial external application of AcCh have inhibitory effects: the membrane is hyperpolarized, presumably by decreasing the K^+ permeability. It has been speculated that these opposite results of stimulation, i.e., depolarizations and hyperpolarizations, may be attributable to different types of neurohumoral transmitters, in spite of the evidence that a compound such as AcCh may have both excitatory and inhibitory effects.

However, an understanding of these seemingly opposite effects requires an understanding of the precise molecular events. We are at present very far from having this information. We do not know why AcCh usually increases, but in some cases apparently decreases, Na^+ permeability or increases the permeability of Cl^- ions. Various factors such as structural organization, small changes in environment, and reactions involving S–S or S–H groups of membrane proteins surrounding the gateways, may readily account for the apparently opposite effects of the same control mechanism. It should be recalled that ATP is required both for muscular contraction and for relaxation.

For many years this phenomenon escaped a satisfactory interpretation until it was found that the effect is controlled by Ca^{2+} ions (e.g., Ebashi and Lipmann, 1962; Hasselbach and Makinose, 1963). In the case of the AcCh-receptor there are S–S groups near the anionic group of the active site (Karlin and Winnik, 1968). Reducing these S–S groups by dithiothreitol (DTT) drastically changes or modifies the effect of compounds acting on the receptor. It was found, for instance, that trimethylbenzene diazonium fluoroborate (TDF) applied to an electroplax forms a covalent bond with the receptor at or near the active site and blocks irreversibly the response to receptor activators. This action of TDF can be prevented by prior application of curare which apparently has a stronger affinity. After exposure of the electroplax to DTT, which reduces the S–S groups in the neighborhood of the active site, TDF at 10^{-6} M becomes a potent reversible receptor depolarizing the membrane (Podleski *et al.*, 1969). Similarly, hexamethonium, a curare-like reversible inhibitor of the receptor of the electroplax, becomes an activator of the receptor after exposure to DTT (Karlin, 1969). Thus, such a minor change as the reduction of S–S groups in the receptor protein may transform an inhibitory into an excitatory action. Until we know more about the specific molecular processes taking place in the membrane and in its proteins the assumption of excitatory and inhibitory transmitters based simply on changes of electrical parameters or release or external application of metabolites such as amino acids, hormones, etc., seems premature. On the basis of the present knowledge there appears to be little justification for the proposal of various transmitters acting independently of the specific AcCh control mechanism (in the pre- and postsynaptic membranes).

It may be briefly mentioned in this connection that the action of some compounds, such as, e.g., veratridine, and of certain neurotoxins, such as tetrodotoxin or batrachotoxin, used in recent years to affect ion permeabilities is also still not understood in molecular terms.

IV. CONCLUDING REMARKS

One of the fundamental notions of biochemical thinking is the biochemical unity of life, a notion which has played an essential role in the development of biochemistry since the time of Pasteur. Nature has shown little imagination in modifying chemical mechanisms associated with a given function. In the hundred millions of years of evolution the cytochromes have changed their essential amino acid composition to a very limited degree (Smith, 1968; Margoliash, 1972). The specific chemical processes associated with various cellular mechanisms are remarkably similar from the most primitive cells to those in the most highly developed organisms. Thus, it appears pertinent to assume that such a specific system as the AcCh cycle is associated with the ion permeability changes during electrical activity and is ubiquitously required for all form of bioexcitability, one of the universal and most vital characteristics of living organisms. The adaptation of the same specific system to the great diversity of needs of a variety of organisms may be achieved by many additional factors, such as changes of form and shape, structure, and organization, as well as by additional chemical factors: Altogether they may modify the effects of specific chemical reactions in any way required.

Typical of chemical processes in cells is their cyclic character. This important principle was first recognized by Otto Meyerhof half a century ago. He found that only about one-sixth of the lactic acid formed anaerobically from glycogen during muscular contraction by a series of intermediary steps is oxidized, as we know today via the citric acid cycle, but five-sixths of the lactic acid formed is resynthesized during recovery to glycogen by the energy derived from the oxidation. His observations provided the basis for the explanation of the well-known Pasteur effect. Meyerhof was the first to recognize the far-reaching general implications of these findings. The series of cycles associated with most cellular functions apparently prevent unnecessary waste, by minimum energy dissipation, using the excess energy for reversing most of the processes and thereby restoring the initial state with maximum efficiency. The series of cyclic processes suggested to be associated with the control of transient changes of electrical properties and the maintenance of nerve excitability as depicted in Fig. 7 may well have a similarly essential bioenergetic function.

Thales, the father of Greek philosophy, was the first to recognize, in the sixth century B.C., the necessity of the principle of explaining the multiplicity of phenomena in which a given material manifests itself by a minimum of assumptions (see Sambursky, 1965). The statement of Katchalsky quoted at the beginning of this chapter reveals that his way of thinking went in a similar direction. It must be stressed that the suggestions made in the framework of our integral model of excitability, as in models in general, do not claim to provide definite answers; on the contrary, they raise many questions. J. J. Thompson once said "a theory is a tool and not a creed." A hypothesis derives its value by stimulating experiments. The results in turn may be used for correcting, modifying, or even discarding a working hypothesis. The elements on which the integral model is based may be subdivided into three categories: (i) well-established biophysical and biochemical data; (ii) indirect evidence which requires further experimental support; and (iii) postulates which, however sound the reasoning may be, present a challenge for experimental tests.

But in contrast to a theory, there are certain axioms, basic notions, without which no field of science can develop, be it biology or physics. At the foundation of all sciences there are fundamental assumptions which cannot be further reduced but have to be accepted by an act of faith (Born, 1949). One axiom in biochemistry is that expressed by Justus von Liebig, in 1847, in the introduction to "Thierchemie," that no manifestation of life is conceivable without molecular changes, i.e., without chemical processes. A second axiom is the basic similarity of the elementary processes underlying a specific cellular mechanism. Finally, the third axiom is the paramount role of proteins, including enzymes, in all mechanisms of living cells. The value of the integral model suggested for the control of bioelectricity will eventually be judged by its ability to inspire experiments providing new important insights, a function characteristic of all endeavors in life as expressed in the words of Goethe: "Was fruchtbar ist, allein ist wahr."

V. SUMMARY

Although numerous experimental data have been accumulated in the various fields of research on bioelectricity, the mechanism of nerve excitability is still an unsolved problem. Many mechanistic interpretations of nerve behavior cover only a part of the facts, are thus selective and unsatisfactory.

An attempt at an integral interpretation of basic data well-established by electrophysiological, biochemical, and biophysical investigations was inspired by the late Aharon Katchalsky and a first attempt had been made previously (Neumann *et al.*, 1973). The present account is a further step toward a quantitative physiochemical theory of bioelectricity. We have further explored the previously introduced notion of a basic excitation unit in excitable membranes. This notion is of fundamental importance for modeling details of sub- and suprathreshold responses, such as threshold behavior and strength–duration curves, in terms of kinetic parameters for specific membrane processes.

Our integral model of excitability is based on the original chemical hypothesis for the control of bioelectricity (Nachmansohn, 1959, 1971b). This specific approach includes some frequently ignored experimental facts on acetylcholine-processing proteins in excitable membranes. According to the integral model, acetylcholine ions are continuously processed through the basic excitation units within excitable membranes: axonal, presynaptic, and postsynaptic parts. Excitability, i.e., the generation and propagation of nerve impulses, is due to a cooperative increase in the rate of AcCh translocation through the cholinergic control system.

At the present stage of the model, the cholinergic control is restricted to the rapidly operating, normally Na^+-ion carrying permeation sites. The variations in the electric field of the membrane, caused by the cholinergically controlled rapid gateway, in turn affect the permeability of the slower, normally K^+-ion carrying permeation sites in the excitable membrane.

The basic biochemical data suggesting a cyclic cholinergic control (AcCh cycle) of the ion movements have been presented and some of the controversial interpretations of biochemical and electrophysiological data on excitability have been discussed.

ACKNOWLEDGMENTS

The authors wish to express their thanks to Prof. Manfred Eigen for his continuous interest. The work was supported in part by the U.S. Public Health Service Grant No. NS-03304, by the National Science Foundation Grant NSF-GB-31122X, by gifts from the New York Heart Association and the Alfred P. Sloan Foundation, and by a grant from the Volkswagen-Stiftung to one of us (E.N.).

NOTE ADDED TO MANUSCRIPT

Recently, another protein which also binds AcCh, α-bungarotoxin (and other cholinergic ligands) has been isolated by Chang (1974) from the electric organ of *Electrophorus electricus*. We recall that the receptor protein (R) can be operationally defined as the membrane component that upon binding of AcCh and other agonists causes a membrane permeability change and that upon binding of α-bungarotoxin and other antagonists prevents AcCh- (or electrically) induced permeability changes. To account for specificity and efficiency, the receptor protein should bind AcCh with a high binding constant and with a binding stoichiometry of not more than one or a few AcCh molecules per receptor protein. The other protein (called by Chang AcCh-R II) would not match this operational definition, because it has a high binding capacity for AcCh associated with a relatively low binding constant. This protein would, however, meet the properties required for a membrane storage site for AcCh and may thus be called S-protein. Furthermore, it has been found that acetylcholine induces a conformational change in the isolated acetylcholine-receptor protein (from *Electrophorus electricus*). This configurational change controls the binding of calcium ions to the polyelectrolytic macromolecule. The kinetic analysis of this fundamentally important biochemical reaction [see Eq. (17)] results in values for apparent rate constants and equilibrium parameters of the participating elementary processes, but also reveals the stoichiometry of the interactions between receptor, acetylcholine, and calcium ions (Chang and Neumann, 1975).

REFERENCES

Abbot, B. C., Hill, A. V., and Howarth, J. V., 1958, The positive and negative heat production associated with a single impulse, *Proc. Roy. Soc. B* **148**:149–187.

Adam, G., 1970, Theory of nerve excitation as a cooperative cation exchange in a two-dimensional lattice, *in* "Physical Principles of Biological Membranes" (F. Snell, J. Wolken, G. Iverson, and J. Lam, eds.), pp. 35–67, Gordon and Breach, New York.

Agin, D., 1967, Electroneutrality and electrodiffusion in the squid axon, *Proc. Nat. Acad. Sci. U. S.* **57**:1232–1238.

Agin, D., 1972, Excitability phenomena in membranes, *in* "Foundations of Mathematical Biology" (R. Rosen, ed.), pp. 253–277, Academic Press, New York.

Armett, J., and Ritchie, J. M., 1960, The action of acetylchloline on conduction in mammalian non-myelinated fibers and its prevention by anticholinesterase, *J. Physiol.* **152**:141–158.

Ascher, L., 1925, Ueber die chemischen Wirkungen der Herznervenreizung, *Arch. Ges. Physiol (Pflueger's)* **210**:689–696.

Barrnett, R. J., 1962, The fine structural localization of acetylcholinesterase at the myoneural junction, *J. Cell Biol.* **12**:247–262.

Bartels, E., 1965, Relationship between acetylcholine and local anesthetics, *Biochim. Biophys. Acta* **109**:194–203.

Bergami, G., 1936a, Liberazione di una sostanza acetilcholine-simile dalla superficie di taglio del nervo durant l'eccitamento fisiologico, *Atti Acad. naz Lincei* **VI**:23:518–521.

Bergami, G., 1936b, Liberazione di una sostanza acetilcolinosimile da un tronco nervoso sopravvivente durante la stimolazione elettrica in vitro, *Boll. Soc. Ital. Biol. Sper.* **11**:275–277.

Bergami, G., Cantoni, G., and Gualtierotti, T., 1936, Sulla liberazione di sostanze biologicamento attive dalla superficie di taglio di nervi durante l'eccitamento fisiologico o provocato. I. La loro azione sul preparato di muscolo dorsale di sanguisuga, *Arch. Ist. Biochim. Ital.* **8**:267–298.

Biesecker, G., 1973, Molecular properties of the cholinergic receptor purified from *Electrophorus electricus*, *Biochemistry* **12**:4403–4409.

Blumenthal, R., Changeux, J.-P., and Lefevre, R., 1970, Membrane excitability and dissipative instabilities, *J. Membrane Biol.* **2**:351–374.

Born, M., 1949, "Natural Philosophy of Cause and Chance," Oxford University Press, London and New York.

Brecht, K., and Corsten, M., 1941, Acetylcholin in sensiblen Nerven, *Arch. Ges. Physiol. (Pflueger's)* **245**:160–169.

Brink, F., 1954, The role of calcium ions in neural processes, *Pharmacol. Rev.* **6**:243–298.

Brzin, M., 1966, The localization of acetylcholinesterase in axonal membranes of frog nerve fibers, *Proc. Nat. Acad. Sci. U. S.* **56**:1560–1563.

Brzin, M., and Dettbarn, W.-D., 1967, Cholinesterase activity of nodal and internodal regions of myelinated nerve fibers of frog, *J. Cell Biol.* **32**:577–583.

Brzin, M., Dettbarn, W.-D., Rosenberg, Ph., and Nachmansohn, D., 1965, Cholinesterase activity per unit surface area of conducting membranes, *J. Cell Biol.* **26**:353–364.

Bullock, T. H., Nachmansohn, D., and Rothenberg, M. A., 1946, Effects of inhibitors of choline esterase on the nerve action potential, *J. Neurophysiol.* **9**:9–22.

Calabro, W., 1933, Sulla regolazione neuro-umorale cardiaca, *Riv. Biol.* **15**:299–320.

Carnay, L. D., and Tasaki, I. 1971, Ion exchange properties and excitability of the squid giant axon, *in* "Biophysics and Physiology of Excitable Membranes" (W. J. Adelman, Jr., ed.), pp. 379–422, Van Nostrand Reinhold, New York.

Chang, H. W., 1974, Purification and characterization of acetylcholine receptor-I from *Electrophorus electricus*, *Proc. Nat. Acad. Sci. U. S.* **71**:213.

Chang, H. W., and Neumann, E., 1975, *Proc. Nat. Acad. Sci. U. S.*, in press.

Changeux, J.-P., Gautron, J., Israel, M., and Podleski, T., 1969, Neurobiologie moléculaire. Séparation de membranes excitables à partir de l'organe électrique d'*Electrophorus electricus*, *C. R. Acad. Sci. Paris D* **269**:1788–1791.

Cole, K. S., 1965, Electrodiffusion models for the membrane of squid giant axon, *Physiol. Rev.* **45**:340–379.

Cole, K. S., 1968, *in* "Membranes, Ions and Impulses" (C. A. Tobias, ed.), University of California Press, Berkeley, California.

Cole, K. S., 1970, Dielectric properties of living membranes, *in* "Physical Principles of Biological Membranes" (F. Snell, J. Wolken, G. Iverson, and J. Lam, eds.), pp. 1–15, Gordon and Breach, New York.

Cowan, S. L., 1934, The action of potassium and other ions on the injury potential and action current in *Maja* nerve, *Proc. Roy. Soc. B* **115**:216–260.

Dale, H. H., Feldberg, W., and Vogt, M., 1936, Release of acetylcholine at voluntary motor nerve endings, *J. Physiol.* **86**:353–380.

Davis, F. A., and Nachmansohn, D., 1964, Acetylcholine formation in lobster in sensory axons, *Biochim. Biophys. Acta* **88**:384–389.

DeLorenzo, A. J. D., Dettbarn, W.-D., and Brzin, M., 1968, Fine structure and organization of nerve fibers and giant axons in lobster *Homarus americanus*, *J. Ultrastruc. Res.* **24**:367–384.

Denburg, J. L., Eldefrawi, M. E., and O'Brien, R. D., 1972, Macromolecules from lobster axon membranes that bind cholinergic ligands and local anesthetics, *Proc. Nat. Acad. Sci. U. S.* **69**:177–181.

Dettbarn, W.-D., 1960a, The effect of curare on conduction in myelinated, isolated nerve fibers of the frog, *Nature* **186**:891–892.

Dettbarn, W.-D., 1960b, New evidence for the role of acetylcholine in conduction, *Biochim. Biophys. Acta* **41**:377–386.

Dettbarn, W.-D., 1961, New evidence for the role of acetylcholine in bioelectrogenesis, *in* "Bioelectrogenesis" (Proc. Symp. on Comparative Bioelectrogenesis) (C. Chagas and A. Paes de Carvalho, eds.), pp. 237–261, Elsevier.

Dettbarn, W.-D., 1967, The acetylcholine system in peripheral nerve, *Ann. N. Y. Acad. Sci.* **144**:483–503.

Dettbarn, W.-D., and Davis, F. A., 1963, Effects of acetylcholine on axonal conduction of lobster nerve, *Biochim. Biophys. Acta* **66**:397–405.

Dettbarn, W.-D., and Rosenberg, P., 1966, Effect of ions on the efflux of acetylcholine from peripheral nerve, *J. Gen. Physiol.* **50**:447–460.

Ebashi, E., and Lipman, F., 1962, Adenosine triphosphate-linked concentration of calcium ions in a particulate fraction of rabbit muscle, *J. Cell Biol.* **14**:389–400.

Eccles, J. C., 1935, After-discharge from the superior cervical ganglion, *J. Physiol.* **84**: 50p–52p.

Eigen, M., 1967, Dynamic aspects of information transfer and reaction control in biomolecular systems, *in* "The Neurosciences" (G. C. Quarton, T. Melnechuk, and F. O. Schmitt, eds.), pp. 130–142, The Rockefeller University Press, New York.

Eigen, M., and DeMaeyer, L., 1963, Relaxation methods, *in* "Technique of Organic Chemistry" (S. L. Friess, E. S. Lewis, and A. Weissberger, eds.), Vol. 8, p. 895, Interscience, New York.

Eldefrawi, M. E., and Eldefrawi, A. T., 1973, Purification and molecular properties of the acetylcholine receptor from *Torpedo* electroplax, *Arch. Biochem. Biophys.* **159**:362–373.

Erlanger, J., 1939, The initiation of impulses in axons, *J. Neurophysiol.* **2**:370–379.

Evans, M. H., 1972, Tetrodotoxin and saxitoxin in neurobiology, *Intern. Rev. Neutobiol.* **15**:83–166.

Feldberg, W., and Vartiainen, A., 1934, Further observations on the physiology and pharmacology of a sympathetic ganglion, *J. Physiol.* **83**:103–128.

Fox, J. M., 1972, Veraenderungen der spezifischen Ionenleitfaehigkeiten der Nervenmembran durch ultraviolette Strahlung, Dissertation, Homburg–Saarbruecken.

Fulton, J. F., 1938, "Physiology of the Nervous System," Oxford University Press, New York (1938, 1943).

Gruber, H., and Zenker, W., 1973. Acetylcholinesterase: histochemical differentiation between motor and sensory nerve fibres, *Brain Res.* **51**:207–214.

Guggenheim, E. A., 1949, "Thermodynamics," Interscience, New York.

Hamprecht, B., 1974, Cell cultures as model systems for studying the biochemistry of differentiated functions of nerve cells, *Hoppe-Seyler's Z. Physiol. Chem.* **355**:109–110.

Harris, A. J., and Dennis, M. J., 1970, Acetylcholine sensitivity and distribution on mouse neuroblastoma cells, *Science* **167**:1253–1255.

Hasselbach, W., and Makinose, M., 1963, Ueber den Mechanismus des Calciumtransportes durch die Membranen des sarkoplasmatischen Reticulums, *Biochem. Z.* **339**:96–111.

Hebb, C. O., 1957, Biochemical evidence for the neural function of acetylcholine, *Physiol. Rev.* **37**:196–220.

Higman, H. B., and Bartels, E., 1961, The competitive nature of the action of acetylcholine and local anesthetics, *Biochim. Biophys. Acta* **54**:543–554.

Higman, H. B., and Bartels, E., 1962, New method for recording electrical characteristics of the monocellular electroplax, *Biochim. Biophys. Acta* **57**:77–82.

Higman, H. B., Podleski, T. R., and Bartels, E., 1963, Apparent dissociation constants between carbamylcholine, d-tubocurarine and the receptor, *Biochim. Biophys. Acta* **75**:187–193.

Higman, H. B., Podleski, T. R., and Bartels, E., 1964, Correlation of membrane potential and K flux in the electroplax of *Electrophorus*, *Biochim. Biophys. Acta* **79**:138–150.

Hill, A. V., 1960, The heat production of muscle, *in* "Molecular Biology. Elementary Processes of Nerve Conduction and Muscle Contraction" (D. Nachmansohn, ed.), pp. 17–24, Academic Press, New York.

Hodgkin, A. L., 1964, "The Conduction of the Nervous Impulse," C. C. Thomas, Springfield, Illinois.

Hodgkin, A. L., and Keynes, R. D., 1955, The potassium permeability of a giant nerve fibre, *J. Physiol.* **128**:61.

Hoskin, F. C. G., and Rosenberg, P., 1964, Alteration of acetylcholine penetration into, and effects on venom-treated squid axons by physostigmine and related compounds, *J. Gen. Physiol.* **46**:1117–1127.

Hoskin, F. C. G., Kremzner, L. T., and Rosenberg, P., 1969, Effects of some cholinesterase inhibitors on the squid giant axon, *Biochem. Pharmacol.* **18**:1727–1737.

Hoskin, F. C. G., Rosenberg, P., and Brzin, M., 1966, Re-examination of the effect of DFP on electrical and cholinesterase activity of squid giant axon, *Proc. Nat. Acad. Sci. U. S.* **55**:1231–1235.

Howarth, J. V., Keynes, R. D., and Ritchie, J. M., 1968, The origin of the initial heat associated with a single impulse in mammalian non-myelinated nerve fibres, *J. Physiol.* **194**:745–793.

Hubbard, J. I., and Schmidt, R. F., 1963, An electrophysiological investigation of mammalian motor nerve terminals, *J. Physiol.* **166**:145–167.

Huneeus-Cox, F., Fernandez, H. L., and Smith, B. H., 1966, Effects of redox and sulfhydryl reagents on the bioelectric properties of the giant axon of the squid, *Biophys. J.* **6**:675–689.

Husain, S. S., and Mautner, H. G., 1973, The purification of choline acetyltransferase of squid-head ganglia, *Proc. Nat. Acad. Sci. U. S.* **70**:3749–3753.

Julian, F. J., and Goldman, D. E., 1962, The effects of mechanical stimulation on some electrical properties of axons, *J. Gen. Physiol.* **46**:197–313.

Karlin, A., 1969, Chemical modification of the active site of the acetylcholine receptor, *J. Gen. Physiol.* **54**:245s–264s.

Karlin, A., and Bartels, E., 1966, Effects of blocking sulfhydryl groups and or reducing disulfide bonds on the acetylcholine-activated permeability system of the electroplax, *Biochim. Biophys. Acta* **126**:525–535.

Karlin, A., and Cowburn, D., 1973, The affinity-labeling of partially purified acetylcholine receptor from electric tissue of *Electrophorus*, *Proc. Nat. Acad. Sci. U. S.* **70**:3636–3640.

Karlin, A., and Winnik, M., 1968, Reduction and specific alkylation of the receptor for acetylcholine, *Proc. Nat. Acad. Sci. U. S.* **60**:668–674.

Karlin, A., Prives, J., Deal, W., and Winnik, M., 1970, Counting acetylcholine receptors in the electroplax, *in* "Ciba Foundation Symposium on Molecular Properties of Drug Receptors" (R. Porter and M. O'Connor, eds.), pp. 247–261, J. & A. Churchill, London.

Katchalsky, A., 1964, Polyelectrolytes and their biological interactions, *in* "Connective Tissue: Intercellular Macromolecules," pp. 9–41, Little, Brown and Co., Boston, Massachusetts.

Katchalsky, A., 1967, Membrane thermodynamics, *in* "The Neurosciences" (G. C. Quarton, T. Melnechuk, and F. O. Schmitt, eds.), pp. 326–343, The Rockefeller University Press, New York.

Katchalsky, A., 1969, Chemical dynamics of macromolecules and its cybernetic significance, *in* "Biology and the Physical Sciences" (S. Devons, ed.), pp. 267–298, Columbia University Press, New York.

Katchalsky, A., and Spangler, R., 1968, Dynamics of membrane processes, *Quart. Rev. Biophys.* **1**:127–175.

Katz, B., 1966, "Nerve, Muscle and Synapse," McGraw-Hill, New York.

Katz, B., and Miledi, R., 1965, Propagation of electrical activity in motor nerve terminals, *Proc. Roy. Soc. B* **161**:453–482.

Koelle, G. B., 1963, Cytological distribution and physiological functions of cholinesterases, *in* "Cholinesterases and Anticholinesterase Agents" (G. B. Koelle, ed.), "Handb. d. Exp. Pharmakol.," Ergw. XV, pp. 187–298, Springer-Verlag, Berlin and Heidelberg.

Koelle, G. B., 1971, Current concepts of synaptic structure and function, *Ann. N. Y. Acad. Sci.* **183**:5–20.

Landowne, D., 1973, Movement of sodium ions associated with the nerve impulse, *Nature* **242**:457–459.

Leuzinger, W., Baker, A. L., and Cauvin, E., 1968, Acetylcholinesterase II. Crystallization, absorption spectra, isoionic point, *Proc. Nat. Acad. Sci. U. S.* **59**:620–623.

Lewis, P. R., and Shute, C. C. D., 1966, The distribution of cholinesterase in cholinergic neurons demonstrated with the electron microscope, *J. Cell. Sci.* **1**:381–390.

Lissak, K., 1939, Liberation of acetylcholine and adrenaline by stimulating isolated nerves, *Am. J. Physiol.* **127**:263–271.

Loewenstein, W. R. (ed.), 1966, Biological membranes: Recent progress, *Ann. N. Y. Acad. Sci.* **137**:403–1048.

Loewi, O., 1921, Ueber humorale Uebertragbarkeit der Nerznervenwirkung. I. Mitteilung, *Arch. Ges. Physiol. (Pflueger's)* **189**:239–242.

Lorente de Nó, R., 1938, Liberation of acetylcholine by the superior cervical sympathetic ganglion and the nodosum ganglion of the vagus, *Am. J. Physiol.* **121**:331–349.

Manson, L. A. (ed.), 1971, "Biomembranes," Vol. I, Plenum Press, New York.

Margoliash, E., 1972, The molecular variations of cytochrome C as a function of the evolution of species, *in* "Harvey Lectures 1970/1971," pp. 171–247, Academic Press, New York.

Martin, R., and Rosenberg, P., 1968, Fine structural alterations associated with venom action on squid giant nerve fibers, *J. Cell Biol.* **36**:341–353.

Masland, R. L., and Wigton, R. S., 1940, Nerve activity accompanying fasciculation produced by Prostigmine, *J. Neurophysiol.* **3**:269–275.

McIntyre, A. R., 1959, Neuromuscular transmission and normal and denervated muscle-sensitivity to curare and acetylcholine, *in* "Curare and Curare-Like Agents" (D. Bovet, F. Bovet-Nitti, and G. B. Marini-Bettolo, eds.), pp. 211–218, Elsevier, Amsterdam.

Meunier, J.-C., Huchet, M., Boquet, P., and Changeux, J.-P., 1971, Separation de la proteine receptrice de l'acetylcholine et de l'acetylcholinesterase, *C. R. Acad. Sci. Paris D* **272**:117–120.

Muralt, A. v., and Stämpfli, R., 1953, Die photochemische Wirkung von Ultraviolettlicht auf den erregten Ranvierschen Knoten der einzelnen Nervenfaser, *Helv. Physiol. Acta* **11**:182–193.

Nachmansohn, D., 1955, Metabolism and function of the nerve cell, *in* "Harvey Lectures 1953/1954," pp. 57–99, Academic Press, New York.

Nachmansohn, D., 1959, "Chemical and Molecular Basis of Nerve Activity," Academic Press, New York.

Nachmansohn, D., 1963, (1) Actions on axons and the evidence for the role of acetylcholine in axonal conduction, pp. 701–740; (2) Choline acetylase, pp. 40–45, *in* "Cholinesterases and Anticholinesterase Agents" (G. B. Koelle, ed.), "Handb. d. Exp. Pharmakol.," Ergw. XV, Springer-Verlag, Berlin and Heidelberg.

Nachmansohn, D., 1966a, Properties of the acetylcholine receptor protein analyzed on the excitable membrane of the monocellular electroplax preparation, *in* "Current Aspects of Biochemical Energetics: Lipmann Dedicatory Volume" (N. O. Kaplan and E. P. Kennedy, eds.), pp. 145–172, Academic Press, New York.

Nachmansohn, D., 1966b, Chemical control of the permeability cycle in excitable membranes during electrical activity, *in* "Biological Membranes: Recent Progress" (W. Loewenstein, ed.), *Ann. N. Y. Acad. Sci.* **137**:877–900.

Nachmansohn, D., 1968, Proteins in bioelectricity: control of ion movements across excitable membranes, *Proc. Nat. Acad. Sci. U. S.* **61**:1034–1041.

Nachmansohn, D., 1969, Proteins of excitable membranes, *J. Gen. Physiol.* **54**:187s–224s.

Nachmansohn, D., 1970, Proteins in excitable membranes. Their properties and function in bioelectricity, *Science* **168**:1059–1066.

Nachmansohn, D., 1971a, Proteins in bioelectricity. Acetylcholine-esterase and -receptor, *in* "Handbook of Sensory Physiology" (W. R. Loewenstein, ed.), Vol. I, "Principles of Receptor Physiology," pp. 18–102, Springer-Verlag, Berlin and Heidelberg.

Nachmansohn, D., 1971b, Chemical events in conducting and synaptic membranes during electrical activity, *Proc. Nat. Acad. Sci. U. S.* **68**:3170–3174.

Nachmansohn, D., 1973, The neuromuscular junction. The role of acetylcholine in excitable membranes, *in* "The Structure and Function of Muscle" (G. H. Bourne, ed.), Vol. III, "Physiology and Biochemistry," pp. 32–117, Academic Press, New York.

Nachmansohn, D., and Lederer, E., 1939, Sur la biochimie de la cholinesterase, *Bull. Soc. Chim. Biol. Paris* **21**:797–808.

Nachmansohn, D., and Machado, A. L., 1943, The formation of acetylcholine. A new enzyme "choline acetylase," *J. Neurophysiol.* **6**:397–404.

Nachmansohn, D., and Meyerhof, B., 1941, Relation between electrical changes during nerve activity and concentration of choline esterase, *J. Neurophysiol.* **4**:348–361.

Nastuk, W. L., 1954, Relation between extracellular Na^+ and the depolarization of Ach^+ on the c.p. nerve, *Fed. Proc.* **13**:104.

Neher, E., and Lux, H. D., 1973, Rapid changes of potassium concentration at the outer surface of exposed single neurons during membrane current flow, *J. Gen. Physiol.* **61**:385–399.

Nelson, P. G., Peacock, J. H., and Amano, T., 1971, Responses of neuroblastoma cells to iontophoretically applied acetylcholine, *J. Cell Physiol.* **77**:353–362.

Neumann, E., 1973, Molecular hysteresis and its cybernetic significance, *Angew. Chem., Int. Ed.* **12**:356–369.

Neumann, E., 1974, An integral physico-chemical model for bio-excitability, *in* "Physics and Mathematics of the Nervous System" (M. Conrad *et al.*, eds.). Lecture Notes in Biomathematics **4**:42–81.

Neumann, E., and Katchalsky, A., 1972, Long-lived conformation changes induced by electric impulses in biopolymers, *Proc. Nat. Acad. Sci. U. S.* **69**:993–997.

Neumann, E., and Rosenheck, K., 1972, Permeability changes induced by electric impulses in vesicular membranes, *J. Membrane Biol.* **10**:279–290.

Neumann, E., Nachmansohn, D., and Katchalsky, A., 1973, An attempt at an integral interpretation of nerve excitability, *Proc. Nat. Acad. Sci. U. S.* **70**:727–731.

Olsen, R. W., Meunier, J.-C., and Changeux, J.-P., 1972, Progress in the purification of the cholinergic receptor protein from *Electrophorus electricus* by affinity chromatography, *FEBS Lett.* **28**:96–100.

Oster, G. F., Perelson, A. S., and Katchalsky, A., 1973, Network thermodynamics: dynamic modelling of biophysical systems, *Quart. Rev. Biophys.* **6**:1–134.

Plonsey, R., 1969, "Bioelectric Phenomena," McGraw-Hill, New York.

Podleski, T. R., 1966, Effects of quaternary ammonium ions on the membrane potential of electroplax, Doctoral Diss., Columbia University, New York.

Podleski, T. R., 1967, Dixtinction between the active sites of acetylcholine-receptor and -esterase, *Proc. Nat. Acad. Sci. U. S.* **58**:268–273.

Podleski, T. R., 1969, Molecular forces acting between ammonium ions and acetylcholine receptor protein, *Biochem. Pharmacol.* **18**:211–226.

Podleski, T. R., and Bartels, E., 1963, Difference between tetracaine and d-tubocurarine in the competition with carbamylcholine, *Biochim. Biophys. Acta* **75**:387–396.

Podleski, T. R., and Nachmansohn, D., 1966, Similarities between active sites of acetylcholinereceptor and acetylcholinesterase with quinolinium ions, *Proc. Nat. Acad. Sci. U. S.* **56**:1034–1039.

Podleski, T., Meunier, J.-C., and Changeux, J.-P., 1969, Compared effects of dithiothreitol on the interaction of an affinity-labeling reagent with acetylcholinesterase and the excitable membrane of the electroplax, *Proc. Nat. Acad. Sci. U. S.* **63**:1239–1246.

Porter, K. R., and Bonneville, M. A., 1964, "An Introduction to the Fine Structure of Cells and Tissues," Lea and Febiger, Philadelphia.

Porter, C. W., Chiu, T. H., Wieckowski, J., and Barnard, E. A., 1973, Types and locations of cholinergic receptor-like molecules in muscle fibres, *Nature, New Biol.* **241**:3–7.

Prigogine, I., 1968, "Thermodynamics of Irreversible Processes," 3rd ed., C. C. Thomas, Springfield, Illinois.

Prince, A. K., 1966, A sensitive fluorometric procedure for the determination of small quantities of acetylcholinesterase, *Biochem. Pharmacol.* **15**:411–417.

Racker, E., 1970, "Membranes of Mitochondria and Chloroplasts" (E. Racker, ed.), Van Nostrand Reinhold, New York.

Rawlings, P. K., and Neumann, E., 1975. *Proc. Nat. Acad. Sci. U. S.*, in press.

Revzin, A., and Neumann, E., 1974, Conformational changes in rRNA induced by electric impulses, *Biophys. Chem.*, **2**:144–150.

Riker, W. F., Jr., Werner, G., Roberts, J., and Kuperman, A., 1959, The presynaptic element in neuromuscular transmission, *Ann. N. Y. Acad. Sci.* **81**:328–344.

Ritchie, J. M., 1963, The action of acetylcholine and related drugs on mammalian non-myelinated nerve fibres, *Biochem. Pharmacol.* **12**(S):3.

Robertson, J. D., 1960, The molecular structure and contact relationships of cell membranes, *in* "Progress in Biophysics" (B. Katz and J. A. V. Butler, eds.), pp. 343–418, Pergamon Press, New York.

Robertson, J. D., 1970, The ultrastructure of synapses, *in* "The Neurosciences" (F. O. Schmitt, ed.), Vol. 2, pp. 715–728, The Rockefeller University Press, New York.

Rosenberg, P., 1966, Use of venoms in studies of nerve excitation, *Mem. Inst. Butantan Simp. Internac.* **33**:477–508.

Rosenberg, P., and Hoskin, F. C. G., 1963, Demonstration of increased permeability as a factor in the effect of acetylcholine on the electrical activity of venom-treated axons, *J. Gen. Physiol.* **46**:1065–1073.

Rothenberg, M. A., Sprinson, D. B., and Nachmansohn, D., 1948, Site of action of acetylcholine, *J. Neurophysiol.* **11**:111–116.

Rothfield, L. I. (ed.), 1971, "Structure and Function of Biological Membranes," Academic Press, New York.

Sambursky, S., 1965, "Das physikalische Weltbild der Antike," Artemis Verlag, Zurich.

Schmidt, J., and Raftery, M. A., 1973, Purification of acetylcholine receptor from *Torpedo californica* electroplax by affinity chromatography, *Biochemistry* **12**:852–856.

Schoffeniels, E., 1957, An isolated single electroplax preparation. II. Improved preparation for studying ion flux, *Biochim. Biophys. Acta* **26**:585–596.

Schoffeniels, E., and Nachmansohn, D., 1957, An isolated single electroplax preparation. I. New data on the effect of acetylcholine and related compounds, *Biochim. Biophys. Acta* **26**:1–15.

Seeman, P., 1972, The membrane actions of anesthetics and tranquilizers, *Pharmacol. Rev.* **24**:583–655.

Segal, J. R., 1968, Surface charge of giant axons of squid and lobster, *Biophys. J.* **8**:470–489.

Sjoestrand, F. S., and Barajas, L., 1968, Effect of modifications in conformation of protein molecules on structure of mitochondrial membranes, *J. Ultrastructure Res.* **25**:121–1255.

Sjoestrand, F. S., and Barajas, L., 1970, A new model for mitochondrial membranes based on structural and on biochemical information, *J. Ultrastructure Res.* **32**:293–306.

Smith, E. L., 1968, The evolution of proteins, *in* "Harvey Lectures 1966/1967," pp. 231–256, Academic Press, New York.

Spector, I., Kimhi, Y., and Nelson, P. G., 1973, Tetrodotoxin and cobalt blockage of Neuroblastoma action potentials, *Nature, New Biol.* **246**:124–126.

Takeuchi, A., and Takeuchi, N., 1972, Actions of transmitter substances on the neuromuscular junctions of vertebrates and invertebrates, *in* "Advances in Biophysics" (M. Kotani, ed.), Vol. 3, pp. 45–95, University Park Press, Baltimore, Maryland.

Tasaki, I., 1968, "Nerve Excitation," Charles C. Thomas, Springfield, Illinois.

Tasaki, I., and Singer, I., 1966, Membrane macromolecules and nerve excitability: a physicochemical interpretation of excitation in squid giant axons, *Ann. N. Y. Acad. Sci.* **137**: 793–806.

Tasaki, I., and Takenada, T., 1964, Ion fluxes and excitability in squid giant axon, *in* "The Cellular Functions of Membrane Transport" (J. F. Hoffman, ed.), Prentice-Hall, Englewood Cliffs, New Jersey.

Tasaki, I., Singer, I., and Takenaka, T., 1965, Effects of internal and external ionic environment on excitability of squid giant axon, *J. Gen. Physiol.* **48**:1095–1123.

Traeuble, H., and Eibl, H., 1974, Electrostatic effects on lipid phase transitions: membrane structure and ionic environment, *Proc. Nat. Acad. Sci. U. S.* **71**:214–219.

Webb, G. D., 1965, Affinity of benzoquinonium and ambenonium derivatives for the acetylcholine receptor, tested on the electroplax, and for acetylcholinesterase in solution, *Biochim. Biophys. Acta* **102**:172–184.

Werner, G., and Kuperman, A. S., 1963, Actions at the neuromuscular junctions, *in* "Cholinesterases and Anticholinesterase Agents" (G. B. Koelle, ed.), "Handb. d. Exp. Pharmakol.," Ergew. XV, pp. 570–678, Springer-Verlag, Berlin and Heidelberg.

Whittaker, V. P., 1973, The biochemistry of synaptic transmission, *Naturwiss.* **60**:281–289.

Zelman, A., and Shih, H. H., 1972, The constant field approximation: numerical evaluation for monovalent ions migrating across a homogeneous membrane, *J. Theor. Biol.* **37**:373–383.

Chapter 7

Peptide Transport

Ze'ev Barak and Charles Gilvarg

Department of Biochemical Sciences
Frick Chemical Laboratory
Princeton University
Princeton, New Jersey

I. INTRODUCTION

Many roles are being ascribed to peptides in nature. These include hormonal activity, control of pituitary tropic hormone secretion, translocation of ions across membranes, regulation of cell growth, memory transmission, carcinogenesis, and antimicrobial and nutritional activity. Some of these activities have been reported to be extracellular, i.e., activation of an endogenous system as a result of the peptide binding to a specific receptor on the membrane; however, others might require the entrance of the peptide into the cell. It is therefore of great interest to determine whether peptides as such can cross membranes and to study this process of peptide transport.

Several reviews have recently appeared of peptide transport and peptide metabolism in bacteria (Payne and Gilvarg, 1971; Sussman and Gilvarg, 1971; Gilvarg, 1972; Payne, 1972a; Simmonds, 1970, 1972) and mammals (Matthews, 1971a, 1971b, 1972a, 1972b; Milne, 1971, 1972; Ugolev, 1972). The main conclusions of these reports are summarized below.

In addition, we shall discuss in this review the contributions of recent publications to the understanding of peptide transport, and offer some comments concerning the thrust of future investigations.

II. METHODS OF STUDYING PEPTIDE TRANSPORT

Transport of peptides into cells can be investigated directly by following the uptake of a labeled peptide (radioactive, fluorescent, or any other specific determinant), or indirectly by observing a specific biological response resulting from the entrance of the peptide into the cell.

A. Direct Methods

Direct measurements of peptide uptake can be carried out by chemical determination of the peptide or its constituent amino acids inside the cell. Alternatively, peptides can be followed when they are labeled with radioactive isotopes or fluorescent groups. The latter are very sensitive methods; they permit facile recognition of the newly penetrated components and the detection of small changes in their concentration. Unfortunately, fluorescent permeable peptides exist only in theory, and radioactive peptides are not commercially available. Therefore investigators have largely used the indirect methods, or when these are inappropriate, the chemical detection system.

Direct measurements isolate the transport action from many nonspecific factors that affect growth, which is the basic assay for uptake by most indirect methods (see Section IIB). Thus direct measurement permits the demonstration of peptide transport in the absence of any biological response. All of the studies of peptide uptake in mammals have utilized the direct method. In most of these studies the detection of peptides was carried out using chemical techniques.

On the other hand, peptide transport in bacteria can easily be demonstrated by indirect methods. The few experiments that have been performed by direct methods measured the uptake of radioactive peptide. (Chemical methods are generally not sensitive enough for peptide uptake measurements in bacteria.) The use of radioactive peptides permitted the quantitation of the peptide transport systems in *Escherichia coli* (De Felice *et al.*, 1973), *Leuconostoc mesenteroides* (Yoder *et al.*, 1965b; Mayshak *et al.*, 1966), and *Lactobacillus casei* (Leach and Snell, 1960). There are, in addition, several advantages of the direct system that have not been thoroughly exploited. Radioactive peptides might be used as markers to trace specific molecular components of their transport system; they enable measurement of peptide uptake in osmotically shocked cells, spheroplasts, or even membrane vesicles. These kinds of experiments should provide a means for the identification of the factors responsible for peptide transport in bacteria.

Lastly, we would like to draw attention to the possibility of labeling peptides by coupling them with fluorescent compounds. Attachment to the carboxy terminal end of an oligopeptide might not affect its transport properties [see Section IVA4(c)]. Alternatively, the probe could be incorporated into an amino acid side chain [see Section IVA2(a)]. The fluorescent peptide derivative, if obtained, can then be useful in studies of peptide uptake by individual mammalian cells within a tissue, using the fluoresence microscope.

B. Indirect Methods

Because of the difficulties in obtaining radioactive oligopeptides with high specific activity, most of the studies in oligopeptide transport in bacteria have used the indirect methods, in which utilization of peptides for growth or their toxic effects on internal cellular systems serve as indicators for peptide transport.

Utilization of a peptide as a nutritional source by an organism requires the existence of an enzyme (or enzymes) capable of liberating the required amino acid from the peptide. This might be an extracellular process in which the peptide is split outside the cell and only the hydrolytic products are transported into the cell as free amino acids. However, utilization of a peptide in the absence of an extracellular peptidase requires its penetration into the cells and thus serves as an indicator of peptide transport. This type of experiment has often been carried out in studies with microorganisms (bacteria and yeast). Frequently, the nutritional activity of the peptide was unequivocally demonstrated by its ability to support the growth of an amino acid auxotroph when it was supplied as the sole source of the required amino acid.

Another indirect way to study peptide transport takes advantage of the existence of toxic peptides. In this case it is the inhibition of growth which indicates peptide transport. This method is of special importance when studying peptide transport in microorganisms that do not require external amino acids for growth, and utilization of peptides as source of an amino acid cannot easily be demonstrated (Barak and Gilvarg, 1974). Toxic peptides can also be used to select for peptide transport deficient mutants (see Section IVA6).

Several synthetic toxic di- and oligopeptides are known, some of which, like Gly-Leu (Simmonds *et al.*, 1951; Vonder Haar and Umbarger, 1972), Lys-Leu (Payne and Gilvarg, 1968a), and tri-L-ornithine (Gilvarg and Levin, 1972; Barak *et al.*, 1973a) in *E. coli* and tri-L-lysine in *Salmonella typhimurium* (Payne, 1968; Sussman and Gilvarg, 1970; B. N. Ames *et al.*, 1973a; B. N. Ames, personal communication) are toxic in their peptide form only, whereas others require cleavage with consequent liberation of the toxic amino acid or amino acid derivative before they can exert their effect. Peptides containing L-valine are toxic to *E. coli* K-12 after intracellular liberation of the L-valine (Sussman and Gilvarg, unpublished observation; Payne, 1971a, 1971b; Vonder Haar and Umbarger, 1972; De Felice *et al.*, 1973). Similarly, peptides containing the toxic amino acid *p*-fluorophenylalanine inhibit the growth of *E. coli* K-12 and *E. coli* W (Barak and Gilvarg, 1974). Toxic amino acid derivatives with low permeability, like norvaline, norleucine, and ethionine, can be brought into *S. typhimurium* cells causing inhibition of growth after attaching them to the C-terminus of peptides (B. N. Ames *et al.*, 1973a).

Most of the studies with toxic peptides have been carried out with *E. coli* and *S. typhimurium*; however, we believe that toxic peptides, especially those containing toxic amino acid derivatives, can be useful in studies of peptide transport in other organisms as well. They should facilitate the isolation of peptide transport deficient mutants, which should be helpful in the investigation of these complex systems.

III. PEPTIDE TRANSPORT AND PEPTIDASES

A major difficulty in studying peptide transport is the existence of extensive peptidase activities in organisms. Because of these activities, peptides originally added to cells are usually found within the cell in the form of their constituent amino acids (Leach and Snell, 1960; Brock and Woolley, 1964; Yoder *et al.*, 1965b; Rubino *et al.*, 1971; Buston *et al.*, 1972). These amino acids might be the products of extra- or intracellular reactions, representing the different mechanisms of peptide utilization (Fig. 1). The demonstration of intact peptide transport (Fig. 1A) requires, therefore, the elimination of extracellular peptidase activity. This argument is particularly valid when measuring utilization of peptide using growth as an indicator of peptide transport. Growth in such experiments can result from external cleavage of the peptides and transport of their degradation products, the free amino acids, into cells. Moreover, the toxicity of those peptides containing inhibitory components might alternatively be explained by extracellular hydrolysis of the peptide and transport of the toxic derivative in its free form.

A. Bacteria

1. *Location of Peptidase Activity*

The existence of nonutilizable peptides capable of being cleaved by cell extracts from *E. coli* (Gilvarg and Katchalski, 1965; Payne and Gilvarg,

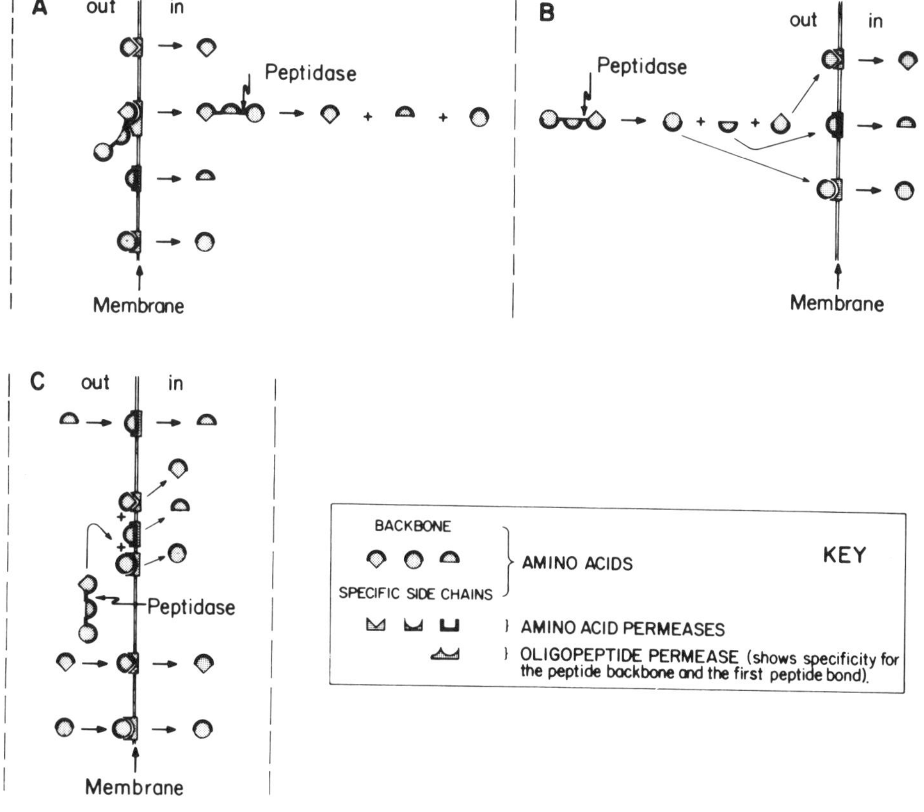

Fig. 1. Hypothetical models for possible mechanisms of peptide utilization.

1968b) provides indirect evidence for the absence of exopeptidases in growing cultures of this microorganism. This notion is supported by the failure to detect any peptidase activity in the supernatants prepared from whole cell suspensions. Moreover, even periplasmic peptidases, as tested in the osmotic shock fluid (Heppel *et al.*, 1972), appear to be absent in *E. coli* (Simmonds and Toye, 1966; Matheson and Murayama, 1966; Van Lenten and Simmonds, 1967). Spheroplast formation by treatment with EDTA and lysozyme also did not result in significant release of peptidase activity. Neither whole spheroplasts (Van Lenten and Simmonds, 1967; Simmonds and Toye, 1966) nor membrane pellets (Sussman and Gilvarg, 1970) showed major peptidase activity, indicating the lack of membrane-bound peptidases in this organism. These negative findings stand in striking contrast to the ease with which a large number of active peptidases have been demonstrated in cytoplasmic extracts derived from *E. coli*. It seems, therefore, that in *E. coli* the peptidases are intracellular. Uptake of radioactivity from peptides, their utilization as nutrilites, or their toxic action must therefore be ascribed to transport as intact peptides.

2. Independence of Peptide Transport and Peptidase Activity

In addition to the direct studies on peptidase location, several indirect approaches were used to identify peptide transport independent of peptidase activity in general and of extracellular peptidase activity in particular. Peptides were found that resist peptidase activities. Such peptides, if transported across the membrane, can accumulate inside the cell in their intact form.

For example, several dipeptides are known with bacteriostatic properties toward *E. coli* (Simmonds *et al.*, 1951; Meisler and Simmonds, 1963; Payne and Gilvarg, 1968a; Gilvarg and Levin, 1972; Payne, 1972c). Since none of their degradation products is toxic, it seems that the intact peptides are responsible for their inhibitory effect. In the case of the bacteriostatic dipeptide Gly-Leu, it was even shown that the peptide is accumulated as such and is very slowly split by cells sensitive to the dipeptide (Meisler and Simmonds, 1963). Tri-L-ornithine is another example of intact-peptide toxicity. This peptide inhibits the growth of several strains of *E. coli* and *S. typhimurium* (Payne and Gilvarg, 1968a; Payne, 1968; Gilvarg and Levin, 1972; Barak *et al.*, 1973a; Fickel and Gilvarg, 1973; De Felice *et al.*, 1973; Barak and Gilvarg, 1974; Sussman and Gilvarg, 1970; B. N. Ames, personal communication), whereas higher concentrations of its degradation products di- and mono-ornithine do not (Gilvarg and Levin, 1972; Barak *et al.*, 1973a). In addition, the tripeptide was found to be relatively resistant to cleavage by peptidases in cell extracts of *E. coli* W (Payne and Gilvarg, 1968a) and *E. coli* B (Barak, 1972; Barak *et al.*, 1973a). Since it was found to affect protein synthesis, an internal cellular function (Barak *et al.*, 1973b, 1970, 1973a; Gilvarg and Levin, 1972), and because mutants of *E. coli* that have lost their ability to transport oligopeptides in general are triornithine resistant [for review see Payne and Gilvarg (1971), Gilvarg (1972), and Section IVA7], it can be concluded that the peptide must be accumulated intracellularly in order to exert its toxic activity.

Intact nontoxic dipeptides have also been detected inside cells. This happens under abnormal conditions ("aged" cells; 168 hr culture) when *E. coli* K-12 cells become cryptic. The peptidases of these cells are practically inactive, whereas the cells can still (actively) accumulate several peptides (Meisler and Simmonds, 1963; Simmonds, 1970).

Peptides containing certain unnatural amino acids are not usually hydrolyzed by peptidases. Some of them can still be absorbed into cells. Indeed, indirect evidence was obtained with *E. coli* indicating that several β-alanyl and sarcosyl peptides can enter the cells without being hydrolyzed (Payne, 1972b, 1973). Recently, Payne (1972a) used radioactive ϵ-acetylated-trilysine to show intracellular accumulation of the intact tripeptide derivative by *E. coli* W. He also demonstrated the identity of its uptake system with that of the general oligopeptide transport system of *E. coli* [see Section IVA2(a)], by showing that an oligopeptide transport deficient (Opt^-)[1] mutant lost over 90% of its ability to import this peptide. The ability of "abnormal" peptides to be accumulated intact by *E. coli* through use of the uptake system used by ordinary peptides would suggest that all peptides, both "abnormal" and "normal," are transported in their intact form.

Further evidence to support the view that peptide transport is independent of cleavage in *E. coli* has come from the genetic field. In 1963 Kessel and Lubin isolated a mutant of *E. coli* W that lacked the ability to hydrolyze diglycine, but was still capable of concentrating this dipeptide. Sussman and Gilvarg (1970) isolated a mutant of *E. coli* K-12 that had lost one of the peptidases that is capable of cleaving trilysine, but could still normally transport this tripeptide inside the cell. Intracellular accumulation of trilysine, as in the case of triornithine, is toxic to *E. coli*. Therefore this mutant, in contrast to its parental strain, was sensitive to trilysine.

Lastly, it should be mentioned that all the dipeptide and oligopeptide transport deficient mutants that were tested showed undiminished peptidase activity in their extracts (Kessel and Lubin, 1963; Payne and Gilvarg, 1968a; Barak, 1972).

It should be recognized that the evidence cited above, while establishing a solid case for the transport of intact peptides in some microbial species, does not rule out the existence of systems in which peptide transport is obligatorily coupled to peptide hydrolysis (Fig. 1C). However, at this point the evidence for the alternative mode of uptake is not convincing. Perhaps the *Mycoplasma*, which apparently contain membrane-bound amino peptidase activity, would be a good place to look for such evidence.

---

[1] The symbols used in this chapter for oligopeptide transport system are Opt and *opt* for the phenotype and genotype, respectively. We have chosen not to use Opp and *opp* (oligopeptide permease) as in previous publications (B. N. Ames *et al.*, 1973a; De Felice *et al.*, 1973; Barak and Gilvarg, 1974), because of the indications that binding proteins are involved in oligopeptide transport (IVA7). In this event, an oligopeptide transport deficient mutant might be altered in its binding protein and not in its permease, although in some cases these might be one and the same. Similarly, Dpt and *dpt* have been used instead of Dpp and *dpp* (De Delice *et al.*, 1973). In this context the word permease is being reserved for the membrane-bound components of the oligopeptide transport system.

3. *Distinction between Amino Acid and Peptide Transport*

If peptides could not penetrate into cells in their intact form, their nutritional activity or intracellularly manifested toxicity would be dependent on extracellular cleavage, followed by amino acid uptake. In this case, the biological activities of peptides should be strictly correlated with their amino acid composition and the presence of the requisite amino acid transport systems. However, studies with several bacterial species revealed that peptides can serve as better amino acid sources than equivalent concentrations of their constituent amino acids (Gale, 1945; Kihara and Snell, 1952; Peters *et al.*, 1953; Miller *et al.*, 1955; Shelton and Nutter, 1964; Meinhart and Simmonds, 1955). In addition, direct measurements sometimes showed faster absorption of amino acids when introduced as peptides than in their free form (Leach and Snell, 1960; Yoder *et al.*, 1965b; Hauschild, 1965). These findings are most easily explained by postulating distinct amino acid and peptide transport systems with similar kinetic properties. (Rapid intracellular hydrolysis of peptides provides more internal amino acid per mole of absorbed compound. Alternatively, absence of competition between the different amino acids when they penetrate as peptides may explain the greater efficiency of peptide feeding.)

An extreme case in which certain amino acids can be integrated into the cells only when administered in their peptide form is represented by the *Bacteroidaceae*. Studies with *Bacteroides ruminicola* have shown its inability to absorb ^{14}C-proline and ^{14}C-glutamic acid despite its ability to accumulate these same amino acids when present in oligopeptides with molecular weights of up to about 2000 (Pittman *et al.*, 1967). Similarly, *B. melaninogenicus* has only a limited ability to ferment amino acids, whereas peptides are readily used (Wahren and Gibbons, 1970). Recently (Wahren and Holme, 1973) a similar phenomenon has been demonstrated in *Fusiformis necrophorus*. This bacterium can neither synthesize proline nor transport it into the cells in its free form. The proline requirement is met by the addition of polyproline (mean molecular weight 2000). In addition, preliminary findings reveal that several species of *Mycoplasma* are unable to accumulate the free amino acids alanine (Pecht *et al.*, 1972), leucine, valine, aspartic acid, and glutamic acid (Cirillo, personal communication). On the other hand, these organisms require many amino acids for growth and it seems plausible that they normally utilize peptides as their amino acid source. However, as was mentioned above, *Mycoplasma* contain membrane-bound amino peptidase activity (Choules and Gray, 1971; Pecht *et al.*, 1972), and might utilize peptides according to the model of Ugolev (see Fig. 1C).

Competition experiments also support the view that there are distinct amino acid and peptide transport systems. Competition for entrance among amino acids sharing the same transport system has often been demonstrated. However, competition occurs only between free amino acids. Interference is usually not detected between amino acids that share a common transport system, when one is introduced in its peptide form (Brock and Woolley, 1964; Mayshak *et al.*, 1966; Yoder *et al.*, 1965b; Shelton and Nutter, 1964). Similarly, peptide uptake is not inhibited by amino acids that interfere with uptake of the constituent amino acids (Leach and Snell, 1959, 1960; Levine and Simmonds, 1960; Kihara and Snell, 1952; Prescott *et al.*, 1953; Brock and Woolley, 1964; Mayshak *et al.*, 1966; Yoder *et al.*, 1965b; Shelton and Nutter, 1964; Kessel and Lubin, 1963). It should be mentioned, however, that peptides are themselves subject to uptake competition (Payne and Gilvarg, 1971).

The presence of separate peptide and amino acid transport systems is further confirmed by the existence of bacterial mutants (Levine and Simmonds, 1960; Guardiola and Iaccarino, 1971) in which cells have lost a specific amino acid transport system, but still retain the ability to absorb the amino acid when administered in peptide form. Experiments with *E. coli* mutants also showed that peptide transport systems can be altered without affecting amino acid uptake as measured either directly with radioactive amino acids (Barak and Gilvarg, unpublished observation), or indirectly by the ability of free amino acids to support normal growth rates in Opt^- mutants (Payne and Gilvarg, 1968a; Payne, 1968; Gilvarg and Levin, 1972; Fickel and Gilvarg 1973; Barak and Gilvarg, 1974).

The ability of *E. coli* (Fickel and Gilvarg, 1973) and *S. typhimurium* (B. N. Ames *et al.*, 1973a) to bring in normally impermeable amino acid derivatives when these are coupled to peptides also clearly distinguishes between amino acid and oligopeptide transport systems.

B. Yeast

Peptide transport has recently been demonstrated in the yeast *Saccharomyces cerevisiae* G1333 (Becker *et al.*, 1973; Naider *et al.*, 1974). These studies utilize the indirect method in which growth response to peptides serves as the indicator for their uptake (Section IIB). The existence of exopeptidases is eliminated by direct measurements on the supernatant of whole cell suspensions and on the osmotic shock fluid of this microorganism (Becker *et al.*, 1973). The absence of extracellular peptidase is further substantiated by the existence of several peptides that are readily cleaved by cell extracts without being utilized as an amino acid source for the yeast auxotroph. It can be concluded, therefore, that peptides are utilized by *S. cerevisiae* after being transported into the cells in their intact form

(Fig. 1A). However, since the variety of peptides tested with the yeast is limited (only peptides containing methionine and glycine residues), one cannot exclude completely, for this microorganism, the alternative or additional use of one of the other models proposed for the mechanism of peptide utilization (Fig. 1B, C).

C. Mammals

In contrast to the findings in *E. coli* (Section IIIA1) and *S. cerevisiae* (Section IIIB), evidence for the existence of proteolytic activity in the extracellular fluid of the intestinal lumen of mammalian organisms already appeared at the beginning of the 20th century (Cohnheim, 1901; Van Slyke and Meyer, 1912). More recently, *in vivo* and *in vitro* investigations confirmed this observation and showed peptidase activity toward triglycine, tetraglycine, and several dipeptides containing glycine in lumen fluid (Agar *et al.*, 1953; Johnston and Wiggans, 1958; Wiggans and Johnston, 1959; Newey and Smyth, 1957, 1959).

This extracellular peptidase activity made it unnecessary to postulate the existence of peptide transport in mammalian organisms. Therefore in the 1940's standard textbooks often simply stated that proteins were absorbed as amino acids (Best and Taylor, 1950) (Fig. 1B). Later, the demonstration of active transport mechanisms for free amino acids in the small intestine (Gibson and Wiseman, 1951; Wiseman, 1953) and the finding that dipeptides could not pass intact across the wall of everted sacs of small intestine *in vitro*, except in traces (Agar *et al.*, 1953), appeared to bring additional support to the hypothesis that proteins were completely hydrolyzed to free amino acids before absorption.

However, this hypothesis ignored statements of early physiologists who found incomplete hydrolysis of proteins in the lumen, and pointed therefore to the possibility that small peptides might be taken up by the epithelial cells of the small intestine where the proteolytic process would be finished by intracellular peptidase activity (Van Slyke and Meyer, 1913–1914). This possibility was further supported by the observation that the extracellular peptidase activity was much lower than the cellular activity tested in extracts (Starling, 1906; Cajori, 1933; Newey and Smyth, 1960). Newey and Smyth (1959, 1960) even calculated that the peptidase activity in the lumen was insufficient to account for the amounts of glycyl peptides capable of disappearing from the mucosal side of the intestine. They were in fact the first to explicitly draw the scientific world's attention to the significance of peptide transport in the process of protein absorption.

Ugolev and co-workers [for review see Ugolev (1972), Ugolev and

DeLaey (1973)] have focused on determining the location of the peptidase activity attributable to the cells of the intestinal wall. Dipeptidase and tripeptidase activity was found to be associated with the glycocalyx (the filaments that cover the membrane of the microvillus). In particular it was shown that under normal conditions dipeptidases are strongly bound to the cell surface. It was assumed therefore that peptide hydrolysis is completed on the external surface of the microvillus membrane (Fig. 1C).

However, the notion of independent peptide transport in the mammalian intestine was supported by the findings that some amino acids are absorbed faster when introduced as peptides than in their free form (Messerli, 1913; Craft and Matthews, 1968; Craft *et al.*, 1968; Matthews *et al.*, 1969; Gangolli *et al.*, 1970; Hellier *et al.*, 1970; Cheng and Matthews, 1970; Crampton *et al.*, 1971; Lis *et al.*, 1971; Burston *et al.*, 1972). Moreover, amino acid absorption from peptides is not inhibited by amino acids that interfere with the uptake of the corresponding amino acid constituents, whereas peptides compete with one another for uptake (Rubino *et al.*, 1971; Addison *et al.*, 1973, 1974a, b, c). The presence of separate peptide and amino acid transport systems is further confirmed by the existence of several inborn physiological disorders in man in which cells have lost a specific amino acid transport system, but still retain the ability to absorb this amino acid when administered as peptides. Specifically, in Hartnup disease intestinal absorption of histidine, phenylalanine, tryptophan, and tyrosine in their free form is negligible, whereas absorption of these "affected" amino acids from dipeptides (β-Ala-His, Gly-His, Phe-Phe, Gly-Trp and Gly-Tyr, respectively) is normal (Navab and Asatoor, 1970; Tarlow *et al.*, 1970; Asatoor *et al.*, 1970b). In cystinuria, kidney and intestinal cells have been shown to lose their ability to absorb free arginine and lysine. Arginine uptake from arginine-containing peptides (Arg-Asp and mixed peptides from protein digestion) is normal in the gut (Asatoor *et al.*, 1971, 1972). Corresponding studies with lysine-containing peptides (Gly-Lys, Lys-Lys) have been more variable and less conclusive, but showed at least some improvement in absorption from the peptides (Hellier *et al.*, 1970, 1971, 1972; Asatoor *et al.*, 1971, 1972). These observations proved that an appreciable fraction of the peptide transport in the intestine of man is independent of amino acid transport.

In addition, several reports have mentioned the appearance, from the lumen, of traces of dipeptides in the intracellular fluid of the enterocyte (Fern *et al.*, 1969; Addison *et al.*, 1974c). The dipeptide Pro-Hyp was even found intact in the bloodstream (Hueckel and Rogers, 1972). Intact-peptide absorption is especially observable with "abnormal" peptides that are resistant to peptidase activity. Indeed, Ford and Shorrock (1971) showed that rats fed with fish cooked by a process producing "abnormal" peptides excreted urine with an increased concentration of such peptides. These observations were recently confirmed *in vitro* using peptides containing β-alanine and sarcosine (Sar). The peptides carnosine (β-Ala-His) and Gly-Sar (which is similar to Gly-Gly except that the N of the peptide bond is methylated) were used (Matthews *et al.*, 1974; Addison *et al.*, 1972, 1973, 1974a, b; Burston *et al.*, 1972) to investigate kinetics of peptide uptake in intestinal cells. These peptides had previously been shown to be peptidase resistant (Payne, 1972b, 1973; Matthews *et al.*, 1974). The cells were able to concentrate carnosine and Gly-Sar over threefold. In addition, the peptides Gly-Gly, Gly-Gly-Gly, Gly-Sar, Gly-Pro, Pro-Hyp, and Met-Met could compete against carnosine uptake, indicating that they shared a transport mechanism.

D. Conclusions

Evidence has been presented for the existence of peptide transport in several microorganisms and in the intestine of mammals. The only systems that have been studied in detail, and on which, therefore, most of our knowledge on peptide transport is based, are those that occur in *E. coli* and, to lesser extent, in the intestinal cells of mammals.

Since *E. coli* is free of extracellular, periplasmic, and membrane-bound peptidases, and peptide uptake processes do not show any correlation to amino acid uptake, it can be concluded that peptides are transported into these cells in their intact form. Only at a later stage, inside the cell, are peptides cleaved to yield their constituent amino acids (Fig. 1A). The same mechanism appears to operate for several species of *Lactobacilli* and for the uptake of methionine peptides by yeast (*S. cerevisiae*).

An interesting group of organisms are those that have to utilize peptides as the source of certain amino acids that cannot penetrate the cell in free form. These microorganisms might "represent the ultimate evolutionary response to a selective advantage of peptide over amino acid transport" (Gilvarg, 1972). For if the energy required for peptide uptake is similar to amino acid uptake, the effort to bring in amino acids in their free form should be higher than that required to accumulate the same amount of amino acids when administered in peptides. Evidence is available to include several species of *Bacteroidaceae* and *Fusiformis necrophorus* and possibly some species of *Mycoplasma* into this group.

In the mammalian intestine minor peptidase activity exists in the fluid of the lumen and therefore a certain amount of peptide might be cleaved to free amino acids before absorption (Fig. 1B). The presence of membrane-bound peptidases on the intestinal cells might account for a peptide transport mechanism that requires surface hydrolysis (Fig. 1C). Since according to this model (Ugolev, 1972) the degradation products of the peptides, the free amino acids, are not released to the lumen, but are immediately transported by the related amino acid transport systems, it might explain many of the findings that show distinction of amino acid and peptide transport. Others have suggested that the postulated peptidase-carrier membrane complex might indeed be defined as the peptide transport system of the mammalian organisms (Kornberg, 1972). We are bothered, however, by the practical difficulty of associating a peptidase unit in the membrane with each of the many different amino acid transport systems. In addition, this model is unable to explain the uptake in the intestinal system of certain intact peptides. It is clear that at least some peptides penetrate into the intestinal cells in their intact form, representing a true peptide transport system (Fig. 1A). However, the complex membrane digestion model as postulated by Ugolev (Fig. 1C) or a modification of it might also exist in the intestine of mammals. The significance and the specificity of each of the latter systems in the overall mechanism of peptide utilization is unknown and therefore still controversial. It should be mentioned, however, that most of the studies in mammalian intestine tested dipeptide uptake. Studies with tri- and tetrapeptides were also performed, but because of the existence of extracellular peptidases, it was not clear whether or not these oligopeptides are taken up into the cells in their intact form. Tripeptide uptake was demonstrated only recently with the use of the peptidase-resistant tripeptides β-Ala-Gly-Gly (Addison *et al.*, 1974a) and Gly-Sar-Sar (Addison *et al.*, 1974b).

IV. THE PROPERTIES OF THE PEPTIDE TRANSPORT SYSTEMS

The peptide transport system is the overall mechanism responsible for transferring an intact peptide across the palsma membrane into the cytoplasm (Fig. 1A). As indicated above, such a mechanism exists in *E. coli* cells and several species of *Lactobacilli*, in *S. cerevisiae*, and at least in certain cases in the mammalian intestine. Most of the properties of the peptide transport systems have been worked out in studies with *E. coli*. In the following sections we shall discuss these properties of *E. coli* and compare them with findings in other organisms.

A. Bacteria

1. Distinction between Dipeptide and Oligopeptide Transports

Studies with *E. coli* W strains by Gilvarg and co-workers [for review see Payne and Gilvarg (1971)] revealed that oligopeptides (peptides containing more than two amino acid residues) are transported via a unique system which is described in the next section. Dipeptides, however, can be taken into *E. coli* cells via the dipeptide transport system or the oligopeptide transport system. Part of the evidence for this conclusion was the finding that dipeptides can compete with tripeptide uptake either when measured by the effect on the utilization of tripeptide for growth (Payne, 1968), or reducing the toxicity of an inhibitory tripeptide (Payne, 1968, 1971a). Tripeptides, however, are unable to interfere completely with the ability of the dipeptides to support growth (Payne, 1968) or with the direct uptake of certain radioactive dipeptides (Kessel and Lubin, 1963). Oligopeptides can interfere, however, with the transport of one another (Payne, 1968, 1971a, b, 1972b). These findings are in accord with the observations that oligopeptide transport deficient (Opt$^-$) mutants retain the normal activity of their dipeptide transport systems. Thus, dipeptides, in contrast to oligopeptides, can meet the amino acid requirement of an Opt$^-$ auxotroph and result in a normal growth response (Payne, 1968, 1971b, 1973). In addition, there is a difference between the two transport systems with regard to structural requirements. Whereas a free terminal carboxyl is needed to allow uptake of a peptide by the dipeptide transport system, no such requirement exists for oligopeptides [see Section IVA4(c)]. This distinction in structural specificities confirms the existence of separate transport systems. Dipeptides that lack their terminal carboxyl and cannot be transported through the dipeptide system are able to penetrate the cell via the oligopeptide transport system, indicating again that dipeptides can use both transport systems.

These findings with *E. coli* W were recently confirmed in *E. coli* K-12 and *S. typhimurium*. Several Opt$^-$ strains of *E. coli* K-12 were found to utilize normally Pro-Phe as proline source and to be sensitive to the toxic dipeptide lysyl-*p*-fluorophenylalanine (Lys-*p*-F-Phe), whereas they were unable to utilize Pro-Phe-Lys and were resistant toward the tripeptides dilysyl-*p*-fluorophenylalanine (Lys-Lys-*p*-F-Phe) and triornithine (Barak and Gilvarg, 1974). Another Opt$^-$ strain of *E. coli* K-12 was found to retain its sensitivity toward the toxic dipeptide Gly-Val. This strain also failed to absorb ^{14}C-triglycine (De Felice *et al.*, 1973). In addition, di- and trilysine competed with the ability of Pro-Gly and Pro-Gly-Gly, respectively, to support the growth of a proline auxotroph of *E. coli* K-12 (Payne, 1971b). These competitive inhibitions were specific; for, as expected, trilysine did not interfere with the dipeptide uptake and the concentration of the dilysine tested was not high enough to affect the oligopeptide transport system. Similarly, Gly-Gly and Gly-Gly-Gly specifically relieved the inhibition caused by Gly-Val and Val-Gly-Gly, respectively (Payne, 1971a). As was the case in *E. coli* K-12, several Opt$^-$ mutants of *S.typhimurium* were inhibited by a toxic dipeptide (ethionylalanine) but not by a number of toxic tripeptides (Ames *et al.*, 1973a).

The inability of De Felice *et al.* (1973) to isolate a spontaneous dipeptide transport deficient (Dpt$^-$) mutant from *E. coli* K-12 also indirectly supported the view that dipeptides are transported through the dipeptide and the oligopeptide transport systems, since only the highly improbable simultaneous mutation in both systems could result in a Dpt$^-$ mutant. Their use of an Opt$^-$ mutant as parental strain eliminated one of the alternative uptake systems and allowed selection for a Dpt$^-$ mutant that could then be created by a single gene mutation.

The mapping of the genes that are responsible for dipeptide and oligopeptide transport in *E. coli* K-12 clearly demonstrated the distinction between these two uptake systems. The *opt* gene (or operon) is the closest marker to the *trp* operon (27 min) on its cysB side (Barak and Gilvarg, 1974; De Felice *et al.*, 1973), whereas the *dpt* locus is separated from *opt* and is located between *pro* C (10 min) and *opt* (De Felice *et al.*, 1973).

Some evidence is available concerning the distinction between dipeptide and oligopeptide transport systems in other bacteria. Thus in *Leuconostoc mesenteroides* (Shelton and Nutter, 1964) and *Lactobacillus arabinosus* (Dunn *et al.*, 1957) tripeptides failed to compete with dipeptide uptake, whereas dipeptides could compete with one another. Several dipeptides were also able to partially reverse the inhibition caused by the tripeptide Val-Val-D-Val in *L. arabinosus* (Shankman *et al.*, 1962), suggesting that as in *E. coli*, dipeptides are capable of utilizing the oligopeptide transport system for their entry. However, there is insufficient evidence in organisms other than *E. coli* to allow a firm decision as to whether or not the distinction between dipeptide and oligopeptide transport systems is a general property of peptide transport in bacteria.

2. *Side-Chain Specificity of Peptide Transport Systems*

a. The Oligopeptide Transport System. All the oligopeptides tested in various mutants of *E. coli* W, except in two cases, have been found to be transported by the same transport system regardless of their amino acid composition. A single gene mutation can result in an oligopeptide transport deficient strain which is unable to transport peptides containing more than two amino acid residues (Payne, 1968, 1971a, b, 1972a, b, 1973; Payne and Gilvarg, 1968a; Fickel and Gilvarg, 1973). Also, oligopeptides containing different amino acids can compete with the transport of one another (Payne, 1968, 1971a, b, 1972a, b). The peptides that have been tested and shown to share the oligopeptide transport system by their failure to be transported into Opt$^-$ mutants or by competition experiments include some with basic amino acid residues (Lys, Orn, Arg), neutral residues (Gly, Ala, Leu, Val), aromatic residues (Tyr, Trp, Phe), hydroxy residues (Ser, Thr), and sulfur residues (Met). Even oligopeptides containing unusual amino acids or amines like cadaverine (Payne and Gilvarg, 1968a), norleucine, norvaline (Payne, 1971a), sarcosine (Payne, 1971a, 1972b), several other alkyl amino acids (Payne, 1974), ϵ-acetylated lysine (Gilvarg and Katchalski, 1965; Payne, 1972a), β-alanine (Payne, 1973), and homoserine-phosphate (Fickel and Gilvarg, 1973) were found to utilize the oligopeptide transport system, since Opt^- mutants failed to take in peptides containing these derivatives. Some of these results were also supported by competition experiments (Payne, 1971a, 1972a, b). It seems, therefore, that a single oligopeptide transport system in *E. coli* W is able to accommodate all peptides and does not show side chain specificity. This observation stands in marked contrast to the high degree of specificity that is found in amino acid and sugar transport systems. However, this lack of specificity enables the system to cope with the enormous diversity of substrates inherent in having 20 different amino acids from which peptides are constructed.

Although the oligopeptide transport system in *E. coli* W does not exclude peptides according to their amino acid constitution, one should not jump to the extreme conclusion that the side chains of the peptides are of no importance to their transport capability. Significant differences in the affinity of peptides to the oligopeptide transport system were demonstrated by Payne (1968) in a set of competition experiments. It appears that highly positively charged oligopeptides like trilysine and triornithine have greater affinity to the oligopeptide transport system than peptides with neutral (glycyl) or aromatic (tyrosyl) residues. This preference in transport might be derived from a different attraction, dependent on structural compatibility, of the peptides to the molecules (permease and/or binding protein) that are directly responsible for uptake. On the other hand, the differences in the transport properties might result from variations in the abilities of the peptides to reach the "transport molecules," whereas the affinities of the peptides to these specific molecules are similar.

The broad specificity of the oligopeptide transport system was also confirmed with *E. coli* K-12. Several Opt^- mutants were isolated from a number of *E. coli* K-12 Hfr and F^- strains, by means of their resistance toward the cytotoxic tripeptide triornithine. They all were unable, therefore, to accumulate this toxic tripeptide. A lysine auxotroph in this group was unable to utilize lysine oligopeptides (Sussman and Gilvarg, unpublished observation) and some of the other Opt^- mutants failed to use Pro-Phe-Lys as proline source (Barak and Gilvarg, 1974). The different Opt^- strains also showed cross-resistance with other cytotoxic tripeptides like trivaline (Sussman and Gilvarg, unpublished observations; De Felice *et al.*, 1973; Barak and Gilvarg, in preparation) and Lys-Lys-*p*-F-Phe (Barak and Gilvarg, 1974). In one case (De Felice *et al.*, 1973), the Opt^- strain was shown to lose its ability to accumulate ^{14}C-triglycine. Although the number of peptides tested is not sufficient to draw a general conclusion, it seems that *E. coli* K-12, like *E. coli* W, possesses an oligopeptide transport system that is able to accumulate oligopeptides with a variety of amino acid residues.

Of course, the demonstration of the existence of a general peptide transport system does not automatically exclude the presence of specialized systems. In that regard it is noteworthy that recently we have some indications (Barak and Gilvarg, unpublished observation) that *E. coli* K-12 TD-V (Barak and Gilvarg, 1974) contains an additional oligopeptide transport system. Thus, Opt$^-$ mutants of this strain were still able to accumulate trithreonine and also gave a normal exponential growth curve when trileucine was supplied as the sole source of leucine to these leucine auxotrophic strains. Moreover, it was possible to demonstrate competition for entry between these two tripeptides in the Opt$^-$ strains. The relationship between trithreonine and trileucine uptake was further established by the isolation from one of the Opt$^-$ strains of a spontaneous mutant that was altered in this new transport system. The substrain failed to take in trithreonine and its growth on high concentrations of trileucine was linear and very poor. This linear growth response correlated with the concentration of the tripeptide, indicating uptake by diffusion. The existence of this additional peptide transport system was also apparent in *E. coli* W TL3, since an Opt$^-$ mutant of this threonine auxotroph (Fickel and Gilvarg, 1973) could utilize trithreonine for growth as well as its parental strain. Recently, Naider and Becker extended these findings and showed (personal communication) that Opt$^-$ mutants of *E. coli* K-12 4212 and *E. coli* B163 were still able to grow normally on low concentrations of trimethionine and trileucine as sources for methionine and leucine, respectively. Trimethionine also competitively inhibited the growth of both TOR strains on trileucine. This Met$_3$-Leu$_3$ transport system was further shown not to admit tetrapeptides containing methionine, since those were unable to support the growth of the Opt$^-$ mutants. In addition, trimethionine was shown by competition experiments to utilize the Thr$_3$-Leu$_3$ transport system observed by us (Barak and Gilvarg, unpublished observation). It appears, therefore, that an additional transport system exists in *E. coli* that can take in trithreonine, trileucine, and trimethionine. It should be mentioned, however, that at least trithreonine and trimethionine and possibly trileucine, to a lesser extent, are also transported by the general oligopeptide transport system described by Gilvarg and his co-workers [for review see Payne and Gilvarg (1971), Gilvarg (1972), Payne (1972a)]. Thus it is reasonable that the oligopeptide transport system serves as a general transport system for peptides and in addition a more specific system (or systems), such as the one mentioned above, also exists. This situation resembles that found for several amino acids that may be taken in by both general and specific transport systems.

Finally, we would like to mention that Payne (1968) reported that the hydrophobic tripeptide Gly-Leu-Gly was able to meet the amino acid requirement of auxotrophic Opt$^-$ strains of *E. coli* W. However, the concentrations of the peptides required for growth were relatively high. Moreover, the growth rate increased in correlation to the concentrations of the peptides but remained essentially linear. This growth response could not be affected by trilysine, indicating a distinct peptide uptake system. The mechanism of uptake through this system is probably by diffusion. A similar phenomenon was demonstrated by the failure of tripeptides to completely reverse the inhibition obtained in *E. coli* W by high concentration of Gly-Gly-norleucine (Payne, 1971a). It appears that although the major peptide uptake in *E. coli* seems to be carried out via a general and nonspecific system, which has been discussed above and will be referred to as the "oligopeptide transport system," some additional systems might exist. Our knowledge about these transport systems is still preliminary.

The only other bacterium in which Opt$^-$ mutants have been identified is *S. typhimurium* (B. N. Ames *et al.*, 1973a). Several spontaneous Opt$^-$ mutants have been isolated by selecting for resistance to one of several inhibitory tripeptides: trilysine, norleucyl-Gly-Gly, or high concentrations of Gly-Gly-histidinol phosphate ester. Each of the mutants is cross-resistant to all the other mentioned toxic tripeptides and in addition to Gly-Gly-norleucine and Gly-Gly-norvaline. These mutants are also unable to use Gly-His-Gly as a histidine source. Since the peptides tested contain a variety of amino acid side chains, including basic (Lys), neutral (Gly, norleu., norval.), and aromatic (His) residues, it would appear that *S. typhimurium*, like *E. coli*, contains a general oligopeptide transport system.

Little evidence exists that bears on the competition among oligopeptides for entry in other bacteria (see Payne and Gilvarg, 1971). However, the variety of peptides and organisms tested is very limited and does not allow a clear decision concerning the generality of the broad specificity and the distribution of the oligopeptide transport system in nature.

b. Dipeptide Transport System. Dipeptides were found to utilize at least two distinct transport systems in *E. coli*: the general oligopeptide transport system and the dipeptide transport system or systems (see Section IVA1). It is difficult, therefore, to interpret competition experiments with dipeptides. For example, the partial inhibition of uptake or growth that is frequently obtained in competition experiments might represent a complete inhibition of the nonspecific uptake through the oligopeptide transport system with no effect on the specific dipeptide transport system. It is therefore important in studies with dipeptides to eliminate the possibility of uptake through the oligopeptide transport system by the use of oligopeptide transport deficient mutants of *E. coli*. Unfortunately, experiments of this type have not been performed.

Competition for entry between dipeptides was demonstrated in normal cells of *E. coli* W. Kessel and Lubin (1963) showed inhibition of ^{14}C-diglycine uptake by all the dipeptides tested that contained two L-amino acids or glycine with one L-amino acid. They checked a variety of dipeptides containing glycine at their N-terminal position, some dipeptides containing other neutral amino acids, and dihistidine. In other experiments the ability of diglycine to support the growth of a glycine auxotroph was shown to be antagonized by dilysine (Payne, 1968) and Gly-Sar (Payne 1972a). The nonutilizable dipeptide Gly-Sar also inhibited the utilization of Gly-Pro as glycine source (Payne, 1972b). Competition for entry was also shown between Gly-Leu and Leu-Gly (Levine and Simmonds, 1962). In *E. coli* K-12, dilysine inhibited growth of a proline auxotroph on Pro-Gly (Payne, 1971b) and Gly-Gly and Sar-Gly competed with Gly-Val toxicity (Payne 1971a).

In addition, a dipeptide transport deficient (Dpt$^-$) mutant of *E. coli* W that failed to accumulate ^{14}C-Gly-Gly was also unable to utilize Gly-Leu or Leu-Gly as glycine sources (Kessel and Lubin, 1963). Another Dpt$^-$ mutant was recently isolated from an Opt$^-$ strain of *E. coli* K-12 (De Felice *et al.* 1973). This mutant was unable to accumulate ^{14}C-Gly-Gly and was resistant to normally toxic dipeptides containing valine: Val-Val, Gly-Val, Val-Leu, and Leu-Val. Several Gly-Leu resistant-strains of *E. coli* K-12 were also found to be defective in their dipeptide transport system and cross-resistant, therefore, to Gly-Val (Vonder Haar and Umbarger, 1972).

It appears that in *E. coli*, dipeptides, much like the oligopeptides, are transported by a nonstringent system. However, several exceptional cases in which a special uptake system was utilized are mentioned in the literature. These include carnosine (β-Ala-His) uptake by *E. coli* K-12 (Payne, 1973) and the uptake of Gly-Norleu and Norleu-Norval by *E. coli* W (Payne, 1971a).

Utilization of a common transport system was also demonstrated for Gly-Ala and Ala-Gly in *Lactobacillus casei* (Leach and Snell, 1960), and for Gly-Val, Gly-Leu, Gly-Gly, Gly-DL-Ser, and DL-Ala-DL-Phe in *Leuconostoc mesenteroides* (Shelton and Nutter, 1964; Mayshak *et al.*, 1966). On the other hand, Pro-Phe and His-His failed to inhibit ^{14}C-Gly-Ala uptake by *L. casei* (Leach and Snell, 1960), possibly indicating independent transport systems.

3. Size Restriction in Peptide Transport

The dipeptide transport system in *E. coli* is a distinct mechanism limited to peptides with two amino acid residues only. This size restriction is probably carried out at the level of the permease, i.e., the molecule that is responsible for the transport of the dipeptide across the cytoplasmic membrane. This permease establishes its specificity for dipeptides by having recognition sites, demanding free N and C termini of the peptide molecule [see Sections IVA4(b, c)]. Since the distance between these two groups is relatively constant and is not greatly affected by differences in side chains, the termini requirements ensure a distinct dipeptide transport system. Peptides that contain more than two residues are transported via the oligopeptide transport system. This has its recognition sites only at the N-terminus of the peptide [see Section IVA4(b)]. The lack of the requirement for a free C-terminal carboxyl enables this permease to transport peptides of different sizes.

The early observation of Gilvarg and Katchalski (1965) and later work by Payne and Gilvarg (1968b) revealed that there is a distinct size limit to oligopeptide uptake. This sharp cutoff in the nutritional effectiveness of peptides is related to their hydrodynamic volume, as measured by the speed of filtration through a Sephadex G-15 column (Payne and Gilvarg, 1968b). It was also established that this limit is independent of the composition of the peptides, since the same cutoff point was observed for a diverse mixture of neopeptone peptides. Heterologous peptides larger than pentalysine failed to meet the lysine and glycine requirements of the respective auxotrophs of *E. coli* W, whereas smaller peptides could support bacterial growth. It should be emphasized that the size restriction for oligopeptide uptake in *E. coli* W, unlike that for the dipeptide transport system, is not determined by the chain length of the peptide backbone. The cutoff point varies in different peptide series. For example, the tetramers: Lys-Lys-Lys-Homoserine-phosphate (Fickel, 1973) and lys-lys-lys-cadavarine (Payne and Gilvarg, 1968a, b) are excluded, but the hexamer (hexaglycine; Payne and Gilvarg, 1968b) is not. Moreover, large peptides like pentalysine and other similar sized oligopeptides cannot interfere with the entry of smaller peptides (Payne and Gilvarg, 1968b, 1971), indicating that the large peptide is unable to reach the oligopeptide permease. It seems, therefore, that the process of elimination of large oligopeptides is not carried out at the stage of binding to the permease. An accessory permeability barrier external to the oligopeptide transport system was postulated to be responsible for the size restriction phenomenon. Because of the location and the special porous structure of the peptidoglycan in bacteria, it was suggested as this "external barrier" (Gilvarg, 1972). However, others have suggested that the lipopolysaccharide (LPS) membrane is responsible for the size discrimination in Gram-negative bacteria [for review see Costerton *et al.* (1974)]. An alternative explanation for the observed findings is that the "external barrier" responsible for size restriction is the microenvironment of the oligopeptide transport system. The structural restraints provided by the location of this transport system in the cytoplasmic membrane might prevent attachment of large peptides to the permease (Fig. 2). One of the predictions of this model is that large peptides should be able to reach the cytoplasmic membrane and be accessible to periplasmic enzymatic activities. Periplasmic enzymes are osmotic-shock-releasable and are believed to be located within the bacterial cell envelope (Heppel *et al.*, 1972).

Recently, experiments were carried out in our laboratory (Fickel, 1973) in which attempts were made to correlate the location of the oligopeptide transport system with that of the periplasmic enzymes. Such correlation could have been helpful in the identification of the size restriction barrier of the oligopeptide transport. Lys-Lys-Lys-Homoserine-phosphate provides a suitable compound for experiments of this type. The peptide, in contrast to the lower homologue Lys-Lys-Homoserine-phosphate, was unable to meet the lysine or threonine requirements of an *E. coli* auxotroph TL3 when the organism was grown on a high-phosphate medium to prevent induction of alkaline phosphatase (Fickel and Gilvarg, 1973). These findings were easily reconciled with previous observations, since the peptide had a larger hydrodynamic volume than pentalysine (Fickel, 1973). Lys-Lys-Lys-Homoserine-phosphate was unable, therefore, to cross the "external barrier" and reach the oligopeptide transport system. However, the peptide could reach the periplasmic enzyme, alkaline phosphatase. The rate of phosphate release from a whole cell preparation, induced for alkaline phosphatase, was roughly the same for the small peptides Lys-Homoserine-phosphate and Lys-Lys-Homoserine-phosphate as for the large peptide Lys-Lys-Lys-Homoserine-phosphate. Lys-Lys-Lys-Homoserine-phosphate was also able to support the lysine requirement of *E. coli* TL3 when the alkaline phosphatase in the bacteria was derepressed. The large peptide was dephosphorylated by the alkaline phosphatase to yield Lys-Lys-Lys-Homoserine. This reduced the hydrodynamic volume enough to allow it to cross through the "external barrier" and be transported into the cell. However, this peptide could not

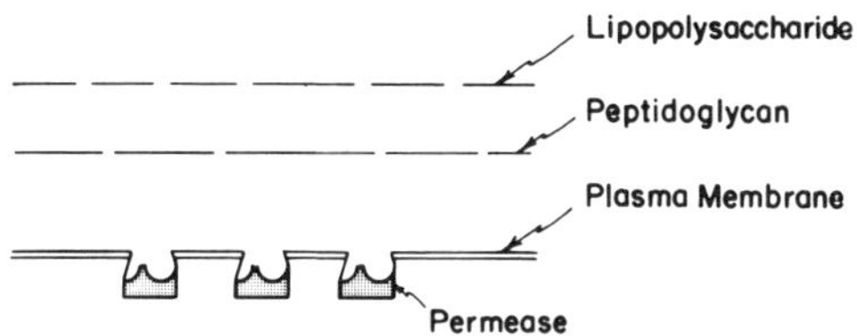

Fig. 2. A model for peptide exclusion at the plasma membrane.

supply homoserine-phosphate to the cells and was unable, therefore, to meet the threonine requirement of the bacteria, since TL3 is blocked at homoserine kinase (Homoserine $\not\rightarrow$ Homoserine-phosphate → threonine).

It was concluded that the periplasmic enzyme alkaline phosphatase and the oligopeptide permease could not occupy the same compartment in the cell envelope of *E. coli.* The barrier between the external medium and the alkaline phosphatase compartment admits larger molecules than the "external barrier" of the oligopeptide transport system. The two ways in which two different barriers can be place with respect to one another are in the "parallel" or "series" arrangement (Fig. 3). As can be seen from Fig. 3, the essential difference between the two is that for a series placement of barriers a peptide must always pass by the alkaline phosphatase to reach the oligopeptide permease, whereas for a parallel placement of the barriers, a peptide need not enter the space containing the alkaline phosphatase in order to reach the permease. The parallel model requires a mosaic arrangement of the cell envelope with small-sized pores in the regions that contain the oligopeptide transport system compartments and large-sized pores in the portions of the envelope that contain the alkaline phosphatase. Asymmetric localization of the alkaline phosphatase at cell termini has been observed (Wetzel *et al.*, 1970; Dvorak *et al.*, 1970). On the other hand, others showed that alkaline phosphatase is evenly distributed throughout the periplasmic space of *E. coli* (MacAlister *et al.*, 1972), so that this point is uncertain.

However, neither the parallel model nor the series model can identify the size restriction barrier of the oligopeptide transport system. Little is known concerning the spatial organization of the macromolecular barriers in the cell envelope of *E. coli.* However, one would assume that the islands of oligopeptide transport suggested by the parallel model contain the normal layers of mucopeptide and LPS. Thus, according to the parallel model, any of the cell envelope layers, including the microenvironment of the oligopeptide permease on the cytoplasmic membrane, might serve as the size restriction barrier of the oligopeptide transport system. The series model could

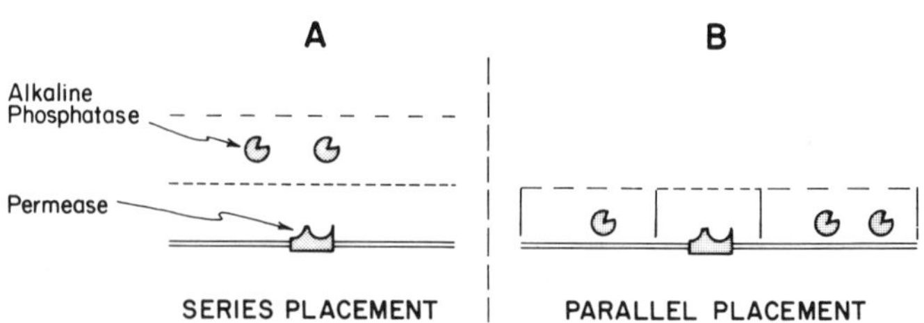

Fig. 3. Possible models for the placement of barriers to alkaline phosphatase and oligopeptide permease.

have been of more help in the process of identification of the size restriction barriers were it not for our ignorance of the precise location of the periplasmic enzyme alkaline phosphatase. If the alkaline phosphatase is located between the LPS membrane and the peptidoglycan, the LPS should serve as the periplasmic barrier and the peptidoglycan or the microenvironment of the system as the oligopeptide transport barrier. On the other hand, the alkaline phosphatase might be located in the space between the cytoplasmic membrane and the peptidoglycan and thus the LPS–peptidoglycan complex serves as the periplasmic barrier, whereas the microenvironment of the permease serves as the oligopeptide transport barrier. Assuming the series model is correct, it is clear that once the exact location of alkaline phosphatase within the cell envelope of *E. coli* is determined this will identify the size restriction barrier of the oligopeptide transport system.

Another approach to the problem is to look for specific changes in the size restriction of oligopeptide transport in *E. coli* cells that have undergone modifications in their envelope. Correlation between an envelope modification and a change in the size restriction of the oligopeptide transport might identify the barrier responsible for this phenomenon. Modifications in a specific layer of the cell have been obtained by mutation in *E. coli* and *S. typhimurium*. Several mutants of this type are mentioned in the literature (Schmidt *et al.*, 1969, 1970; Monner *et al.*, 1971; Henning *et al.*, 1972; Rooney and Goldfine, 1972; Wu, 1972; Tamaki *et al.*, 1971; Tamaki and Matsuhashi, 1973; Lehmann *et al.*, 1973; Kadner and Liggins, 1973; White *et al.*, 1973; Lindsay *et al.*, 1973; B. N. Ames *et al.*, 1973b; G. F. Ames *et al.*, 1974; Koplow and Goldfine, 1974; Kamiryo and Strominger, 1974). However, none of these have been used in peptide transport studies. One might also carry out transport studies with spheroplasts that lack their peptidoglycan, or with cytoplasmic membrane vesicles. The maintenance or disappearance of the size restriction barrier in these systems should provide the answer to the role that the peptidoglycan plays in the process of size restriction of oligopeptide transport in *E. coli*. It is of interest that many mutations in the LPS membrane of *E. coli* and *S. typhimurium* are accompanied by changes in the size restriction barrier of the cells toward several drugs, detergents, dies, mutagens, or lysozyme (Wu, 1972; Tamaki *et al.*, 1971; Tamaki and Matsuhashi, 1973; B. N. Ames *et al.*, 1973b; G. F. Ames *et al.*, 1974). Moreover, specific chemical changes in the LPS membrane of *E. coli* caused by short EDTA treatment also result in a heightened sensitivity to actinomycin D, probably by increasing the permeability of the cells for the drug (Leive, 1968; Leive *et al.*, 1968). Therefore it seems plausible that the LPS membrane of *E. coli* serves as a size restriction barrier. It is not clear, however, whether this barrier is identical with the "external barrier" of the oligopeptide transport system. The possibility that a change in the LPS layer might indirectly affect the peptidoglycan layer that is primarily responsible for the size restriction for oligopeptides would have to be considered.

Little evidence is available concerning the size limit of oligopeptide uptake in other organisms. Preliminary findings in our laboratory (Raychaudhuri and Gilvarg, unpublished observation) revealed that the size restriction barrier for *Bacillus megaterium* is similar to that of *E. coli* W. Cells were able to utilize tetralysine as a lysine source, whereas pentalysine failed to support the growth of a lysine auxotroph of *B. megaterium*. This similarity to *E. coli* might be related to the similar percentage of crosslinking in the peptidoglycans of the two species (Fordham and Gilvarg, 1974). Moreover, *B. megaterium*, unlike *E. coli*, is a Gram-positive bacterium that lacks the lipopolysaccharide layer. However, since the number of peptidoglycan layers in *B. megaterium* is different from *E. coli*, and because of the existence of an external teichuronic acid layer in *B. megaterium*, it is difficult to draw the definitive conclusion that the peptidoglycan is the barrier in both species. Different cutoff points were observed in studies of peptide uptake and utilization in *Bacteroides ruminicola* (Pittman *et al.*, 1967) and *Fusiformis necrophorus* (Wahren and Holme, 1973). Both microorganisms were capable of utilizing proline peptides with a molecular weight of up to about 2000. It should be possible to examine this change in the barrier properties and relate them to the different cell envelopes of these microorganisms as compared to *E. coli*.

In summary, it is clear that a barrier exists which prevents the transport of large oligopeptides into *E. coli* and that this barrier does not prevent access of these peptides to the periplasm. The present inability to precisely locate alkaline phosphatase within the cell envelope, and the lack of studies with envelope-modified cells, prevent the identification of the barrier responsible for the size restriction of oligopeptide transport. Evidence from peptide transport studies in Gram-positive microorganisms might support the previous suggestion that the peptidoglycan is the size-restricting barrier of the oligopeptide transport system. However, the possibility that the microenvironment of the oligopeptide transport system or the LPS membrane represents the size restriction barrier is not excluded.

4. Structural Requirements of Peptide Transport

Carrier-type transport utilizes binding of substrates to carriers followed by movement of the complex across the cytoplasmic membrane. The carrier should contain, therefore, an active site which recognizes some structural features of the substrate. Studies on the structural requirements of the peptide transport systems, therefore, may be helpful in identifying transport mechanisms. Knowledge of the structural requirements should also aid in the construction of affinity chromatographic systems to permit the purification of the peptide carriers. This knowledge may also permit the design of affinity labelling compounds, which would also be helpful in the characterization of the active site of the carrier. Moreover, because of the many important roles that certain peptides play in nature, some of which might involve transport into the cells, such studies may provide a means to control the specific biological activities of the peptides.

Peptide transport systems were found to exhibit requirements for structural features in their substrates. These requirements and their significance are now summarized.

a. α-Peptide Bond. The inability of several β, γ-, and ϵ-linked dipeptides to support the growth of an appropriate auxotroph of *E. coli* and their failure to compete with the nutritional effectiveness of α-linked dipeptides led Payne (1972a, 1973) to the conclusion that the α-peptide bond is an essential feature of the dipeptide uptake system. This conclusion is in accord with the requirement for both the N-terminal α-amino group and the C-terminal carboxyl group for dipeptide uptake [see Sections IVA4(b, c)], since only the presence of an α-peptide bond can preserve the fixed spatial arrangement of these two groups in the dipeptide molecule. However, most of the dipeptides that were not utilized were also resistant toward peptidase activities in cell extracts; thus the lack of growth could have been due to the failure of the peptides to yield the required amino acids inside the cells. Moreover, certain unusual dipeptides—carnosine (β-Ala-His) and γ-Glu-ε-Lys—that could be hydrolyzed by intracellular peptidases were able to be utilized and met the amino acid requirement of an appropriate auxotroph of *E. coli* (Payne, 1972a, 1973). The above-mentioned conclusion is therefore based mainly on the negative observation that showed inability of the β-, γ-, and ε-linked dipeptides to compete with the uptake of normal dipeptides (Payne, 1972a). It should also be mentioned that as in *E. coli*, carnosine (β-Ala-His) could be utilized nutritionally by several strains of *Corynebacteriae* (Mueller, 1938) and by *Lactobacillus delbrueckii* (Peters *et al.*, 1953). On the other hand, *Pedicoccus cerevisiae* is unable to utilize carnosine, whereas α-linked histidyl dipeptides support normal growth (Florsheim *et al.*, 1962). Lastly, the above observations with *E. coli* W do not prove a need for a peptide bond for uptake. They might instead be interpreted to indicate a requirement for a certain distance between the N and C termini of the transported molecule. It would therefore be of interest to study uptake of molecules that lack the peptide bond but maintain a distance between an amine and a carboxyl group which is similar to that found in dipeptides.

Studies were also carried out with tripeptides containing β-alanine. In *E. coli* it was shown that Gly-Gly-β-Ala could use the oligopeptide transport system, whereas β-Ala-Gly-Gly could not (Payne, 1973). These results are in agreement with other observations [see Sections IVA4(b, c)], which support the notion that the C-terminal carboxyl group is not essential for oligopeptide transport in *E. coli* (Gly-Gly-β-Ala might be considered as Gly-Gly-Asp devoid of a C-terminal carboxyl), whereas the N-terminal amine is needed. The tripeptide that contains the β amino group on its N-terminal side is not transported, probably because the first peptide bond participates in the binding of peptides to the carrier molecules. The effect of the presence of a β-peptide bond in the second position from the N-terminal end has not yet been studied. Therefore additional experiments will be required to allow a definitive conclusion with respect to the utilizability of such oligopeptides.

b. N-Terminal α-Amino Group. Early observations showed that α-N-terminal-substituted dipeptides are not utilized by *E. coli* (Simmonds *et al.*, 1947; Simmonds and Fruton, 1948). Later, it was demonstrated by Gilvarg and Katchalski (1965) that α-acetylated di- and tetralysine failed to be utilized because they could not penetrate into *E. coli* cells, whereas acetylation in the ϵ position did not affect the uptake. It was concluded, therefore, that a free α-amino group is essential for di- and oligopeptide uptake in *E. coli* W. However, recent experiments by Payne (1971a, b, 1974) indicated that certain substitutions at the N-terminal α-amino group could be tolerated. Thus di- and tripeptides of glycine with methylated α-amino groups (*N*-sarcosyl glycyl peptides) could still penetrate into *E. coli* cells (Payne, 1971a). Moreover, these peptides continued to utilize the normal peptide transport systems of *E. coli.* They were able to compete for entry into the cells with natural di- and tripeptides, respectively, and α-N-methylated triglycine (Sar-Gly-Gly) failed to be utilized by an Opt^- mutant as glycine source. Further investigations were carried out with glycyl peptides that were N-substituted with larger alkyl groups, e.g., ethyl, propyl, isopropyl, butyl, isobutyl (Payne, 1974). All of these peptide derivatives were nutritionally active and capable of inhibiting the uptake of unsubstituted peptides. Additional studies (Payne, 1971b) with peptides containing N-terminal proline demonstrated that the di- and the oligopeptide transport systems of *E. coli* can handle peptides with α-imino groups as well. In contrast to these findings, N-acyl-substituted di- and triglycine derivatives (acetyl, propionyl, succinyl, glutaryl, maleyl, citraconyl) failed to meet the amino acid requirement of a glycine auxotroph of *E. coli* in spite of the existence of peptidases capable of cleaving these peptides in the cell extract (Payne, 1971a). These analogs could not compete with natural peptides for entry. The failure of the N-acyl derivatives to reach the peptide transport systems could have been explained by steric hindrance of the substituent groups. However, this possibility is excluded, since several of the alkyl groups that were larger or equal in size to at least some of these acyl groups were able to substitute for a hydrogen on the α-amino groups without affecting the uptake of the peptides. It should also be mentioned that dimethylglycyl peptides were both nutritionally and competitively inactive (Payne, 1974). It appears, therefore, that the di- and the oligopeptide transport systems of *E. coli* require the existence on their substrate of an N-terminal nitrogen with at least one hydrogen and with a substantially unaltered *pK* in order to be transported.

c. C-Terminal Carboxyl Group. It has been shown that the C-terminal carboxyl is not essential for the uptake of peptides through the oligopeptide transport system of *E. coli* W. Oligopeptides were found to be taken up by the bacterial cells even in the absence of the terminal carboxyl (Payne and Gilvarg, 1968a; Payne, 1973) or when this group was substituted (Payne and Gilvarg, 1968a). However, the dipeptide transport system appears to require the presence of the terminal carboxyl, since a dipeptide (lysyl-cadaverine) that lacks this group was unable to utilize the dipeptide transport system. It should be noted that this carboxyl-deficient dipeptide was transported into the cells via the oligopeptide transport system (Payne and Gilvarg, 1968a). In addition, various amides and esters of dipeptides were found to lose most of their biological effectiveness (Simmonds and Griffith, 1962; Kessel and Lubin, 1963).

The lack of a requirement for the C-terminal carboxyl for oligopeptide uptake was recently supported by the observation that peptides containing impermeant substances at their C-termini were readily transported into *E. coli* (Fickel and Gilvarg, 1973) and *S. typhimurium* (B. N. Ames *et al.*, 1973a) (see Section IVA5). The transport was carried out through the oligopeptide transport system, since mutants deficient in this system (Opt$^-$) failed to take in these peptide derivatives.

A requirement for the C-terminus for dipeptide uptake and lack of such requirement for oligopeptide uptake was also demonstrated in *Lactobacilli* (Woolley *et al.*, 1955; Merrifield and Woolley, 1956; Shankman *et al.*, 1962).

d. Stereospecificity. Dipeptide transport into *E. coli* was shown to be stereospecific. Dipeptides containing D-amino acids are ineffective in competitively inhibiting the uptake of natural L–L dipeptides (Levine and Simmonds, 1962; Kessel and Lubin, 1963). Examples of similar behavior in *Lactobacilli* have been reported (Shankman *et al.*, 1960; Leach and Snell, 1960; Kihara *et al.*, 1961; Yoder *et al.*, 1965a). In contrast, D-Ala-His served as a good histidine source for *L. delbrueckii* (Peters *et al.*, 1953).

Several experiments were also carried out with tripeptides containing D-amino acids. It was recently demonstrated (Payne, 1972a) that tri-D-alanine is unable to compete with the uptake of the toxic peptides tri-L-valine and tri-L-ornithine into *E. coli.* However, tri-L-alanine at similar concentrations relieved the toxic effect of these peptides. Tri-D-alanine also failed to compete against triglycine utilization and the uptake of radioactive lysyl oligopeptides, whereas the tripeptide of the L-isomer was a good competitor. It seems, therefore, that the oligopeptide transport system in *E. coli* possesses stereospecific properties, preferring the L-isomer. It is not clear, however, how strict these stereospecific requirements are, since no experiments have been carried out with mixed D–L oligopeptides in *E. coli.* Experiments of this type were carried out in other bacteria. In investigations on the eight stereoisomers of L- and D-Val tripeptides, LLD was found to be inhibitory to *Pedicoccus cerevisiae.* LLL-Val_3 was the only isomer that was able to reverse the inhibition (Shankman *et al.*, 1961). Shankman *et al.* (1962) also showed that radioactive LLD-Val_3 was actively transported into this strain. The peptides LLL-Val_3 and Val-Leu-D-Val inhibited the uptake. None of the other stereoisomers of Val_3 affected the uptake. LLD-Val_3, in contrast to the other mixed stereoisomers of Val_3, was also shown to serve as valine source by several lactic acid bacteria (Shankman *et al.*, 1960). These observations correlate with the lack of specific transport site on the C-terminal end of the oligopeptides (Section IVA4(c)]. It should be mentioned, however, that *Lactobacillus casei* and *Streptococcus faecalis* do utilize DLL-Val_3, although not as well as LLD-Val_3 (Shankman *et al.*, 1960).

Thus it appears that peptide transport systems in bacteria have a stereospecific requirement for L-isomers and the addition of a D-isomer strongly reduces the affinity of the peptides to their permease. The degree of flexibility appears to be greater for oligopeptide transport, where it has been shown that some tripeptides containing D-isomers can be transported. It was postulated by Payne and Gilvarg (1971) that the presence of a D-isomer at the C-terminal end of a tripeptide should have little influence on the transport of the peptide. It is not yet clear, however, whether or not a tripeptide that contains D-isomer on its N-terminal is accepted by the oligopeptide transport system.

5. Transport of Impermeant Substances by Way of Oligopeptide Transport System

The lack of a requirement for the free C-terminal carboxyl for oligopeptide transport in *E. coli* [Section IVA4(c)] might be explained by the absence of recognition sites on the permease for this end of the peptide molecule. This characteristic of the oligopeptide transport system, coupled with the established broad specificity of the system [see Section IVA2(a)], provided Fickel and Gilvarg (1973) and B. N. Ames *et al.*, (1973a) with a tool to bring normally impermeant substances into the bacterial cells. This was achieved by attaching the impermeant compounds to peptides through the nonessential terminal carboxyl group. The peptide derivative containing the new compound as substituent of the carboxyl is then transported into the cells to yield the constituent substances intracellularly. Thus the impermeant threonine precursor, homoserine-phosphate, was brought into *E. coli* W TL3 in a peptide form as Lys-Lys-Homoserine-phosphate (Fickel and Gilvarg, 1973). Histidinol-phosphate ester, a histidine biosynthetic intermediate that failed to enter into *S. typhimurium* cells, was also transported into the bacteria as Gly-Gly-Histidinol-phosphate (B. N. Ames *et al.*, 1973a). In both cases, the peptide carriage was found to utilize the oligopeptide transport system, as Opt$^-$ mutants failed to take in these peptide derivatives.

The possibility of bringing impermeant or poorly permeable substances into the bacterial cells through the oligopeptide transport system opens a number of unexplored opportunities. It might be of great help in the construction of new antibiotics by coupling known toxic analogs to peptides. This was already proved to a certain extent by the observation that the toxicity of norleucine, norvaline, and ethionine toward *S. typhimurium* was increased when introduced to the bacteria in their peptide form (B. N. Ames *et al.*, 1973a). This principle might also be useful to increase permeability of known antibiotics, or to allow substances with a potential biological activity that had been established *in vitro* to reach their internal site of action in the intact cell.

6. *Genetic Studies on Peptide Transport*

Genetic analysis helps to identify the steps of any biological process. Comparing different mutations of the same system provides evidence concerning the complexity of the system, e.g., the number of factors responsible for its activity.

Several spontaneous mutants deficient in dipeptide and oligopeptide transport have been isolated from *E. coli* and *S. typhimurium*. In general, the mutants were selected through use of cytotoxic peptides, on the assumption that one of the ways to gain resistance would be loss of transport. However, Kessel and Lubin (1963) did use the penicillin selection method to isolate mutants that are unable to utilize Gly-Gly.

a. Dipeptide Transport Mutants. Dipeptides are transported into *E. coli* through at least two systems, e.g., their specific dipeptide transport system and the oligopeptide transport system (Section IVA1). A Dpt$^-$ mutant should be altered, therefore, in both transport systems. The probability of getting spontaneous mutations in two independent systems is very low (10^{-12}). This is in keeping with the difficulties experienced by De

Felice *et al.*, (1973) in their efforts to isolate a spontaneous Dpt⁻ mutant of *E. coli* K-12, using for selection, resistance to the dipeptide Gly-Val and sensitivity to the monomer valine. It is surprising, therefore, that Kessel and Lubin (1963) and Vonder Haar and Umbarger (1972) were successful in isolating spontaneous Dpt⁻ mutants from *E. coli* W and *E. coli* K-12, respectively. One possibility was that their strains had a single gene mutation in a region responsible for a common factor in the di- and the oligopeptide transport. Such Dpt⁻ mutants should be altered in their oligopeptide transport as well, and would not be able to take in tri- and tetrapeptides. However, this expectation contrasts with our own unsuccessful efforts to directly select for spontaneous mutants of *E. coli* K-12 with modifications in both peptide transport systems. Such a selection pressure was created by exposing *E. coli* to a combination of a toxic tripeptide (triornithine) and a toxic dipeptide (Lys-*p*-F-Phe). Since our attempts failed, it seems either there is no common factor for di- and oligopeptide transport systems in *E. coli*, or that mutations in this factor are lethal.

A possible explanation for the spontaneous Dpt⁻ mutants mentioned above may be that the dipeptides used for selection of the mutants [Gly-Gly (Kessel and Lubin, 1963); Gly-Leu (Vonder Haar and Umbarger, 1972)] have very low or no affinity to the oligopeptide transport system. A Dpt⁻ mutant would, in this case, result from a single gene mutation at the region that is responsible for the uptake of the dipeptide, since no alternative uptake is functioning. According to this postulation, the Dpt⁻ mutants obtained should be altered in their dipeptide transport system only, whereas their oligopeptide transport system should function normally. Unfortunately, the ability of the above-mentioned Dpt⁻ mutants to take in oligopeptides was not tested.

The way De Felice *et al.* (1973) overcame their difficulties in isolating a Dpt⁻ mutant was by blocking the alternative way of taking in dipeptides through the oligopeptide transport system. An Opt⁻ mutant was used as the parental strain to further select for Dpt⁻ mutants. The obtained mutant was therefore an Opt⁻, Dpt⁻ strain. These authors also mapped the *dpt* region and showed its independence from the *opt* marker. The *dpt* locus was located between *pro*C (10 min) and *opt* (27 min).

b. Oligopeptide Transport Mutants. Since oligopeptides are in general not transported into *E. coli* through several pathways simultaneously, the isolation of spontaneous Opt⁻ mutants is easier than the isolation of Dpt⁻ mutants. As a matter of fact, the frequency of Opt⁻ mutants in unselected populations of *E. coli* is often 10^{-4}, indicating that this locus has a mutation rate approximately 100-fold higher than normal (Gilvarg and Levin, 1972; Barak and Gilvarg, 1974).

Spontaneous Opt$^-$ mutants have been isolated from several strains of *E. coli* W (Payne and Gilvarg, 1968a; Payne, 1968; Gilvarg and Levin, 1972; Fickel and Gilvarg, 1973), *E. coli* K-12 (Sussman and Gilvarg, unpublished results; De Felice *et al.*, 1973; Barak and Gilvarg, 1974), *E. coli* B (Barak, 1972), and *S. typhimurium* (B. N. Ames *et al.*, 1973a). The selection for Opt$^-$ mutants was always carried out by resistance to cytotoxic tripeptides. Triornithine was used for the isolation of mutants from the various strains of *E. coli*, trivaline in one case of *E. coli* K-12, and trilysine, Norleucyl-Gly-Gly, and high concentrations fo Gly-Gly-Histidinol-phosphate were used in *S. typhimurium*. The most efficient compound used for the isolation of Opt$^-$ mutants in *E. coli* is triornithine. This is a synthetic basic tripeptide which, when accumulated by *E. coli*, specifically blocks protein biosynthesis (Barak *et al.*, 1970; 1973a, b; Gilvarg and Levin, 1972). Triornithine resistance (TOR) could arise for any of the following reasons: (a) a loss of the ability to transport the tripeptide, (b) increased intracellular peptidase activity, (c) derivatization of the toxic peptide, (d) change in sensitivity of the target site. It is therefore necessary to determine the type of change responsible for the TOR phenotype in any given organism. However, in all the cases examined to date, TOR mutants were found to be defective in their oligopeptide transport system (Opt$^-$). This is demonstrated by cross-resistance with different cytotoxic tripeptides (Sussman, unpublished observation; B. N. Ames *et al.*, 1973a; De Felice *et al.*, 1973; Barak and Gilvarg, 1974), inability to utilize oligopeptides (Payne and Gilvarg, 1968; Payne, 1968; Gilvarg and Levin, 1972; Fickel and Gilvarg, 1973; B. N. Ames *et al.*, 1973a; De Felice *et al.*, 1973; Barak and Gilvarg, 1974), or failure to accumulate radioactive tripeptides (Payne, 1972a; De Felice *et al.*, 1973; Barak and Gilvarg, unpublished observation). None of the other theoretical possibilities to achieve triornithine resistance in *E. coli* has been detected. The failure to detect the other postulated types of TOR mutants might be due to the very high frequency of mutation at the *opt* locus (see above). Another possibility that should be considered is that all the other TOR mutations are lethal and cannot exist under normal growth conditions.

Since the Opt$^-$ mutants isolated are all of spontaneous origin, it is reasonable to assume that they are single gene mutations. However, the basis of the Opt$^-$ phenotype might be due either to a change in the specific oligopeptide transport system or to a general change in the cell envelope. The latter type of mutation, which is selected for by phage resistance, is often shown to be pleiotropic and to affect the permeability of certain components like iron, antibiotic drugs, and colicins to *E. coli* cells (Wang and Newton, 1969; Tamaki *et al.*, 1971; Tamaki and Matsuhashi, 1973; Takagaki *et al.*, 1973). This possibility is ruled out by the following observations: (a) A number of Opt$^-$ mutants tested show the same sensitivity as their parental strains towards the phages ϕ 80 vir, Pl, T_4, and T_1 (Barak and Gilvarg, 1974), (b) no morphological changes in the cells or the colonies of the mutants can be detected (Barak and Gilvarg, 1974); (c) the rate of growth of the mutants is similar to their parental strains (Payne and Gilvarg, 1968a; Payne, 1968; Gilvarg and Levin, 1972; Fickel and Gilvarg, 1973; Barak and Gilvarg, 1974); (d) neither is the dipeptide transport system affected by the Opt⁻ mutation (Payne and Gilvarg, 1968a; Payne, 1968; Fickel and Gilvarg, 1973; B. N. Ames *et al.*, 1973a; De Felice *et al.*, 1973; Barak and Gilvarg, 1974), nor is there a change in the uptake of several radioactive amino acids and sugars (Barak and Gilvarg, unpublished observation). It appears, therefore, that Opt^- strains are single gene mutants with lesions in the gene (or genes) responsible for the oligopeptide transport system. However, the Opt^- might represent a family of strains having defects in different genes that are essential for the normal activity of the oligopeptide transport system. Genetic analysis of different Opt^- strains can be used, therefore, to study the complexity of the oligopeptide transport system. Mapping of the *opt* locus was carried out by De Felice *et al.* (1973) and Barak and Gilvarg (1974) in several Opt^- mutants of various Hfr and F^- strains of *E. coli* K-12. The gene (or genes) was mapped near the *trp* operon on its *cys*B side. All 12 Opt^- mutants mapped were located approximately at the same region, suggesting involvement of only a single gene or that this region might represent an operon that contains several genes that are essential for the oligopeptide transport. Variations in Pl-mediated *trp–opt* cotransduction frequencies were obtained. However, the number of colonies scored is not high enough to define different locations of the *Opt⁻* mutations within this region. A remarkable difference in Pl-mediated linkage to *trp* is observed when comparing our cotransduction frequencies, which vary between 80 and 95%, with those of De Felice *et al.*, who found 54%. This difference might result from mutations in separate genes, or, as is more likely, from variations between the different strains and the different techniques used. (De Felice *et al.* transferred *opt*⁺ into an *opt*⁻ recipient and checked for triornithine sensitivity, whereas in our cases the donor was always *opt*⁻, the recipient *opt*⁺, and we checked for triornithine-resistant colonies). Complementation experiments using episomes or ϕ80-*opt* from the different Opt^- strains should provide the answer concerning the complexity of the oligopeptide transport locus.

The specific location of the *opt* gene(s) near *trp* provides a finding of great potential utility, since this locus is near the attachment site of phage ϕ80. One can construct an HFT phage hybrid ϕ80-*opt* which might yield, after induction, a culture synthesizing large amounts of the gene product (Muller-Hill *et al.*, 1968) and might serve therefore in the isolation of the oligopeptide permease.

It is of interest to mention that several strains of *E. coli* have been found to be resistant to triornithine without prior selection (De Felice *et al.*, 1973; Barak and Gilvarg, 1974). However, the mechanism of resistance has not been studied in any of these cases. The existence of such strains in a laboratory can be explained by the lack of selection pressure to maintain the oligopeptide transport under laboratory conditions, and the high rate of Opt$^-$ mutation.

Though the rate of Opt$^-$ mutation in *E. coli* is high, the rate of spontaneous reversion of *opt*$^-$ to *opt*$^+$ is low (less than 2×10^{-8} in three cases that were examined in our laboratory). This can best be explained by the observation that the Opt$^-$ lesion was a deletion mutation in each of these three strains. These findings were confirmed with five more independent TOR strains that were tested and found not to be point mutations, since no nitrosoguanidine-induced revertant could be detected. These observations of the difficulty of reverting *opt*$^-$ are further supported by the stability of the TOR mutants in the laboratory. We have several Opt$^-$ strains which have undergone more than 50 passages without any selection pressure and still maintain the character of being fully oligopeptide transport deficient. The high rate of Opt$^-$ mutation, which seems to result mainly by deletions, has not yet been explained.

Another type of mutation in which there is a qualitative change in the oligopeptide transport system has been mentioned by Fickel (1973). This spontaneous mutant was able to utilize Lys-Homoserine-phosphate, in contrast to its parental strain *E. coli* TL3. This is an oligopeptide transport mutant since the Opt$^-$ substrain of this mutant failed to take in Lys-Homoserine-phosphate.

7. *Peptide Transport—A Carrier-Type Transport System*

The structural requirements for di- and oligopeptide transport (Section IVA4), the competition for entry between respective peptides (Sections IVA1, 2), and the possibility of obtaining specific deficiencies of di- and oligopeptide transport in *E. coli* by single gene mutations (Section IVA6) provide indirect evidence that define the peptide transport mechanisms as carrier-mediated transport systems.

There are two general transport mechanisms that would involve carriers: facilitated diffusion and active transport. The criteria for active transport include dependence on metabolic energy and the ability to transport intact substrates against a chemical gradient. Because of the presence of extensive peptidase activity in bacteria, it is difficult to demonstrate intracellular accumulation of intact peptides. However, by the use either of peptidase-deficient cells or poorly hydrolyzed peptides as substrates, it was shown that di- and oligopeptide transport is independent of peptidase activity (Sections IIIA2, 3). Moreover, a direct accumulation of radioactive dipeptides was demonstrated in peptidase-deficient cells that were obtained either by mutation (Kessel and Lubin, 1963) or by special physiological conditions (Meisler and Simmonds, 1963). Accumulation of radioactive tripeptide derivatives resistant to peptidase was also recently demonstrated (Payne, 1972a). It can therefore be concluded that peptides are transported by active transport systems. This conclusion is further supported by the energy requirement for peptide uptake. The dependence of peptide transport on energy was demonstrated with dipeptides in several bacterial species (Leach and Snell, 1959, 1960; Meisler and Simmonds, 1963; Kessel and Lubin, 1963; Yoder *et al.*, 1965b; Mayshak *et al.*, 1965; Simmonds, 1966). It was also shown with tripeptides in bacteria (Shankman *et al.*, 1962; Young *et al.*, 1964; Pittman *et al.*, 1967; Smith *et al.*, 1970; Payne, 1972a; De Felice *et al.*, 1973).

Active transport systems in *E. coli* fall into at least two broad categories, those whose activity depends upon periplasmic binding protein, which are shock releasable, and others that are tightly associated with the plasma membrane and are resistant to osmotic shock (Heppel *et al.*, 1972). Evidence has been presented (Berger, 1973) that the binding-protein-mediated amino acid transport systems are driven directly by phosphate-bond energy, formed either by oxidative phosphorylation or glycolysis, whereas the tightly bound systems are coupled to an energized membrane state, which can be generated either by electron transport or ATP hydrolysis.

Dipeptide transport into *E. coli* is significantly reduced by osmotic shock. Moreover, membrane vesicles of *E. coli* free of binding proteins have lost most of their ^{14}C-Gly-Gly uptake capability (Cowell, personal communication). It seems, therefore, that dipeptide uptake in *E. coli* requires the presence of a binding protein. However, such a protein has not yet been detected. It should be mentioned that these observations are in contrast to those of Kaback (personal communication, 1972), who found ^{14}C-Gly-Gly uptake in membrane vesicles.

Preliminary results of ours suggest a dependence of oligopeptide transport on a shock-releasable factor. A sharp decrease in the ^{14}C-Gly-Gly-Gly transport capability after osmotic shock was observed. Moreover, we could not detect any tripeptide uptake in membrane vesicles of *E. coli* W M-123. These are preliminary results only and further investigation is required to characterize the molecular basis of di- and oligopeptide transport in *E. coli.*

B. Yeast

Peptide transport has been recently demonstrated in the yeast *Saccharomyces cerevisiae* G1333 (Becker *et al.*, 1973; Naider *et al.*, 1974). These studies utilized the indirect method in which growth response to peptides serves as the indicator for their uptake (Section IIB). The existence of extracellular peptidases was eliminated, indicating that *S. cerevisiae* is a suitable test organism for measuring peptide uptake in terms of growth response (Section IIIB).

In contrast to the findings with *E. coli*, none of the lysyl oligopeptides tested (Lys-Lys, Lys-Gly, Gly-Lys, trilysine, tetralysine, hexalysine, heptalysine, octalysine, and polylysine) were able to meet the lysine requirement of various lysine auxotrophs of the yeast. These lysine peptides, however, do yield free lysine when incubated in yeast-cell-free extract (Becker *et al.*, 1973). The yeast transport system also shows preference toward peptides containing COOH-terminal methionine (Naider *et al.*, 1974). It seems therefore that *S. cerevisiae* does not possess a general oligopeptide transport system of nonspecific nature, as does *E. coli* (Section IVA2). However, neither competition experiments nor isolations of peptide transport deficient mutants have been carried out, and only a limited number of peptides have been tested. There is also no evidence available concerning a possible distinction between di- and oligopeptide transport in yeast (for comparison to *E. coli* see Section IVA1). It should be mentioned, however, that these differences in behavior toward the different peptides might be due to special secondary changes in the yeast's organization that enable or prevent certain peptides from reaching their transport system or the peptidases inside the cell.

Observations with *S. cerevisiae* G1333 also suggest the existence of a different peptide transport size barrier than that found in *E. coli* (Section IVA3). The yeast methionine auxotroph is capable of utilizing pentamethionine as the sole source of the required amino acid, whereas the *E. coli* auxotroph cannot (Naider *et al.*, 1974). This change in the size restriction property is probably due to the dramatic differences in the cell envelope between *E. coli* and yeast.

The structural requirements of the peptide transport system of the yeast are also different from those of *E. coli*. Experiments with *S. cerevisiae* G1333 showed for some peptides different behavior concerning the N-terminus requirements of the peptide transport systems [Section IVA4(b)]. A number of N-acetylated (Ac-) and *t*-butyloxycarbonyl (Boc-) di- and tripeptides containing methionine and glycine (Ac-Met-Met, Ac-Gly-Met, Boc-Gly-Met, Ac-Met-Met-Met, Ac-Met-Gly-Met, Boc-Gly-Gly-Met, Boc-Gly-Met-Met) could enter the yeast cells and meet the methionine requirement of this auxotroph (Naider *et al.*, 1974). However, Boc-Met-Met and Boc-Met-Met-Met served only as poor substrates for growth, and a number of amino-blocked peptides tested (Ac-Met-Gly, Ac-Met-Gly-Gly, Boc-Met-Met-Gly) failed to support the growth of this strain. In the absence of extracellular peptidase activities in *S. cerevisiae* G1333 (Becker *et al.*, 1973), it is clear that di- and tripeptides with acylated α-amino groups are transported into the yeast cells.

The failure of several peptide derivatives to support growth might result either from their inability to penetrate into the cells or from their resistance to intracellular cleavage. Since no peptidase activities were detected against any of the acetylated peptides (the utilizable and the nonutilizable) in the cell extract, the growth response could not be correlated to the presence or absence of the appropriate hydrolytic activity (Naider *et al.*, 1974). Therefore the possibility that these nonutilizable acetylated peptides are also taken into the yeast without being split cannot be excluded. It should be noted, however, that all the nutritionally ineffective analogs tested were derivatives of peptides containing glycine at their carboxy-terminal end. These peptides in general were utilized poorly as methionine sources by the yeast, and might be transported by a special, less effective, transport system. This transport system, in contrast to the other, might require the presence of a free or positively charged amine in its substrates to allow transportation. Nevertheless, it is clear that *S. cerevisiae* G1333 contains at least one peptide transport system that can handle peptides without free α-amino groups.

Further studies with the yeast strain *S. cerevisiae* G1333 revealed that tri-, tetra-, and pentamethionine methyl esters could support the growth of the methionine auxotroph (Naider *et al.*, 1974). However, the growth response was significantly poorer than that obtained with unsubstituted oligomethionine peptides. This might be due to a reduction in the affinity of the peptide derivatives for their transport system. Such a reduction in affinity as a result of a loss or modification of the C-terminal carboxyl was also observed in *E. coli* (Payne and Gilvarg, 1968a). It is possible, therefore, that the methionine oligopeptide transport system in the yeast resembles the one of *E. coli* in its lack of absolute requirement for C-terminal carboxyl [Section IVA4(c)].

In summary, it seems that *S. cerevisiae* contains a peptide transport system. This peptide transport shows side chain specificity and might even be exclusive to peptides containing methionine residues. The structural requirements for this transport system are different from those of *E. coli* as well as its size restriction barrier. The existence of a distinct dipeptide transport system in the yeast has not been tested. In spite of the limited variety of peptides tested in *S. cerevisiae*, it can be concluded that its peptide transport mechanism is different from that of *E. coli*.

C. Mammals

1. *Distinction between Di- and Oligopeptide Transport Systems*

The distinction between di- and oligopeptide transport systems was demonstrated in *E. coli* by the isolation of mutants deficient in the different transport systems and by the inability of tripeptides to compete with the uptake of dipeptides (Section IVA1).

However, the isolation of pinpointed classes of mutants in mammals is almost impossible. Moreover, no inborn disorder is known in which peptide transport is impaired. For this reason, studies directed at determining the kinds of peptide transport systems present in mammals have relied on the competition technique. Several competition experiments between di- and tripeptides were carried out in mammalian intestinal systems. Uptake of the dipeptide carnosine (β-Ala-His) by everted rings of hamster jejunum was shown to be inhibited by triglycine (Addison *et al.*, 1973, 1974c). However, it was not possible to distinguish inhibition caused by the intact tripeptide from that caused by diglycine liberated from it by extracellular hydrolysis. In addition, the uptake of the hydrolysis-resistant tripeptides β-Ala-Gly-Gly (Addison *et al.*, 1974a) and Gly-Sar-Sar (Addison *et al.*, 1974b) was inhibited by tripeptides as well as by dipeptides, but not by amino acids. This situation is similar to the one observed in *E. coli*, where dipeptides can use the oligopeptide transport system and are therefore able to compete with the entry of oligopeptides into the cells. It is not yet clear, however, whether a special dipeptide transport system or systems exist in the mammalian intestine. Investigation of this problem would be facilitated by the use of tripeptides that are stable in the presence of peptidase, like Gly-Sar-Sar and β-Ala-Gly-Gly, as competitors. These tripeptides, unlike triglycine, are transported intact into the intestinal cells and their effect on dipeptide uptake could then be interpreted in terms of the action of an oligopeptide on dipeptide uptake.

2. *Side Chain Specificity of Peptide Transport Systems*

a. The Oligopeptide Transport System. The existence of extracellular peptidase activity in the mammalian intestine makes it difficult to demonstrate a real competition for entry between peptides in this system. For this reason, little work was done in this area until very recently when several peptides (β-Ala-Gly-Gly and Gly-Sar-Sar) were found to be taken up by hamster jejunum with very little hydrolysis. Gly-Sar-Sar inhibited uptake of β-Ala-Gly-Gly (Addison *et al.*, 1974b), indicating that it shared the same transport mechanism. This is in fact the only experiment that demonstrates competition for transport between two intact tripeptides in the mammalian intestine. Indeed, the uptake of the two poorly hydrolyzed tripeptides was also inhibited by triglycine (Addison *et al.*, 1974a, b). However, since diglycine was also inhibitory to these systems, inhibition could be due to this degradation product of the tripeptide. In any case, competition was discovered only between similar tripeptides containing neutral amino acid side chains and nothing can be concluded concerning the generality of this transport system and its resemblance to the oligopeptide transport system found in *E. coli* (Section IVA2). One should bear in mind that no evidence is as yet available to indicate that a distinct oligopeptide transport system exists in the intestinal cells. Oligo- and dipeptides might be transported by a common uptake system, and the established broadly specific nature of the dipeptide uptake system in mammals (see below) might therefore reflect the situation for the oligopeptides as well.

b. The Dipeptide Transport System. Competition for entry between dipeptides occurs in the mammalian intestine. Rubino *et al.* (1971) measured competition using a 60-sec influx or ^{14}C-Gly-Pro into rabbit ileal mucosa *in vitro*. Met-Pro, Phe-Pro, Leu-Leu, Phe-Gly, Gly-Gly, and Gly-Sar inhibited the uptake. The constituent amino acids of these peptides did not affect the uptake of Gly-Pro, ruling out possible competition by their degradation products. In addition, the uptake of carnosine by hamster jejunum was inhibited by Gly-Gly, Gly-Sar, Gly-Pro, Met-Met, and Pro-Hyp, but not by the equivalent free amino acids (Addison *et al.*, 1973, 1974c). It appears that a wide variety of neutral dipeptides are transported into the small intestine by the same system. It is, therefore, as in *E. coli*, a relatively non-stringent transport system. However, special uptake systems might exist for charged dipeptides, since dilysine and diglutamic acid did not inhibit carnosine uptake in hamster (Addison *et al.*, 1974c). These two peptides and other charged dipeptides were also poorly transported into rat intestine when compared to neutral and mixed dipeptides (Burston *et al.*, 1972). The failure of the charged peptides to compete might reflect their use of an independent transport system. On the other hand, the charged dipeptides might utilize the same transport system as the neutral dipeptides but their affinity for the system might be very low. Low affinity would also explain the lack of competition of charged dipeptides against the neutral dipeptide carnosine observed in hamster.

In summary, it seems that dipeptides utilize a transport system with broad specificity that can take in dipeptides containing neutral aliphatic amino acids, aromatic amino acids, and sulfur amino acids. However, in contrast to the findings with *E. coli*, dipeptides that contain only basic amino acids seem to utilize a specific transport system. It is also possible that additional specific dipeptide transport systems exist in mammals.

3. Structural Requirements for Peptide Transport

a. α-Peptide Bond. One of the structural requirements for di- and oligopeptide transport in *E. coli* is an α-peptide bond between the first two

N-terminal amino acids [Section IVA4(a)]. Carnosine (β-Ala-His) was found to be transported into *E. coli*. This β-alanine-containing dipeptide was also readily transported into mammalian intestinal cells and was even found to share its transport system with many normal α-linked dipeptides (Asatoor *et al.*, 1970a; Navab and Asatoor, 1970; Addison *et al.*, 1973; Matthews *et al.*, 1974). This latter observation is in contrast to Payne's finding concerning the importance of the α-peptide bond for dipeptide uptake and his suggestion that carnosine is transported into *E. coli* using a special transport system (Payne, 1973). These contrasting requirements serve as another example illustrating differences between dipeptide transport systems in *E. coli* and mammals.

It is of interest that β-Ala-Gly-Gly was also found to be taken up by hamster jejunum *in vitro* (Addison *et al.* 1974a). Moreover, this uptake was carried out by the normal peptide transport system, for it was competed against by Gly-Gly-Gly, Gly-Sar-Sar, and Gly-Gly. It is therefore another indication of possible difference in the oligopeptide transport system between *E. coli* and the mammalian intestine.

b. N-Terminal α-Amino Group. Evidence concerning the uptake of several α-N-substituted dipeptides by mammalian intestine have been recorded in the literature. For example, α-N-acetyl-Gly-Gly failed to compete with Gly-Pro (Rubino *et al.*, 1971) or with carnosine (Addison *et al.*, 1974c) for entry, indicating inability of this derivative to utilize the dipeptide transport system (or systems). Moreover, even N-methyl-Gly-Gly (Sar-Gly) was very poorly transported by the mucosa (Burston *et al.*, 1972). It was also postulated that the failure of Pro-Gly to compete with the uptake of Gly-Pro might result from the lack of affinity of the former to the peptide entry mechanism. It was suggested that this lack of affinity could be due to the absence of a free α-amino in the Pro-Gly (Rubino *et al.*, 1971). It appears that any substitution of the α-amino group of a dipeptide, including methylated and imino-N-terminal derivatization, in which the amine retains its positive charge, affects its uptake by intestinal cells. This contrasts with the findings in *E. coli* and in the yeast, as noted above.

c. C-Terminal Carboxyl. In the mammalian intestine, Gly-Gly-amide failed to compete with carnosine uptake, indicating lack of affinity to the peptide transport system (Addison *et al.*, 1974b). This observation correlates with the absolute requirement for free C-terminal carboxyl for dipeptide transport in *E. coli*. However, this is a single experiment, which does not permit a generalized conclusion. In addition, no experiment was carried out concerning the role of the carboxyl in oligopeptide uptake by the intestinal cells.

d. Stereospecificity. As had been observed in *E. coli*. dipeptide entry into mammalian intestine was strongly reduced by the presence of D-amino

Table I

Properties of Peptide Transport System[a]

Property	Organism		Evidence	Remarks
1. Distinction between di- and oligopeptide transport systems	*E. coli*	+	a. Opt^- mutants take in dipeptides but not oligopeptides b. Dpt^- mutants map at a different location than Opt^- c. Oligopeptides will not eliminate dipeptide uptake d. C-terminal carboxyl required only for dipeptide uptake	—
	S. cerevisiae	NT	—	—
	Mammalian intestine	+(?)	Inhibition of dipeptide uptake by oligopeptides	Not clear whether this competition for entry occurs between intact peptides
2. Side chain specificity	*E. coli* A	—	a. Dpt^- mutants are altered in transport of a variety of dipeptides b. Competition for entry between different dipeptides	Special dipeptide transport system(s) exist
	E. coli B	—(P)	a. Dpt^- mutants are altered in transport of a large variety of oligopeptides b. Competition for entry between different oligopeptides c. Uptake of impermeant substances by way of OPTS	a. Special oligopeptide transport system(s) exist b. Affinity to the OPTS varies with different peptides
	S. cerevisiae	+	a. Uptake of peptides containing Met and Gly, but no uptake of lysyl peptides b. Preference toward peptides containing C-terminal Met	—
	Mammalian intestine	—	Competition for entry between different peptides	Special transport system(s) might exist for charged peptides

3. Size restriction	*E. coli*	A	+	Opt⁻ mutants (Dpt⁺) take in peptides containing two amino acid residues only	The restriction is probably determined at the level of the permease
	E. coli	B	+	Existence of a size limit in uptake of peptides which corresponds to *Kd* values of 0.195 with Sephadex G-15	Macromolecular components of carrier not determined
	S. cerevisiae		NT	Size restriction has not been determined	The variety of peptides tested is limited
	Mammalian intestine		NT	Size restriction has not been determined	Experiments with large, hydrolysis-resistant peptides have not been performed
4. Requirement for α-peptide bond at the first position from the N-terminus	*E. coli*	A	+	Failure of β-, γ-, and ϵ-linked dipeptides to compete with α-linked dipeptides	Special dipeptides with abnormal peptide bond might utilize specific transport system(s)
	E. coli	B	+(?)	β-Ala-Gly-Gly is not transported, in contrast to Gly-Gly-β-Ala	a. β-Peptide bond at the C-terminus does not affect uptake of oligopeptides b. The importance of α-linkage in positions other than the N or C-terminus has not been studied c. Additional examples for this requirement have not been reported
	S. cerevisiae		NT	—	—
	Mammalian intestine			a. β-Ala-His and β-Ala-Gly-Gly are transported b. Uptake is inhibited by a variety of α-linked di- and tripeptides	—

[a] The following symbols and abbreviations are used in this table: +, presence of the indicated property; —, absence of the indicated property; NT, not tested; (P), partial requirement; (?), evidence is not clear; (NC), some peptides act differently; A, dipeptide transport; B, oligopeptide transport; OPTS, oligopeptide transport system; DPTS, dipeptide transport system.

(table continued)

Table I *(continued)*

Property	Organism			Evidence	Remarks
5. Requirement for NH_2 terminal with at least one hydrogen with an unaltered *pK*	*E coli*	A	+	N-acylated dipeptides are not transported, in contrast to N-alkylated dipeptides	—
	E. coli	B	+	N-acylated tripeptides are not transported, in contrast to N-alkylated tripeptides	Dialkylated peptides are not transported
	S. cerevisiae		−(NC)	A number of N-acetylated and *t*-Boc-di- and tripeptides are transported	—
	Mammalian intestine		+	Any substitution of the free α-NH_2 of a dipeptide strongly reduces its uptake	Methylation (Sar-Gly) or imino-N-terminal derivatization (Pro-Gly) of a dipeptide affects its uptake
6. Requirement for COOH terminal	*E. coli*	A	+	a. A dipeptide that lacks the terminal carboxyl is unable to utilize the DPTS b. Amides and esters of dipeptides lose most of their biological effectiveness	A dipeptide that lacks the terminal carboxyl is transported through the OPTS
	E. coli	B	−(P)	a. Oligopeptides are transported in the absence of terminal carboxyl b. Oligopeptides are transported when the terminal carboxyl is substituted	Alteration of the terminal carboxyl reduces the affinity to the OPTS
	S. cerevisiae		−(P)	Methyl esters of oligopeptides can be transported	The affinity of the methyl esters to the peptide transport system(s) is reduced
	Mammalian intestine		+(?)	Gly-Gly-amide fails to compete with carnosine uptake	Additional examples have not been reported
7. Stereospecificity requirement for L-configuration	*E. coli*	A	+	Dipeptides containing D-amino acids do not compete with L–L dipeptides for entry nor are they transported	—

	E. coli	B	+	Tri-D-Ala does not compete with the uptake of tripeptides containing L-amino acids	a. Experiments with additional combinations of stereoisomers have not been performed with *E. coli* b. In *Lactobacilli* the OPTS can tolerate D-amino acids at the C-terminus and to a lesser extent at the N-terminus of tripeptides
	S. cerevisiae		NT	—	—
	Mammalian intestine		+	Dipeptide entry is strongly reduced by the presence of D-amino acids in the peptide.	Mixed D–L and L–D isomers are absorbed at intermediate rates as compared to the respective L–L and D–D isomers
8. Energy-dependent uptake, active transport	*E. coli*	A	+	a. Uptake is dependent upon the presence of an energy source	—
	E. coli	B	+	b. Uptake is inhibited by uncoupling reagents c. Radioactive peptides are accumulated against chemical gradient	
	S. cerevisiae		NT	—	—
	Mammalian intestine		+	a. Anoxia reduces uptake of peptides b. Uncoupling reagents inhibit uptake of peptides. c. Intact peptides are accumulated against chemical gradient	—
9. A carrier-type transport	*E. coli*	*A*	+	a. Structural requirements for di- and oligopeptide transport	—
	E. coli	B	+	b. Competition for entry between peptides c. A single gene mutation alters the uptake of peptides	
	S. cerevisiae		+(?)	Structural requirements for peptide transport	—
	Mammalian intestine		+	a. Structural requirements for peptide transport b. Competition for entry between peptides c. Peptide transport is Na^+ dependent	—

acids in the peptides. This was shown *in vitro* (Burston *et al.*, 1972; Cheeseman and Smyth, 1973) and *in vivo* (Asatoor *et al.*, 1973). In most of these cases, the low rate of absorption was found to correlate with the rate of hydrolysis by homogenates of the intestinal mucosa. In spite of the slow entry, it was shown that D-Leu-Gly is taken into the intestinal cells by a mechanism shared with Leu-Ala and is not simple diffusion (Cheeseman and Smyth, 1973). It was also demonstrated with stereoisomers of Ala-Phe, Leu-Leu, and Gly-Trp that mixed D–L and L–D isomers were absorbed at intermediate rates as compared to the respective L–L and D–D dipeptides (Asatoor *et al.*, 1973). These authors suggested that diffusion plays an appreciable part in the uptake of D–D isomers by the jejunum of the rat.

4. Peptide Transport—A Carrier-Type Transport System

Because of the extensive peptidase activity in mammalian intestinal cells, it is difficult to demonstrate intracellular accumulation of intact peptides. However, the ability of certain hydrolysis-resistant peptides to be accumulated inside the intestinal cells (Section IIIC2) does indicate a process of active transport. This notion is further supported by the dependence of peptide transport on energy sources. This was demonstrated with dipeptides (Newey and Smyth, 1962; Cheng *et al.*, 1971; Addison *et al.*, 1972, 1973; Matthews *et al.*, 1974) as well as tripeptides (Addison *et al.*, 1974a, 1974b). In addition, it was shown that dipeptide uptake in this system is Na^+ dependent. Rubino *et al.* (1971) demonstrated this dependence with radioactive Gly-Pro. Accumulation of Gly-Sar and Car was also inhibited by replacement of medium Na^+ by K^+ or Li^+ (Addison *et al.*, 1972, 1973; Matthews *et al.*, 1974).

D. Summary

The properties of the peptide transport system are summarized in Table I.

REFERENCES

Addison, J. M., Burston, D., and Matthews, D. M., 1972, Evidence for active transport of dipeptide glycylsarcosine by hamster jejunum *in vitro*, *Clin. Sci.* **43**:907–911.

Addison, J. M., Burston, D., and Matthews, D. M., 1973, Carnosine transport by hamster jejunum *in vitro* and its inhibition by other di- and tripeptides, *Clin. Sci. and Molec. Med.* **45**:3–4p.

Addison, J. M., Burston, D., and Matthews, D. M., 1974a, Transport of the tripeptide β-alanyl-glycyl-glycine by hamster jejunum *in vitro*, *Clin. Sci. Molec. Med.* **46**:5–6p.

Addison, J. M., Burston, D., Matthews, D. M., Payne, J. W., and Wilkinson, S., 1974b, Evidence for active transport of the tripeptide glycylsarcosylsarcosine by hamster jejunum *in vitro*, *Clin. Sci. Molec. Med.*, **46**:30P.

Addison, J. M., Matthews, D. M., and Burston, D., 1974c, Competition between carnosine and other peptides for transport by hamster jejunum *in vitro*, *Clin. Sci. Molec. Med.* **46**:707–714.

Agar, W. T., Hired, F. J. R., and Sidhu, G. S., 1953, The active absorption of amino acids by the intestine, *J. Physiol.* **121**:255–263.

Ames, B. N., Ames, G. F., Young, J. D., Isuchiya, D., and Lecocq, J., 1973a, Illicin transport, the oligopeptide permease, *Proc. Nat. Acad. Sci. U. S.* **70**:456–458.

Ames, B. N., Lee, F. D., and Durston, W. E., 1973b, An improved bacterial test system for the detection and classification of mutagens and carcinogens, *Proc. Nat. Acad. Sci. U. S.* **70**:782–786.

Ames, G. F., Spudich, E. N., and Nikaido, H., 1974, Protein composition of the outer membrane of *Salmonella typhimurium*: effect of lipopolysaccharide mutations, *J. Bacteriol.* **117**:406–416.

Asatoor, A. M., Bandoch, J. K., Lant, A. F., Milne, M. D., and Navab, F., 1970a, Intestinal absorption of carnosine and its constitutent amino acids in man, *Gut* **11**: 250–254.

Asatoor, A. M., Cheng, B., Edwards, K. D. G., Lant, A. F., Matthews, D. M., Milne, M. D., Naveb, F., and Richards, A. J., 1970b, Intestinal absorption of two dipeptides in Hartnup disease, *Gut* **11**:380–389.

Asatoor, A. M., Crouchman, M. R., Harrison, A. R., Light, F. W., Loughridge, L. W., Milne, M. D., and Richards, A. J., 1971, Intestinal absorption of oligopeptides in cystinuria, *Clin. Sci.* **41**:23–33.

Asatoor, A. M., Harrison, B. D. W., Milne, M. D., and Prosser, D. I., 1972, Intestinal absorption of an arginine-containing peptide in cystinuria, *Gut* **13**:95–98.

Asatoor, A. M., Chadha, A. K., Milne, M. D., and Prosser, D. I., 1973, Intestinal absorption of stereoisomers of dipeptides in the rat, *Brit. J. Nutr.* **28**:417–423.

Barak, Z., 1972, Effect of basic oligopeptides on the biosynthesis of macromolecules, Ph.D. thesis, Weizmann Institute of Science.

Barak, Z., and Gilvarg, C., 1974, Triornithine-resistant strain of *Escherichia coli*: isolation, definition and genetic studies, *J. Biol. Chem.* **249**:143–148.

Barak, Z., Sarid, S., and Katchalski, E., 1970, Effect of tri-L-ornithine on nucleic acid and protein biosynthesis in intact and bacteriophage infected *E. coli* B cells, *Israel J. Chem.* **8**:121.

Barak, Z., Sarid, S., and Katchalski, E., 1973a, Inhibition of protein biosynthesis in *Escherichia coli* B tri-L-ornithine, *Eur. J. Biochem.* **34**:317–324.

Barak, Z., Sarid, S., and Katchalski, E., 1973b, Inhibition of T_4 maturation by tri-L-ornithine, *Eur. J. Biochem.* **34**:325–328.

Becker, J. M., Naider, F., and Katchalski, E., 1973, Peptide utilization in yeast: studies on methionine and lysine auxotrophs of *Saccharomyces cerevisiae*, *Biochim. Biophys. Acta* **291**:388–397.

Berger, E. A., 1973, Different mechanisms of energy coupling for active transport of proline and glutamine in *Escherichia coli*, *Proc. Nat. Acad. Sci. U. S.* **70**:1514–1518.

Best, C. H., and Taylor, N. B., 1950, "The Physiological Basis of Medical Practice," 5th ed., p. 588, Bailliere, Tindall and Cox, London.

Brock, T. D., and Wooley, S. O., 1964, Glycylglycine uptake in *Streptococci* and a possible role of peptides in amino acid transport, *Arch. Biochem. Biophys.* **105**:51–57.

Burston, D., Addison, J. M., and Matthews, D. M., 1972, Uptake of dipeptides containing basic and acidic amino acids by rat small intestine *in vitro*, *Clin. Sci.* **43**:823–837.

Cajori, F. A., 1933, The enzyme activity of dogs' intestinal juice and its relation to intestinal digestion, *Am. J. Physiol.* **104**:659–668.

Cheeseman, C. I., and Smyth, D. H., 1973, Specific transfer process for intestinal absorption of peptides, *J. Physiol.* **229**:45–46P.

Cheng, B., and Matthews, D. M., 1970, Rates of uptake of amino acid from L-methionine and the peptide L-methionyl-L-methionine by rat small intenstine *in vitro*, *J. Physiol.* **210**:37–38P.

Cheng, B., Navab, F., Lis, M. T., Miller, T. N., and Matthews, D. M., 1971, Mechanisms of dipeptide uptake by rat small intenstine *in vitro*, *Clin. Sci.* **40**:247–259.

Choules, G. L., and Gray, W. R., 1971, Peptidase activity in the membranes of *Mycoplasma laidlawii*, *Biochem. Biophys. Res. Commun.* **45**:849–855.

Cohnheim, O., 1901, Die umwandlung des eiweiss durch die darmwand, *Z. Physiol. Chem.* **33**:451–465.

Costerton, J. W., Ingram, J. M., and Cheng, K. J., 1974, Structure and function of the cell envelope of gram-negative bacteria, *Bacteriol. Rev.* **38**:87–110.

Craft, I. L., and Matthews, D. M., 1968, The absorption of glycine and glycylglycine in man, following surgery and in gastrointestinal disorders, *Brit. J. Surg.* **55**:158.

Craft, I. L., Geddes, D., Hydge, C. W., Wise, I. J., and Matthews, D. M., 1968, Absorption and malabsorption of glycine and glycine peptides in man, *Gut* **9**:425–427.

Crampton, R. F., Gangolli, S. D., Simson, P., and Matthews, D. M., 1971, Rates of absorption by rat intestine of pancreatic hydrolysates of proteins and their corresponding amino acid mixtures, *Clin. Sci.* **41**:309–417.

De Felice, M., Guardiola, J., Lamberti, A., and Iaccarino, M., 1973, *Escherichia coli* K-12 mutants altered in the transport systems for oligo- and dipeptides, *J. Bacteriol.* **116**:751–756.

Dunn, F. W., Humphreys, J., and Shive, W., 1957, Utilization of tripeptides, *Arch. Biochem. Biophys.* **71**:475–476.

Dvorak, H. F., Wetzel, B. K., and Heppel, L. A. 1970, Biochemical and cytochemical evidence for the polar concentration of periplasmic enzymes in a minicell strain of *Escherichia coli*, *J. Bacteriol.* **104**:543–548.

Fern, E. B., Hider, R. C., and London, D. R., 1969, The site of hydrolysis of dipeptides containing leucine and glycine by rat jejunum *in vitro*, *Biochem. J.* **114**:855–861.

Fickel, T. E., 1973, The oligopeptide permease of *E. coli* as a vehicle for the transport of impermeant substances and its accessibility to large oligopeptides, Ph.D. thesis, Princeton University.

Fickel, T. E., and Gilvarg, C., 1973, Transport of impermeant substances in *E. coli* by way of oligopeptide permease, *Nature, New Biol.* **241**:161–163.

Florsheim, H. A., Makineni, S., and Shankman, S., 1962, The isolation, identification and synthesis of a peptide growth factor for *P. cerevisiae*, *Arch. Biochem. Biophys.* **97**:243–249.

Ford, J. E., and Shorrock, C., 1971, Metabolism of heat-damaged proteins in the rat. Influence of heat damage on the excretion of amino acids and peptides in the urine, *Brit. J. Nutr.* **26**:311–322.

Fordham, W. D., and Gilvarg, C., 1974, Kinetics of crosslinking of peptidoglycan in *Bacillus megaterium*, *J. Biol. Chem.* **249**:2478–2482.

Gale, E. F., 1945, The arginine, ornithine and carbon dioxide requirements of *Streptococci* (Lancefield group D) and their relation to argine dehydrolase activity, *Brit. J. Exp. Path.* **26**:225–233.

Gangolli, S. D., Simson, P., Lis, M. T., Crampton, R. F., and Matthews, D. M., 1970, Amino acid and peptide uptake in protein absorption, *Clin. Sci.* **39**:18P.

Gibson, Q. H., and Wiseman, G., 1951, Selective absorption of stereoisomers of amino acids from loops of the small intestine of the rat, *Biochem. J.* **48**:426–429.

Gilvarg, C., 1972, Peptide transport in bacteria, *in* "Peptide Transport in Bacteria and Mammalian Gut," p. 11, Ciba Foundation Symposium, Elsevier, Excerpta Medica, North Holland, Associated Scientific Publishers, Amsterdam, London, New York.

Gilvarg, C., and Katchalski, E., 1965, Peptide utilization in *Escherichia coli*, *J. Biol. Chem.* **240**:3093–3098.

Gilvarg, C., and Levin, Y., 1972, Response of *Escherichia coli* to ornithyl peptides, *J. Biol. Chem.* **247**:543–549.

Guardiola, J., and Iaccarino, M., 1971, *Escherichia coli* K-12 mutants altered in the transport of branched-chain amino acids, *J. Bacteriol.* **108**:1034–1044.

Hauschild, A. H. W., 1965, Incorporation of ^{14}C from amino acids and peptides into protein by *clostridium perfringens* type D, *J. Bacteriol.* **90**:1569–1574.

Hellier, M. D., Perret, D., and Holdsworth, C. D., 1970, Dipeptide absorption in cystinuria, *Brit. Med. J.* **4**:782–793.

Hellier, M. D., Perret, D., Holdsworth, C. D., and Thirumalai, 1971, Absorption of dipeptides in normal and cystinuric subjects, *Gut* **12**:496–497.

Hellier, M. D., Holdsworth, C. D., Perrett, D., and Thirumalai, C., 1972, Intestinal dipeptide transport in normal and cystinuric subjects, *Clin. Sci.* **43**:659–668.

Henning, U., Braun, V., Höhn, B., and Schwarz, U., 1972, Cell envelope and shape of *Escherichia coli* K-12, properties of a temperature-sensitive *rod* mutant, *Eur. J. Biochem.* **26**:570–586.

Heppel, L. A., Rosen, B. P., Friedberg, I., Berger, E. A., and Weiner, J. H., 1972, The molecular basis of biological transport, *in* "Miami Winter Symposia," Vol. 3, pp. 133–156.

Hueckel, H. J., and Rogers, Q. R., 1972, Prolylhydroxyproline absorption in hamsters, *Can. J. Biochem.* **50**:782–790.

Johnston, J. M., and Wiggans, D. S., 1958, The absorption *in vitro* of alanylphenylalanine, *Biochim. Biophys. Acta* **27**:224–225.

Kadner, R. J., and Liggins, G. L., 1973, Transport of vitamin B_{12} in *Escherichia coli*: genetic studies, *J. Bacteriol.* **115**:514–521.

Kamiryo, T., and Strominger, J. L., 1974, Penicillin-resistant temperatures ensitive mutants of *Escherichia coli* which synthesize hypo- or hyper-cross-linked peptidoglycan, *J. Bacteriol.* **117**:568–577.

Kessel, D., and Lubin, M., 1963, On the distinction between peptidase activity and peptide transport, *Biochim. Biophys. Acta* **71**:656–663.

Kihara, H., and Snell, E. C., 1952, Peptides and bacterial growth: L-alanine peptides and growth of *Lactobacillus casei*, *J. Biol. Chem.* **197**:791–800.

Kihara, H., Ikawa, M., and Snell, E. E., 1961, Peptides and bacterial growth: relation of uptake and hydrolysis to utilization of D-alanine peptides for growth of *Streptococcus faecalis*, *J. Biol. Chem.* **236**:172–176.

Koplow, J., and Goldfine, H., 1974, Alterations in the outer membrane of the cell envelope of heptose-deficient mutants of *Escherichia coli*, *J. Bacteriol.* **117**:527–543.

Kornberg, H. J., 1972, in Discussion to: Membrane digestion and peptide transport, *in* "Peptide Transport in Bacteria and Mammalian Gut," p. 137, Ciba Foundation Symposium, Elsevier Excerpta Medica, North-Holland, Associated Scientific Publishers, Amsterdam, London, New York.

Leach, F. R., and Snell, E. E., 1959, Occurrence of independent uptake mechanisms for glycine and glycine peptides in *Lactobacillus casei*, *Biochim. Biophys. Acta* **34**:292–293.

Leach, F. R., and Snell, E. E., 1960, The absorption of glycine and alanine and their peptides by *Lactobacillus casei*, *J. Biol. Chem.* **235**:3523–3531.

Lehmann, V., Hammerling, G., Nurminen, M., Ruschmann, E., Luderitz, O., Kuo, T., and Stocker, B. A. D., 1973, A new class of heptose-defective mutant of *Salmonella typhimurium*, *Euro. J. Biochem.* **32**:268–275.

Leive, L., 1968, Studies on the permeability change produced in coliform bacteria by ethylenediaminetetraacetate, *J. Biol. Chem.* **243**:2373–2380.

Leive, L., Shovlin, V. K., and Mergenhagen S. E., 1968, Physical chemical and immunological properties of lipopolysaccharide released from *Escherichia coli* by ethylenediaminetetraacetate, *J. Biol. Chem.* **243**:6384–6391.

Levine, E. M., and Simmonds, S., 1960, Metabolite uptake by serine-glycine auxotrophs of *Escherichia coli*, *J. Biol. Chem.* **235**:2902–2909.

Levine, E. M., and Simmonds, S., 1962, Further studies on metabolite uptake by serine-glycine auxotrophs of *Escherichia coli*, *J. Biol. Chem.* **237**:3718–3724.

Lindsay, S. S., Wheeler, B., Sanderson, K. E., Costerton, J. W., and Cheng, K. J., 1973, The release of alkaline phosphatase and lipopolysaccharide during growth of rough and smooth strains of *Salmonella tiphimurium*, *Can. J. Microbiol.* **19**:333–343.

Lis, M. T., Crampton, R. F., and Matthews, D. M., 1971, Rates of absorption of a dipeptide and the equivalent free amino acid in various mammalian species, *Biochim. Biophys. Acta* **233**:453–455.

MacAlister, T. J., Costerton, J. W., Thompson, L., Thompson, J., and Ingram, J. M., 1972, Distribution of alkaline phosphatase within the periplasmic space of gram-negative bacteria, *J. Bacteriol.* **111**:827–832.

Matheson, A. T., and Murayama, T., 1966, The limited release of ribosomal peptidase during formation of *Escherichia coli* spheroplasts, *Can. J. Biochem.* **44**:1407–1415.

Matthews, D. M., 1971a, Experimental Approach in chemical pathology, *Brit. Med. J.* **3**:659–664.

Matthews D. M. 1971b, Protein absorption, *J. Clin. Path.* 24, *Suppl. Roy Coll. Path.* **5**:29–40.

Matthews, D. M., 1972a, Rates of Peptide uptake by small intestine, *in* "Peptide Transport in Bacteria and Mammalian Gut," pp. 71–88, Ciba Foundation Symposium, Elsevier, Excerpta Medica, North-Holland, Associated Scientific Publishers, Amsterdam, London, New York.

Matthews, D. M., 1972b, Intestinal absorption of amino acids and protein, *Proc. Nutr. Soc.* **31**:171–177.

Matthews, D. M., Lis, M. T., Cheng, B., and Crampton, R. F., 1969, Observations on the intestinal absorption of some oligopeptides of methionine and glycine in the rat, *Clin. Sci.* **37**:751–764.

Matthews, D. M., Addison, J. M., and Burston, D., 1974, Evidence for active transport of the dipeptide carnosine (ß-alanyl-L-histidine) by hamster jejunum *in vitro*, *Clin. Sci. Molec. Med.* **46**:693–705.

Mayshak, J., Yoder, O. C., Beamer, K. C., and Shelton, D. C., 1966, Inhibition and transport kinetic studies involving L-leucine, L-valine and their dipeptides in *Leuconostic mesenteroides*, *Arch. Biochem. Biophys.* **113**:189–194.

Meinhart, J. O., and Simmonds, S., 1955, Metabolism of serine and glycine peptides by mutants of *Escherichia coli* Strain K-12 *J. Biol. Chem.* **216**:51–65.

Meisler, N., and Simmonds, S., 1963, The metabolism of glycyl-L-leucine by *Escherichia coli*, *J. Gen. Microbiol.* **31**:109–123.

Merrifield, R. B., and Woolley, D. W., 1956, The synthesis of L-seryl-L-histidyl-L-leucyl-L-valyl-L-glutamic acid, a peptide with strepogenin activity, *J. Am. Chem. Soc.* **78**: 4646–4649.

Messerli, H., 1913, Über die Resorptiongeschwindigkeit der Eiweisse und ihrer Abbauprodukte in Dünndarm, *Biochem. Z.* **54**:446–473.

Miller, A., Neidle, A., and Welsch, H., 1955, Chemical stability and metabolic utilization of asparagine peptides, *Arch. Biochem. Biophys.* **56**:11–21.

Milne, M. D., 1971, Transport of amino acids and peptides in the gut and the kidney, *Sci. Basis Med.* **1971**:161–177.

Milne, M. D., 1972, Peptides in genetic errors of amino acid transport, *in* "Peptide Transport in Bacteria and Mammalian Gut," p. 93, Ciba Foundation Symposium, Elsevier, Excerpta Medica, North-Holland, Associated Scientific Publishers, Amsterdam, London, New York.

Monner, D. A., Jonsson, S., and Boman, H. G., 1971, Ampicillin-resistant mutants of *Escherichia coli* K-12 with lipoplysaccharide alterations affecting mating ability and susceptibility to sex-specific bacteriophages, *J. Bacteriol.* **107**:420–432.

Mueller, J. H., 1938, The utilization of carnosine by *Diphteria bacillus*, *J. Biol. Chem.* **123**:421–432.

Muller-Hill, B., Crapo, L., and Gilbert, W., 1968, Mutants that make more lac repressor, *Proc. Nat. Acad. Sci. U. S.* **59**:1259–1264.

Naider, F., Becker, J. M., and Katzir-Katchalski, E., 1974, Utilization of methionine-containing peptides and their derivatives by a methionine-requiring auxotroph of *Saccharomyces cerevisiae*, *J. Biol. Chem.* **249**:9–20.

Navab, F., and Asatoor, A. M., 1970, Studies on intestinal absorption of amino acids and a dipeptide in a case of Hartnup disease, *Gut* **11**:373–379.

Neu, H. C., and Heppel, L. A., 1966, The release of enzymes from *Escherichia coli* by osmotic shock and during the formation of spheroplasts, *J. Biol. Chem.* **240**:3605–3692.

Newey, H., and Smyth, D. H., 1957, Intestinal absorption of dipeptides, *J. Physiol.* **135**: 43–44.

Newey, H., and Smyth, D. H., 1959, The intestinal absorption of some dipeptides, *J. Physiol.* **145**:48–56.

Newey, H., and Smyth, D. H., 1960, Intracellular hydrolysis of dipeptides during intestinal absorption, *J. Physiol.* **152**:367–380.

Newey, H., and Smyth, D. H., 1962, Cellular mechanisms in intestinal transport of amino acids, *J. Physiol.* **164**:527–551.

Payne, J. W., 1968, Oligopeptide transport in *Escherichia coli*: specificity with respect to side chain and distinction from dipeptide transport, *J. Biol. Chem.* **243**:3395–3403.

Payne, J. W., 1971a, The requirement for the protonated α-amino group for the transport of peptides in *Escherichia coli*, *Biochem. J.* **123**:245–253.

Payne, J. W., 1971b, The utilization of prolyl peptides by *Escherichia coli*, *Biochem. J.* **123**:255–260.

Payne, J. W., 1972a, Mechanisms of bacterial peptide transport, *in* "Peptide Transport in Bacteria and Mammalian Gut," p. 17, Ciba Foundation Symposium, Elsevier, Excerpta Medica, North-Holland, Associated Scientific Publishers, Amsterdam, London, New York.

Payne, J. W., 1972b, Effects of N-methylpeptide bonds on peptide utilization by *Escherichia coli*, *J. Gen. Microbiol.* **71**:259–265.

Payne, J. W., 1972c, in Discussion to: Mechanisms of bacterial peptide transport, *in* "Peptide Transport in Bacteria and Mammalian Gut," p. 38, Ciba Foundation

Symposium, Elsevier, Excerpta Medica, North-Holland, Associated Scientific Publishers, Amsterdam, London, New York.

Payne, J. W., 1973, Peptide utilization in *Escherichia coli*: Studies with peptides containing β-alanyl residues, *Biochim. Biophys. Acta* **298**:469–478.

Payne, J. W., 1974, Peptide transport in *Escherichia coli*: Permease specificity towards terminal amino group substituents, *J. Gen. Microbiol.* **80**:269–276.

Payne, J. W., and Gilvarg, C., 1968a, The role of terminal carboxyl group in peptide transport in *Escherichia coli*, *J. Biol. Chem.* **243**:335–340.

Payne, J. W., and Gilvarg, C. 1968b, Size restriction on peptide utilization in *Escherichia coli*, *J. Biol. Chem.* **243**:6291–6299.

Payne, J. W., and Gilvarg, C., 1971, Peptide transport, *Advan. Enzymol.* **35**:187–244.

Pecht, M., Giberman, E. Keysary, A., Yariv, J., and Katchalski, E., 1972, Hydrolysis of alanine oligopeptides by an enzyme located in the membrane of *Mycoplasma laidlawii*, *Biochim. Biophys. Acta* **290**:267–273.

Peters, V. J., Prescott, J. M., and Snell, E. E., 1953, Peptides and bacterial growth: Histidine peptides as growth factors for *Lactobacillus delbrueckii* 9649, *J. Biol. Chem.* **202**: 521–532.

Pittman, K. A., Lakshmanan, S., and Bryant, M. P., 1967, Oligopeptide uptake by *Bacteroides ruminicola*, *J. Bacteriol.* **93**:1499–1508.

Prescott, J. M., Peters, V. J., and Snell, E. E., 1953, Peptides and bacterial growth: Serine peptides and growth of *Lactobacillus delbrueckii* 9649, *J. Biol. Chem.* **202**:533–540.

Rooney, S. A., and Goldfine, H., 1972, Isolation and characterization of 2-keto-3-deoxyoctonate-lipid A from a heptose deficient mutant of *Escherichia coli*, *J. Bacteriol.* **111**:531–541.

Rubino, A., Field, M., and Shwachman, H., 1971, Intestinal transport of amino acid residues of dipeptides: Influx of the glycine residue of glycyl-L-proline across mucosal border, *J. Biol. Chem.* **246**:3542–3548.

Schmidt, G., Jann, B., and Jann, K., 1969, Immunochemistry of R lipopolysaccharides of *Escherichia coli*, different core regions in the lipopolysaccharides of 0 group 3, *Eur. J. Biochem.* **10**:501–510.

Schmidt, G., Jann, B., and Jann, K., 1970, Immunochemistry of R lipopolysaccharides of *Escherichia coli*, studies on R mutants with an incomplete core, derived from *E. coli* 08:K27, *Eur. J. Biochem.* **16**:382–392.

Shankman, S., Higa, S., Florsheim, H. A., Schvo, Y., and Gold, V., 1960, Peptide studies: Growth-promoting activity of peptides of L-leucine and L- and D-value for lactic acid bacteria, *Arch. Biochem. Biophys.* **86**:204–209.

Shankman, S., Higa, S., and Gold, V., 1961, Peptide studies: Inhibition of bacterial growth by di- and tripeptides, *Texas Rept. Biol. Med.* **19**:358–369.

Shankman, S., Gold, V., Higa, S., and Squires, R., 1962, On the mode of action of a peptide inhibitor of growth in *P. cerevisiae*, *Biochem. Biophys. Res. Communs.* **9**:25–31.

Shelton, D. C., and Nutter, W. E., 1964, Uptake of valine and glycylvaline by *Leuconostoc mesenteroides*, *J. Bacteriol.* **88**:1175–1184.

Simmonds, S., 1966, The role of dipeptidases in cells of *Escherichia coli* K-12, *J. Biol. Chem.* **241**:2502–2508.

Simmonds, S., 1970, Peptidase activity and peptide metabolism in *Escherichia coli* K-12, *Biochem.* **9**:1–9.

Simmonds, S., 1972, Peptidase activity and peptide metabolism in *Escherichia coli*, *in* "Peptide Transport in Bacteria and Mammalian Gut," p. 43, Ciba Foundation Symposium, Elsevier, Excerpta Medica, North-Holland, Associated Scientific Publishers, Amsterdam, London, New York.

Simmonds, S., and Fruton, J. S., 1948, The utilization of proline derivatives by mutant strains of *Escherichia coli*, *J. Biol. Chem.* **174**:705–715.

Simmonds, S., and Griffith, D. D., 1962, Metabolism of phenylalanine containing peptide amides in *Escherichia coli*, *J. Bacteriol.* **83**:256–263.

Simmonds, S., and Toye, N. O., 1966, Peptidases in spheroplasts of *Escherichia coli* K-12, *J. Biol. Chem.* **241**:3852–3860.

Simmonds, S., Tatum, E. L., and Fruton, J. S., 1947, The utilization of phenylalanine and tyrosine derivatives by mutant strains of *Escherichia coli*, *J. Biol. Chem.* **169**:91–101.

Simmonds, S., Harris, J. I., and Fruton, J. S., 1951, Inhibition of bacterial growth by leucine peptides, *J. Biol. Chem.* **188**:251–262.

Smith, R. L., Archer, E. G., and Dunn, F. W., 1970, Uptake of ^{14}C-labeled tri-, tetra- and pentapeptides of phenylalanine and glycine by *Escherichia coli*, *J. Biol. Chem.* **245**: 2962–2966.

Starling, E. H., 1906, "Recent Advances in the Physiology of Digestion," p. 127, Constable, London.

Sussman, A. J., and Gilvarg, C., 1970, Peptidases in *Escherichia coli* K-12 capable of cleaving lysine homopeptides, *J. Biol. Chem.* **245**:6518–6524.

Sussman, A. J., and Gilvarg, C., 1971, Peptide transport and metabolism in bacteria, *Ann. Rev. Biochem.* **40**:397–408.

Tamaki, S., and Matsuhashi, M., 1973, Increase in sensitivity to antibiotics and lysozyme on deletion of lipopolysaccharides in *Escherichia coli* strains, *J. Bacteriol.* **114**:453–454.

Tamaki, S., Sato, T., and Matsuhashi, M., 1971, Role of lipopolysaccharides in antibiotic resistance and bacteriophage absorption of *Escherichia coli* K-12, *J. Bacteriol.* **105**: 968–975.

Takagaki, Y., Kunugita, K., and Matsuhashi, M., 1973, Evidence for direct action of colicin K on aerobic ^{32}Pi uptake in *Escherichia coli in-vivo* and *in-vitro*, *J. Bacteriol.* **113**:42–50.

Tarlow, M. J., Seakins, J. W. T., Lloyd, J. K., Matthews, D. M., Cheng, B., and Thomas, A. J., 1970, Intestinal absorption and biopsy transport of peptides and amino acids in Hartnup disease, *Clin. Sci.* **39**:18–19p.

Ugolev, A. M., 1972, Membrane digestion and peptide transport, *in* "Peptide Transport in Bacteria and Mammalian Gut," pp. 123–137, Ciba Foundation Symposium, Elsevier, Excerpta Medica, North-Holland, Associated Scientific Publishers, Amsterdam, London, New York.

Ugolev, A. M., and DeLaey, P., 1973, Membrane digestion: a concept of enzymic hydrolysis of cell membranes, *Biochim. Biophys. Acta* **300**:105–128.

Van Lenten, E. J., and Simmonds, S., 1967, Dipeptidases in spheroplasts and osmotically shocked cells prepared from *Escherichia coli* K-12, *J. Biol. Chem.* **242**:1439–1444.

Van Slyke, D. D., and Meyer, G. M., 1912, The amino acid nitrogen of the blood. Preliminary experiments on protein assimilation, *J. Biol. Chem.* **12**:399–410.

Van Slyke, D. D., and Meyer, G. M., 1913–1914, The fate of protein digestion products in the body. The absorption of amino acids from the bood by the tissues, *J. Biol. Chem.* **16**:197–212.

Vonder Haar, R. A., and Umbarger, H. E., 1972, Isoleucine and valine metabolism in *Escherichia coli*, *J. Bacteriol.* **112**:142–147.

Wahren, A., and Gibbons, R. J., 1970, Amino acid fermentation by *Bacteroides melaninogenicus*, *Antonie van Leeuwenhoek J. Microbiol. Serol.* **36**:149–159.

Wahren, A., and Holme, T., 1973, Amino acid and peptide requirement of *Fusiformis necrophorus*, *J. Bacteriol.* **116**:279–284.

Wang, C. C., and Newton, A., 1969, Iron transport in *Escherichia coli*: relation between chromium sensitivity and high iron requirement in mutants of *Escherichia coli*, *J. Bacteriol.* **98**:1135–1141.

Wetzel, B. K., Spicer, S. S., Dvorak, H. F., and Heppel, L. A., 1970, Cytochemical localization of certain phosphatases in *Escherichia coli*, *J. Bacteriol.* **104**:529–542.

White, J. C., Di Girolamo, P. M., Fu, M. L., Preston, Y., and Bradbeer, C., 1973, Transport of vitamin BJJ in *Escherichia coli*, location and properties of the initial B_{12}-binding site, *J. Biol. Chem.* **248**:3978–3986.

Wiggans, D. S., and Johnston, J. M., 1959, The absorption of peptides, *Biochim. Biophys. Acta* **32**:69–73.

Wiseman, G., 1953, Absorption of amino acids using an *in-vitro* technique, *J. Physiol.* **120**:63–72.

Woolley, D. W., Merrifield, R. B., Ressler, C., and Du Vigneaud, V., 1955, Strepogenin activity of synthetic peptides related to oxytocin, *Proc. Soc. Exp. Biol. Med.* **89**: 669–673.

Wu, H. C., 1972, Isolation and characterization of an *Escherichia coli* mutant with alteration in the outer membrane proteins of the cell envelope, *Biochim. Biophys. Acta* **290**: 274–289.

Yoder, O. C., Beamer, K. C., and Shelton, D. C., 1965a, Structural and stereochemical specificity of transport systems for glycine, valine and their dipeptides in *L. mesenteroides*, *Fed. Proc.* **24**:352.

Yoder, O. C., Beamer, K. C., Cipolloni, Jr., P. B., and Shelton, D. C., 1965b, Kinetic studies of L-valine and glycyl-L-valine uptake by *leuconostoc mesenteroides*, *Arch. Biochem. Biophys.* **110**:336–340.

Young, E. A., Bowen, D. O., and Diehl, J. F., 1964, Transport studies with peptides containing unnatural amino acids, *Biochem. Biophys. Res. Commun.* **14**:250–255.

Chapter 8

Factors Influencing the Retention of *K* in a Halobacterium

M. Ginzburg and B. Z. Ginzburg

Botany Department
The Hebrew University
Jerusalem, Israel

I. OUTSIDE KCl CONCENTRATION

A. Introduction

The cells of at least two species of *Halobacterium* contain 3–4 mol of potassium and require for survival a similar concentration of NaCl in the outside medium (Christian and Waltho, 1962; Ginzburg *et al.*, 1970). It has been suggested (Ginzburg *et al.*, 1970, 1971a, b) that the retention of cell potassium is not directly dependent upon metabolism. The principal basis for this conclusion comes from experiments with starved bacteria: These respire at about 1 % of the normal rate and are presumed to have a low rate of metabolism. The cell membrane of these starved bacteria allows the free passage of the K^+, Na^+, and Cl^- ions, yet the amount of potassium is equal to that present in cells metabolizing normally. Thus metabolism cannot be of importance in maintaining large amounts of potassium in the cells, and it is concluded that the cell ions are at equilibrium with those in the outside solution.

The present work includes measurements on the effects of the concentration of external potassium chloride on the amount of cell potassium.

B. Materials and Methods

Methods of culture of the *Halobacterium* species used have already been described (Ginzburg *et al.*, 1970). The methods of analysis and calculation were described in the same paper. For the method of preparation of starved bacteria, see Ginzburg *et al.* (1971b).

For experiments, 4-ml portions of starving bacteria in suspension were centrifuged at 13,000*g* in a Sorvall RC-B centrifuge at room temperature for 10 min. The supernatant was poured off and all remaining droplets removed by suction. Four milliliters of experimental salt solution was poured onto the pellet of bacteria, which were partially resuspended by scraping the tube wall with a spatula. The experimental salt solution contained 3.5 M NaCl, 0.15 M $MgSO_4$, 1.4×10^{-3} M $CaCl_2$, and 2.5×10^{-7} M $MnCl_2$ together with 10 mM *N,N*-bis(2-hydroxyethyl)-2-amino-ethane sulfonic acid buffer (BES) at *p*H 7.0 and KCl at the desired concentration. The contents of the tube were transferred to a 25-ml Erlenmeyer flask which was shaken at the experimental temperature (0–37°C) for 2 hr.

At the end of the incubation period ten 300-μl portions of bacterial suspension were centrifuged in a Beckman Microfuge. Five were used for determination of K and five for protein determination.

C. Results

It became clear at the start of the work that starving bacteria preserve their integrity only in the presence of KCl; there is much cell breakage in solutions without KCl. Not less than 0.1 mM KCl is required for the maintenance of the cell integrity even when the outside NaCl concentration is at

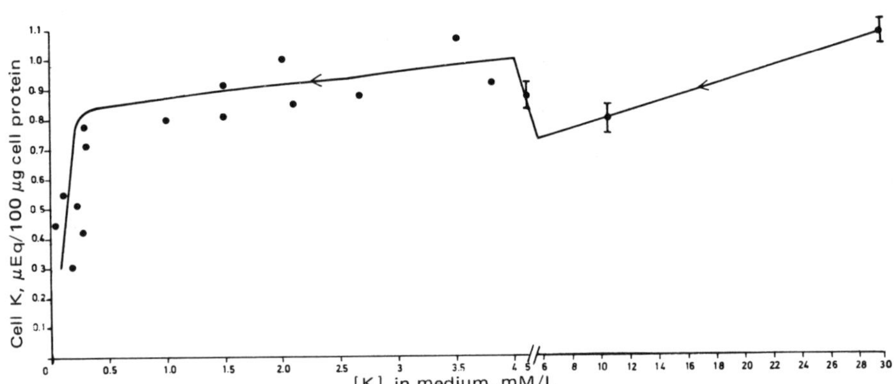

Fig. 1. Effect of outside concentration of KCl on amount of cell K (desorption curve). The bacteria were starved at 30 mM/liter KCl and were equilibrated at the given KCl concentration for 2 hr before measurements were made. Temperature of equilibration was 37°C. K/pro: μequiv cell K per 100 μg cell protein. Mean of four experiments.

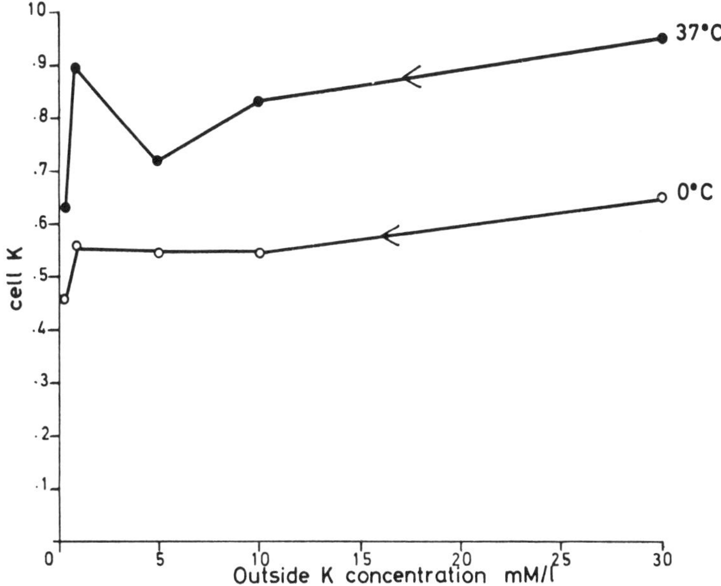

Fig. 2. Desorption of cell K at 37° and 0°C. For other conditions see legend to Fig. 1. Mean of two experiments.

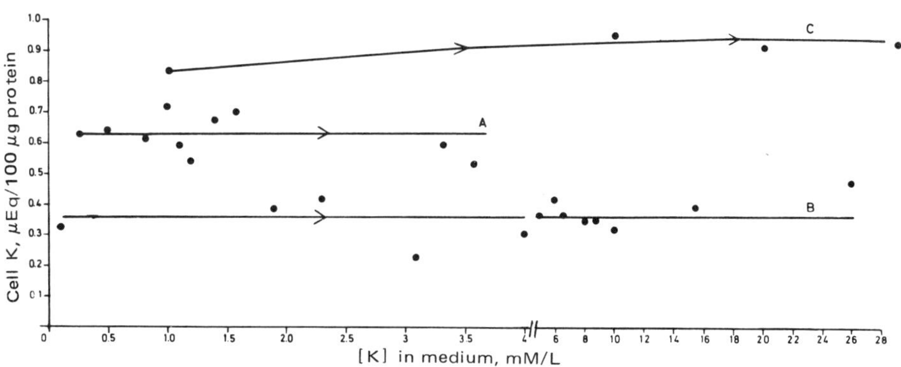

Fig. 3. Effect of outside concentration of KCl on amount of cell K (absorption curve). After starvation in medium with KCl, the bacteria were equilibrated at low KCl concentration for not less than 2 hr; after this period KCl was added to the ambient medium; measurements of cell K and protein were taken 2 hr later. The KCl concentrations during periods of equilibration were (mM/liter): 0.3 (curve A); 0.6 (curve B); 0.8 (curve C). Temperature of experiments, 37°C. K/pro: μequiv cell K per 100 μg cell protein.

its optimum. Thus experiments on the effect of KCl on starving bacteria have a lower limit of 0.1 mM.

When bacteria starved at 30 mM/liter KCl were transferred to a solution containing KCl at some lower concentration, there was a loss of cell K; Fig. 1 represents a desorption curve obtained in this way; it has a peculiar shape with maxima at 4 and 30 mM KCl. The bimodal shape of the curve is temperature dependent (Fig. 2); it is seen clearly at 37°C and not at all at 0°C.

The reverse experiment was also attempted: Bacteria incubated at a low enough KCl concentration to cause partial depletion of cell K were subjected to an increase in the KCl concentration. In no such experiment was there any significant gain in cell K (Fig. 3). No increase whatsoever in cell K occurred at 0 and 17°C (Fig. 4); a very slight increase, of doubtful significance, was detected at 27 and 37°C. Thus it is generally concluded that cell K lost when the outside KCl concentration is lowered cannot be regained by subsequently returning the outside KCl concentration to its initial higher value.

In contrast, potassium lost when cells were cooled was regained when the suspension was warmed. This was noted at several different outside concentrations of KCl (Table I).

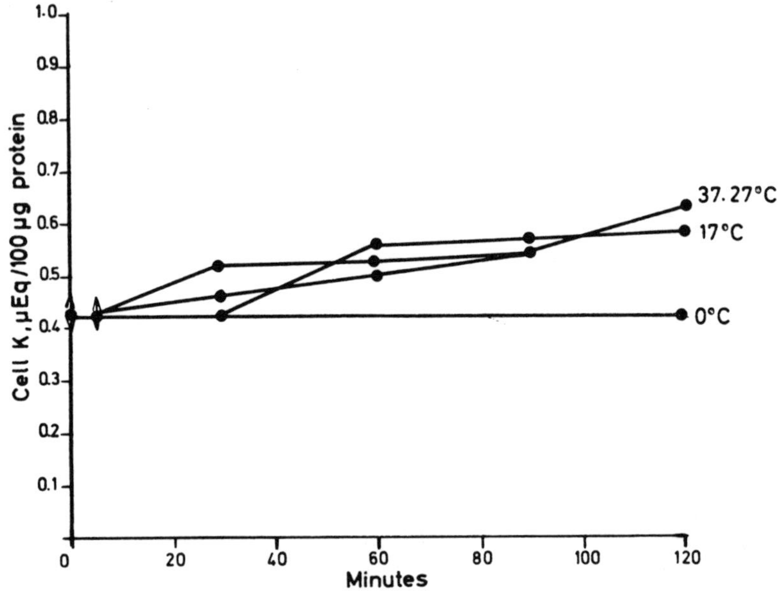

Fig. 4. Cell K in bacteria 2 hr after transference from 0.1 mM/liter KCl to 30 mM/liter KCl at 0, 17, 27, and 37°C. Cell K: µequiv cell K per 100 µg cell protein.

Table I

Effect of a 2-hr Exposure to 0°C on Bacteria in Media Containing Different Concentrations of KCl[a]

Time after start of experiment, min	0	45	120	185
Temperature, °C	37	0	0	37
KCl, mmol/liter	μequiv K/100 μg protein			
0.5	0.90	0.50	0.43	0.65
5	0.72	0.58	0.51	0.79
10	0.85	0.60	0.50	0.80
30	0.93	0.69	0.64	0.89

[a] After the period at 0°C, the bacteria were warmed to the initial temperature of 37°C.

D. Discussion

The major conclusion from this work is that potassium within the cells of *Halobacterium* sp. exists at equilibrium with the outside K^+, the presence of which is essential for the integrity of the cells; the concentration required is below 0.5 mM. In contrast, starving and growing *Halobacterium* cells have requirements for NaCl of not less than 0.5 and 2 M, respectively.

In starving bacteria the cell K lost when the outside KCl concentration was lowered could not be regained by restoring the KCl to its initial value (cf. Figs. 1 and 4). On the other hand, K lost when the bacteria were cooled was regained on warming them (Table I). Similarly, cell K lost by lowering the NaCl concentration of the medium is regained on increasing the NaCl concentration (Section II). Thus loss of cell K is not necessarily irreversible. Two possible reasons can be suggested for the lack of absorption of K by the bacteria when the outside KCl concentration is increased. A first possibility is that loss of K due to lowered KCl concentration is accompanied by an irreversible change in the sites to which the K is normally attached; this possibility seems unlikely in view of the fact that the bacteria do regain K when there is an increase in temperature or in the NaCl concentration of the outside medium. Alternatively, the energy of activation of absorption may be very high, so that a form of hysteresis might be expected, i.e., reabsorption of K might occur at a very high outside concentration of KCl. If so, the required concentration must be very high indeed since suspensions of starved bacteria have been subjected to KCl concentrations of 0.8 M and even at that concentration there was little increase in cell K above the control level (Section II). Unfortunately, very high concentrations of KCl—of the order of 4 M—are not tolerated by the bacteria for more than 5–10 min and the hypothesis cannot be more thoroughly tested. At any rate, there may be an energy barrier preventing bacteria with low cell K to attain equilibrium when supplied with relatively high levels of KCl in the medium (Katchalsky *et al.*, 1964).

Results presented here and in Section II show that there are two major interactions involving ions in these *Halobacterium* cells; there is a strong interaction, as with potassium, and a weak interaction, as with sodium. The strong interaction is demonstrated by the variation of cell K with changes in the outside KCl concentration; the midpoint of saturation occurs at 0.1–0.2 mM KCl and saturation is reached at 0.5 mM. The weak interaction is demonstrated by the variation of cell K with changes in the NaCl concentration in the medium; the midpoint of saturation is at 2 M NaCl and saturation is barely reached at 4 M NaCl. The weak interaction is demonstrated with several chlorides of cations with small crystal radii; with all of them the midpoint of saturation is approximately at 2 M.

A standard free energy change ΔG can be calculated for the two interactions:

$$\Delta G = -RT \ln K$$

where K is the apparent mass action equilibrium constant, assuming all ions are in solution. The value of ΔG is about 5 kg-cal/mol for the cell-K interaction and about 0.5 kg-cal/mol for the cell-Na interaction.

The quantitative expression of the Na–K specificity within the *Halobacterium* cell yields another aspect of the relative strengths of the two ion interactions; on the assumption that all ions are in solution, $K^+_{in}Na^+_{out}/K^+_{out}Na^+_{in} = 10{,}000$ in starved bacteria, i.e., the cells show a 10,000-fold preference for K^+ over Na^+.

Comparison of kinetic data yields similar ratios of K^+ relative to Na^+; the half-time of exchange of Na^+ is less than 1 min (Ginzburg *et al.*, 1971a), whereas the half-time of exchange for the cell K is about 900 min at 37°C (unpublished observations).

Thus both thermodynamic and kinetic data show that there is a difference of about four orders of magnitude between the cell-K and the cell-Na interactions. The data strongly support the hypothesis that the cell K is bound within the cell whereas Na serves as modulator of the K-binding system.

E. Summary

The organism used for this work was a species of *Halobacterium* which grows optimally at NaCl concentrations of not less than 3.5 M. The cells of this organism contain 3–4 M potassium, which is retained when the rate of metabolism is reduced to its lowest level, even though the outer membrane is permeable to K^+; cell K is therefore apparently bound. Measurements were made of the variation of cell potassium in starving bacteria with change in external [KCl], the NaCl concentration being kept constant at 3.5 M. Cell K was calculated relative to content of cell protein. An average value of 1 μequiv of K per 100 μg of cell protein was obtained from bacteria at 2 mM KCl; cell K was present at half the amount at 0.1–0.2 mM KCl. Cell K was lost when the outside [KCl] was lowered but could not be regained when the [KCl] was raised to 100 times the original concentration. Cell K was also lost when the temperature or NaCl concentration of the ambient medium was lowered; the losses induced by these conditions, were, however, reversible. It is postulated that the reaction controlling uptake of K from KCl in solution has a very high energy of activation.

II. pH AND NATURE AND CONCENTRATION OF SALT IN OUTSIDE MEDIUM

A. Introduction

The cells of a species of *Halobacterium* isolated some years ago from the Dead Sea contain 4 M potassium (Ginzburg *et al.*, 1970). Similar concentrations have been detected in at least one other species of this genus (Christian and Waltho 1962). The growth medium for species of this genus contains at least 3.5 M NaCl. The purpose of this Section is to explain the state of potassium within the cells of the *Halobacterium* sp. from the Dead Sea and to contrast potassium with sodium within the cell, thus emphasizing the extraordinary specificity that exists within the bacteria with regard to the two ions (Ginzburg *et al.*, 1970). It has been argued on thermodynamic grounds that the cell potassium must be bound whereas the cell sodium is free (Ginzburg *et al.*, 1971a). Unfortunately, the term "binding" is ambiguous; its meaning depends upon the technique of measurement used. For instance, Berendsen (1974) distinguished among three different aspects of binding: energetic, dynamic, and structural. Adamson (1967) prefers to describe the interaction of small molecules as "sorption" rather than binding. He distinguishes between physical or chemical sorption, according as to whether the time of residence of the sorbate on the absorbing surface is shorter or longer than 10^{-7} sec. According to this definition, potassium is chemically absorbed within the *Halobacterium* cell.

There are, however, difficulties in trying to explain the state of the cell potassium in terms of binding. First, the potassium is released into the medium in freely diffusible form when the cells are broken; thus retention of cell K is dependent upon the integrity of the cell. Second there is not sufficient organic material within the cell to provide binding sites for all the K that is there: The maximum ratio of K to cell protein is 1500–1600 mol K per 10^5 g protein; if K were bound to protein, there would be one order of magnitude more of ion bound than in any other case of cation–protein complex known (Steinhardt and Reynolds, 1969). It follows that, even though the absorption of K within the bacterial cell is chemical, the nature of the bond involved in the K interaction is difficult to suggest.

It has already been shown that the sum of the inorganic cations is greater than that of the inorganic anions within the *Halobacterium* cell (Christian and Waltho, 1962; Ginzburg *et al.*, 1970). It was argued that the cell K is electrostatically balanced partly by Cl and partly by negative organic ions; these two K fractions are referred to as K_{Cl} and K_x, respectively. Results presented in this section show that the amount of K_{Cl} is controlled by the NaCl concentration in the ambient medium, while K_x depends upon the *p*H of the medium. Hypotheses to explain these effects are given and a unified model is presented in Section IID.

B. Materials and Methods

1. *Methods of Culture*

Methods of culture of the *Halobacterium* sp. used have already been described (Ginzburg *et al.*, 1970). The methods of analysis and calculation were described in the same paper. For the preparation of starved bacteria see Ginzburg *et al.*, (1971b).

2. *pH Experiments with Starving Bacteria*

Bacteria were suspended in saline solution consisting of 3.5 M Nacl, 0.15 M $MgSO_4$, 1.4×10^{-3} M $CaCl_2$, 2.5×10^{-7} M $MnCl_2$, and 5×10^{-3} M KCl. This solution is referred to as Complete Salt Solution (CSS). Later experiments were performed with bacteria suspended in 3.5 M NaCl and 2 mM/liter KCl. The nature of the suspending solution is indicated in Section IIC. The *p*H was monitored with a joint glass electrode (Radiometer GK 264 B, which is largley insensitive to high salt concentrations below *p*H 10). The *p*H was maintained at the desired value by addition of 0.01 N NaOH or 0.01 N HCl in 3.5 M NaCl. The temperature was maintained at 37°C. Samples for the measurement of ions or protein were taken at intervals of time, as required.

3. Ion Experiments with Starving Cells

Four-milliter portions of starving bacteria suspension were centrifuged at 13,000 × *g* in a Sorvall RC 2-B centrifuge at room temperature for 10 min. The supernatant was poured off and all remaining droplets removed by suction. Four milliliters of the experimental salt solution was poured onto the pellet of bacteria, which were partially resuspended by scraping the tube wall with a spatula. The experimental salt solution contained the major salt at the desired concentration together with 2 mM/liter KCl and 10 mM *N*,*N*-bis(2-hydroxyethyl)-2-aminoethane sulfonic acid (BES) buffer at *p*H 7.0. The tube contents were transferred to a 25-ml Erlenmeyer flask, which was shaken at the experimental temperature (0–50°C) for 2 hr.

At the end of the incubation period ten 300-μl portions of bacterial suspension were centrifuged in a Beckman Microfuge. Five were used for determination of K and five for protein determination.

In certain experiments in which the range of ion concentration to be tested was relatively narrow (e.g., Figs. 16 and 18) bacterial suspensions in 4 M NaCl were diluted with 4 M solutions of the second salt. The effect of dilution per se was checked by performing a parallel control experiment in which 4 M NaCl was used as diluent. All solutions contained 10 mM/liter BES and 2 mM/liter KCl.

All experiments were compared with a parallel set of suspensions tested in NaCl at the same concentration as the unknown salt. Thus all results shown are relative to NaCl.

4. Analysis of Ions and Protein

Methods are described in detail by Ginzburg *et al.* (1970). The measurement of K with the Eppendorf flame photometer, which was used routinely in these experiments, is affected by the presence of other ions in the medium (e.g., Na^+, Mg^{2+}, Ca^{2+}, Cs^+, Rb^+, Cl^-, Br^-).The K concentration of the experimental samples was therefore measured by standard solutions containing, in addition to K, the same additional ions at the same concentrations as were present in the sample.

5. Calculation of Results

Cell potassium was calculated relative to the amount of cell protein, as measured by the Lowry test.

In order to distinguish between K_{Cl} and K_x, it was necessary to measure the Na and Cl contents of pellets on which measurements of K had been made. Sodium is known to be entirely balanced by Cl, and K partly so (Ginzburg *et al.*, 1971b). Thus the difference between the measured Cl and Na represents that portion of the Cl that balances K, and is numerically equal to K_{Cl}. The difference between total cell K and K_{Cl} is referred to as K_x.

C. Results

1. *Effect of pH on Cell Potassium*

The amount of K in *Halobacterium* sp. cells was measured as function of external *p*H both in growing and starving cultures (Fig. 5). Measurements were made 2 hr after the *p*H of a given suspension had been set at a new

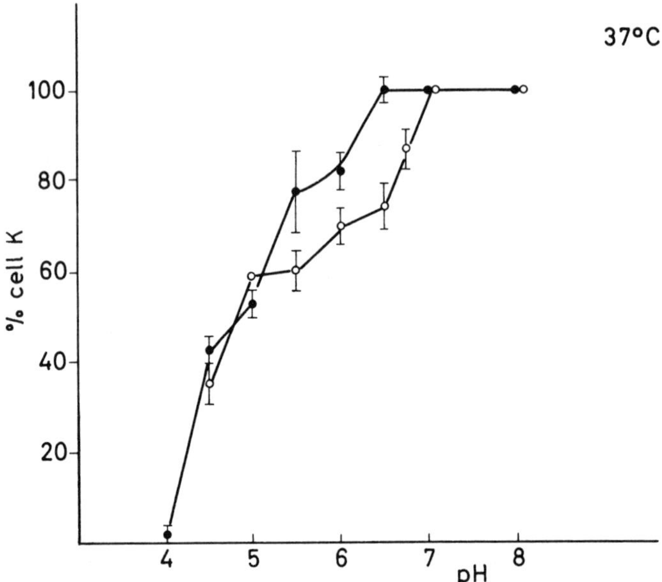

Fig. 5. Cell K content of *Halobacterium* sp. as a function of *p*H of suspending medium, expressed as percentage of amount of cell K at *p*H 7. Suspending medium consisted of Complete Salt Solution (see Section IIB2) plus 10% yeast autolyzate for the growing cultures. The *p*H of the medium was controlled by addition of small volumes of acid. Measurements were made 2 hr after adjustment of *p*H. Temperature, 37°C. All points are average ±SE of not less than four separate experiments. ○, growing bacteria; ●, starving bacteria.

value; at this time a state close to equilibrium had been reached. Figure 5 shows that at 37°C the amount of cell K, as proportion of that present at *p*H 7, falls as the *p*H is lowered until at *p*H 4, none is left at all. In the upper *p*H range the amount of K present differed according to the state of the cells; at *p*H 6.5 growing bacteria had lost 25% of the total K, while starving bacteria had lost none at all. However, the K of both types of bacteria was affected in the same way at *p*H values under 5.

Figure 6 shows the result of exposing starving bacteria to *p*H 4 at temperatures from 37 to 0°C. The rate of loss of cell K is shown to be strongly temperature dependent and to consist of two phases, an initial, faster one lasting 25 min and a subsequent, slower one. The amount of K lost during the initial phase was also temperature dependent. In Fig. 7 the logarithm of the rates of loss are plotted against $1/T$. Since straight-line relationships were obtained, the slopes can be used to calculate the enthalpies of activation;

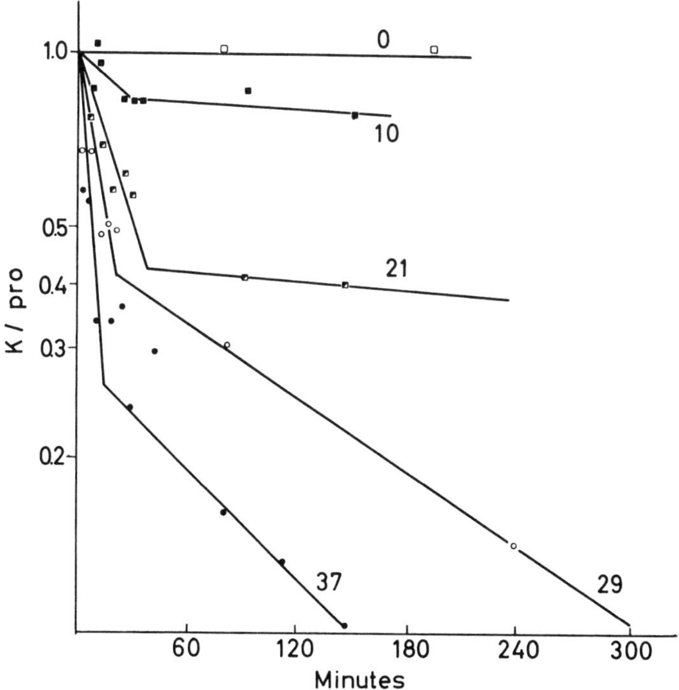

Fig. 6. Rate of K loss from starving bacteria maintained at *p*H 4. The temperature at which loss was measured is indicated next to each curve. Bacteria were suspended in Complete Salt Solution. The results are from one of two similar experiments. K/pro: μequiv cell K per 100 μg cell protein.

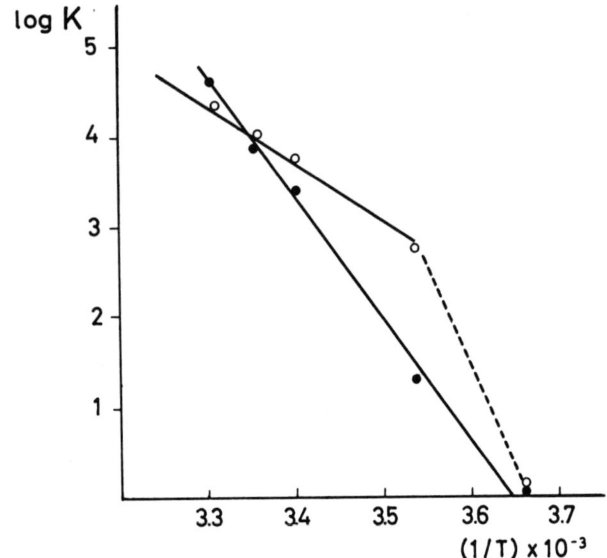

Fig. 7. Arrhenius plot of $\log_{10}$ (rate of loss of cell potassium), log *K*, vs. the reciprocal of the absolute temperature, $1/T$. ○, refers to initial phase; ●, refers to later phase of K loss.

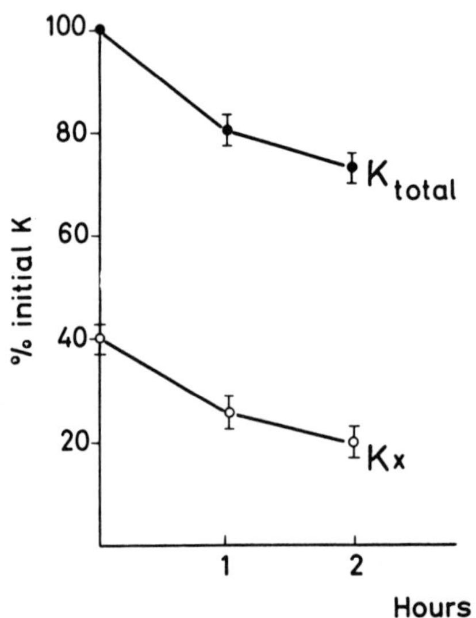

Fig. 8. Loss of cell K from starving bacteria maintained at H 5.5 at 37°C. K_x refers to that part of the cell K not balanced by Cl. Mean of three separate experiments performed in 3.5 M NaCl plus 2 mM/liter KCl. Mean total K/pro was 0.85 μequiv K/100 μg cell protein.

these amount to −7000 cal/mol for the initial phase and −18,000 cal/mol for the later phase.

It was of interest to discover whether potassium was lost in exchange for Na^+ or H^+, or in conjunction with Cl^-. The problem proved to be difficult experimentally since at *p*H values of 5 or below, loss of K was accompanied by rapid gains of water and NaCl, whereas at higher *p*H values the loss of K is small. Finally, the following method was adopted: starved cells were incubated at *p*H 5.5 at 37°; at this *p*H there is significant loss of K (Fig. 8) and uptake of NaCl is slight. The figure shows that during the 2 hr of the experiment total cell K fell by 26%; since K_x fell by 20%, it is assumed that this portion of the K was exchanged for H^+.

It can be surmised from Fig. 8 that at some *p*H value below 5.5 (say 4.5–5.0) all of the K_x is lost. It is concluded that at lower *p*H the remaining cell K is lost in the form of K_{Cl}.

2. *Effect of* [*NaCl*] *in Medium on Cell K in Growing* Halobacterium

Figure 9, curve M, shows the cell K, per unit of bacterial protein, in cultures of *Halobacterium* grown at NaCl concentrations from 2 to 5 M. Measurements were made on 48-hr cultures, when growth, as measured by bacterial protein per milliter of suspension, had ceased. The species of *Halobacterium* used does not grow at NaCl concentrations below 2 M; it grows at close to its optimum rate at 5.2 M NaCl (the maximum solubility of NaCl).

The figure shows that the cell potassium content increases steeply with increase in NaCl concentration of the medium; there is a 13-fold increase in cell K as the [Nacl] is increased from 2 to 4.8 M.

3. *Effect of* [*NaCl*] *in Medium on Cell K in Starving* Halobacterium *sp.*

In these experiments suspensions of *Halobacterium* sp. at 3.5 M NaCl were transferred to solutions containing salt at some other concentration (Fig. 9). The greater part of the curve represents transfer of bacteria from a higher to a lower [NaCl] in the medium and loss of K from the bacteria. This loss is not fully reversible (see Section IIC5), although it is partially so, and the curve may be said to represent a state of equilibrium. It will be noticed that at 4 M NaCl cell K content is the same in growing and starving bacteria. However, starving bacteria retained their integrity at concentrations as low as 0.25 M NaCl. Cell K content was proportional to the outside [NaCl] between 0.25 and 4 M.

The effect of K concentration in the medium on cell K was examined in experiments in which [K] and [NaCl] were varied simultaneously. In the

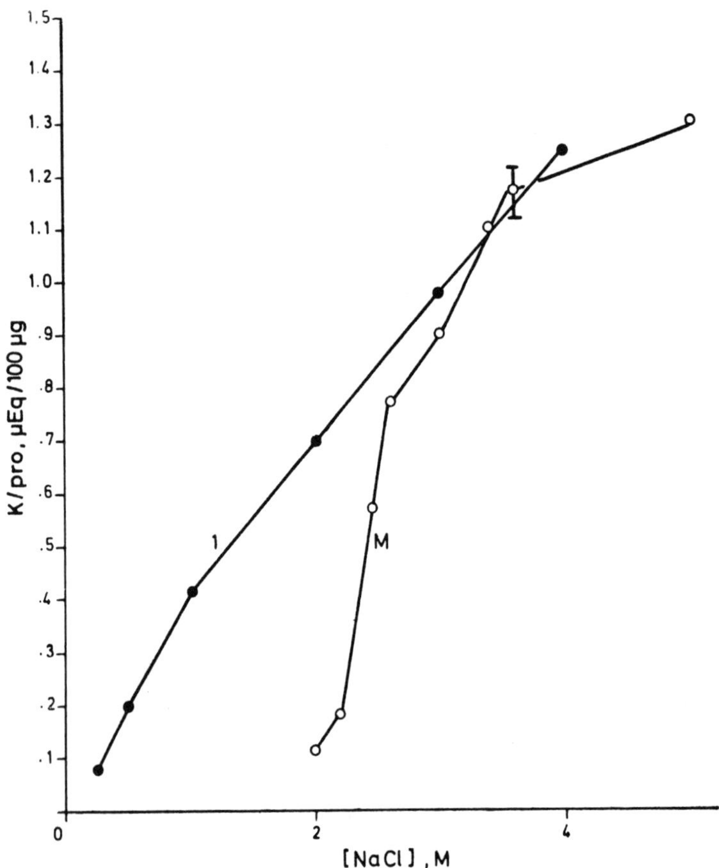

Fig. 9. Effect of NaCl concentration in medium on cell K content of growing (M) and starved bacteria (Ginzburg *et al.*, 1970). Each point is the average of six determinations. The growing bacteria were at 37°C and the starved bacteria at 20°C. K/pro represents cell K in μequiv K per 100 μg cell protein.

range tested the outside [K] was found to have little influence on cell K (Fig. 10). The relatively low cell K at 4 mM/liter KCl has been observed repeatedly (Section I, Fig. 1). It should be noted that the [K] in the medium does not affect the capacity of the binding system: At 4 M NaCl the cell K content was the same at all [K^+] tested.

By making measurements of cell Na and Cl at the same time as cell K, it was possible to determine the form in which the cell K was lost when the medium [NaCl] was decreased. Results of these measurements are shown in Fig. 11. The upper curve shows the total cell K as a function of [NaCl] in the

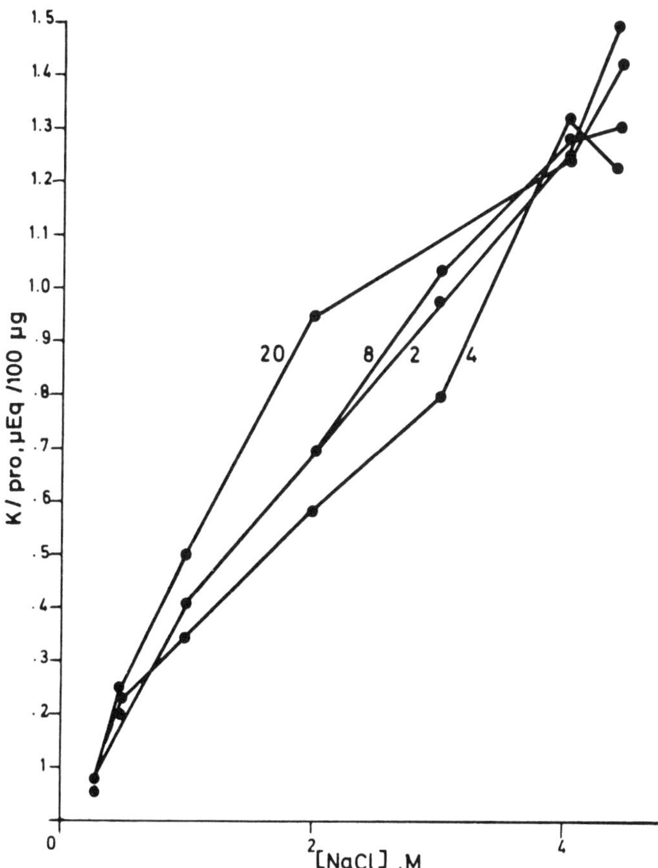

Fig. 10. Effect of variation of NaCl and KCl concentrations in medium on the cell K content of starved bacteria. The KCl concentration is indicated by the figure next to each curve. All the measurements were made at 20°C.

medium. K_{Cl} refers to that part of the cell K that is balanced by Cl. The fall of K_{Cl} was parallel to the fall of total cell K. Thus the loss of cell K was accounted for fully by loss of K_{Cl} and the K_x component remained unchanged except at the lowest [NaCl], measured where K_x is below the average value of 0.26 μequiv/100 μg. In other words, K is lost together with Cl; there is no uptake of Na in exchange for the K except at the lowest NaCl concentration.

4. Rate of Loss of K on Lowering [NaCl] in Medium

Loss of cell K is complete not more than 1 min after the [NaCl] is lowered (Table II). The precise rate of fall of K has not been measured.

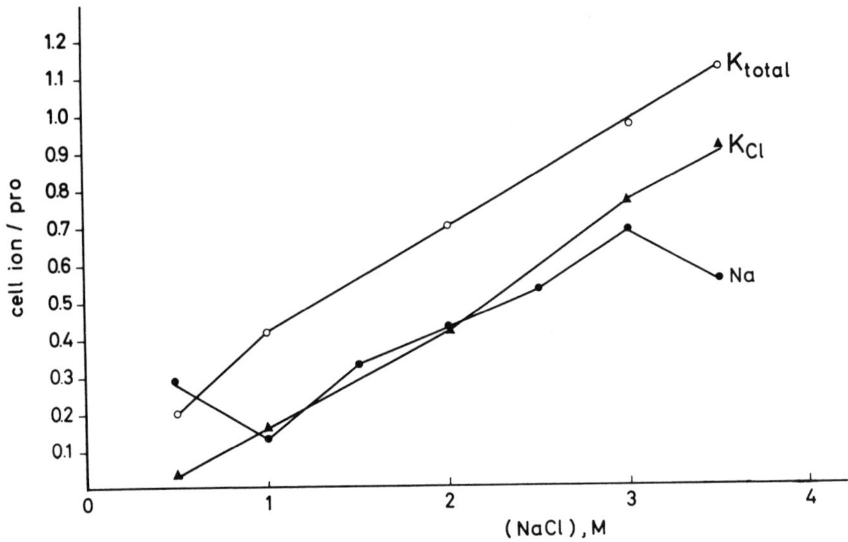

Fig. 11. Effect of the NaCl concentration in the medium on the cell K and Na content of starved bacteria equilibrated at 20°C. K_{total}, total cell K content; K_{Cl}, that part of the cell K balanced by Cl. Cell ion/pro represents cell K or Na in μequiv per 100 μg cell protein.

5. *Test of Uptake of K by Starving* Halobacterium

It has already been established that growing bacteria take up potassium [e.g., Fig. 5 in Ginzburg *et al.* (1971a)]. It was also shown that starving bacteria take up K when provided with yeast autolyzate [Fig. 14 in Ginzburg *et al.* (1971a)]. It seemed desirable to discover under which other conditions, if any, cells of *Halobacterium* sp. could take up potassium. As starving

Table II

Rate of Loss of Cell K When NaCl Concentration in Medium is Lowered[a]

Time	NaCl, M	Percent cell K	NaCl, M	Percent cell K
Initial	4	100	3.5	100
0	3.5	—	2	—
1 min	3.5	76	2	53
90 min	3.5	76	2	53

[a] Dilution of NaCl took place at time zero. Results are mean of not less than three separate determinations at each concentration.

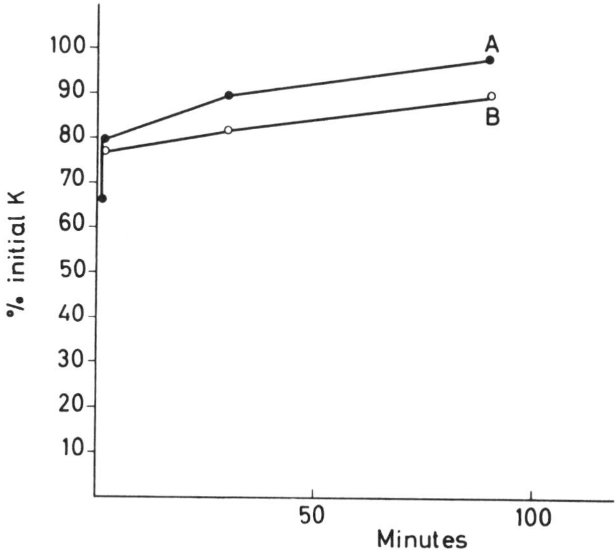

Fig. 12. Gain in cell K by starving bacteria due to increase in [NaCl] of medium. Group A cells spent 1 min at 2 M NaCl, and immediately after time zero they were returned to 4 M NaCl. Group B had spent 2 hr at 2 M NaCl. "Initial K" on the ordinate refers to the amount of K in bacteria before the [NaCl] had been lowered to 2 M; at this latter concentration the cell K fell to 66% of its value at 4 M. Temperature, 20°C. Mean of three experiments.

bacteria lost *K* when the [NaCl] of the ambient medium was lowered, it was wondered whether a subsequent raising of the [NaCl] could bring about gain of cell K. On investigation, it was found that a gain of this type depended upon the duration of exposure to the lower concentration. Figure 12, curve A, shows the K content of bacteria in 4 M NaCl after a 1-min exposure to 2 M NaCl and demonstrates that in the time interval of 90 min the cell K content returns to 97% of its initial value before the exposure to 2 M Nacl. Curve B refers to cells left for 2 hr at 2 M NaCl and only then returned to higher [NaCl] of 4 M. Although regain of cell potassium was considerable, it lagged behind at every step. The increase in cell K is clearly not coupled to metabolism, since it is dependent upon the time that the bacteria spent at the lower NaCl concentration. It is suggested that during this time some irreversible conformational change takes place within the bacteria.

6. *Effect of Temperature on Cell K in Starved* Halobacterium *sp.*

Suspensions of starving bacteria were incubated at different concentrations of NaCl, at temperatures varying from 0 to 40°C. At 4 M NaCl, significantly less cell K was found at 0 and 10°C (Fig. 13). We took 20°C as the temperature of choice for experiments since slight, but persistent reductions of cell K were found at higher temperatures.

It should be noted that the optimum temperature for growth of *Halobacterium* sp. is around 40°C. This is distinct from the optimum temperature for maintenance of cell K, i.e., 20°C. The difference between those two temperatures is yet another indication of the lack of dependence of cell K content on metabolism.

7. *Ion Replacement Experiments*

It might be possible to explain the role of NaCl in maintaining cell K in *Halobacterium* sp. by attempting to replace the NaCl with some substance of equal effectiveness. It has been known for years that the effect of NaCl is not due to its osmotic pressure (Larsen, 1967). It follows that it is the ionic nature of the substance which is of importance to the bacterial cell.

Preliminary experiments demonstrated that starving bacteria would have to be used for these experiments since growing cultures were unsuitable: Table III shows that in *Halobacterium* sp. cultures provided with the minimum concentration of 2 M NaCl to permit growth, the addition of 1 M LiCl brought about slight growth stimulation, but an additional 2 M LiCl inhibited growth (as measured by bacterial protein per milliliter of suspension), cell expansion, and K uptake.

It was decided to study two series of cations, the alkali cations Li^+, Na^+, K^+, Rb^+, Cs^+ and the alkaline earth cations Mg^{2+}, Ca^{2+}, Sr^{2+}, Ba^{2+}. It was quickly found that starved bacteria could withstand immersion in solutions of only a few of these cation chloride salts, namely Li^+, Na^+, Mg^{2+}, Ca^{2+}. In solutions of the others, at all concentrations from the lowest tried to 4 M, the bacteria disintegrated and suspensions became transparent. The effects on

Table III

Effect of LiCl on Growth of *Halobacterium*[a]

Major salts in medium	Bacterial protein, μg/ml	Volume/protein, mm³/100 μg	K/protein, μequiv/100 μg
3.5 M NaCl	560	0.82	1.04
2 M NaCl	200	0.24	0.24
2 M NaCl + 1 M LiCl	290	0.45	0.58
2 M NaCl + 2 M LiCl	Disintegration	—	—

[a] Measurements made on 48-hr-old cultures.

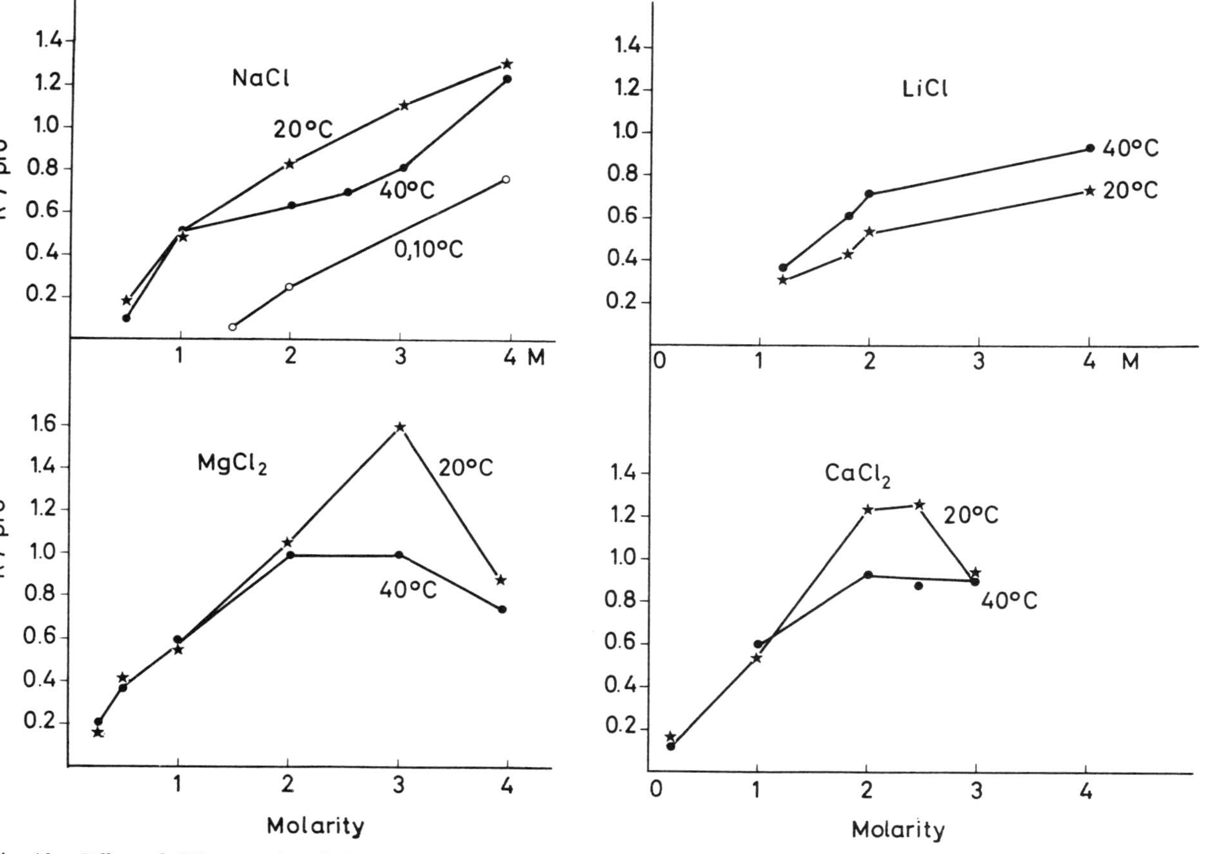

Fig. 13. Effect of different salt solutions on cell K content of starved *Halobacterium* sp. K/pro: μequiv cell K per 100 μg cell protein. Temperature of experiment indicated on each curve.

cell K content of incubations in solutions of LiCl, NaCl, $MgCl_2$, and $CaCl_2$ are shown in Fig. 13. Higher concentrations than those shown bring about coagulation. LiCl is shown to be somewhat less effective than NaCl in maintaining cell K, while $MgCl_2$ and $CaCl_2$ are actually more effective than NaCl. As the bacteria were transferred from solutions of NaCl to $MgCl_2$, the transfer must have been accompanied by an uptake of K. This uptake, entirely nonmetabolic, was brought about by the difference in nature of the cations. In Fig. 14 cell K is plotted against the activity of the major salt in the solution; it is clear that $CaCl_2$ and NaCl are of equal effectiveness.

The changes in cell stability and K content with size of crystal radius of the cation are shown in Fig. 15. It seems then that the only factor that cations of similar effectiveness have in common is their size; the charge on the ion does not appear to be of importance. The significance of this is discussed later (p. 243).

$MgCl_2$ and $CaCl_2$ are similar to NaCl in that their effectiveness in maintaining cell K is lower at 40°C than at 20°C (Fig. 13). This temperature

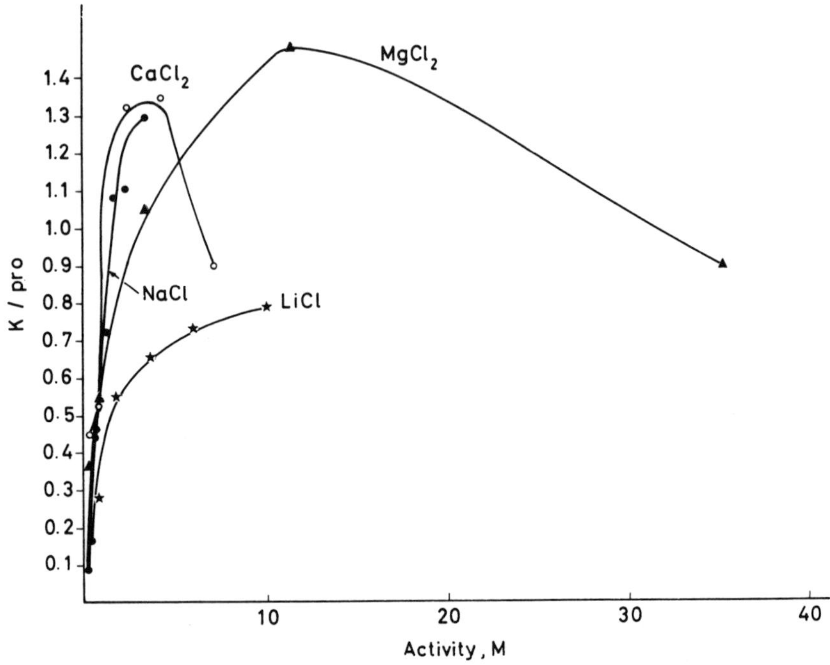

Fig. 14. Effect of activity of salt solution on cell K content in starved *Halobacterium* sp. K/pro represents cell K in μequiv K per 100 μg cell protein. ●, NaCl; ○, $CaCl_2$; ▲, $MgCl_2$; ★, LiCl.

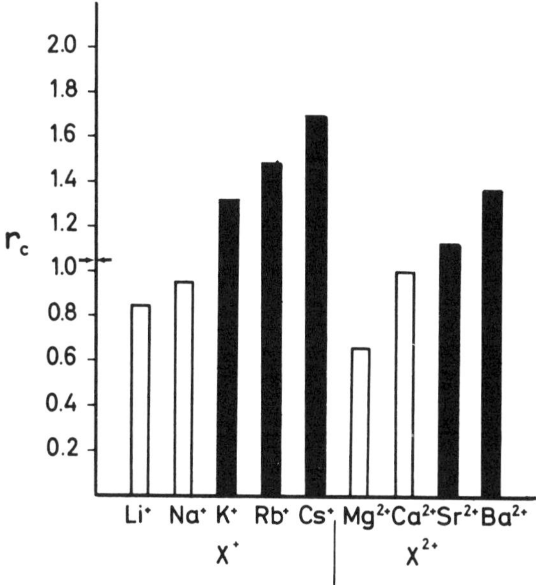

Fig. 15. Crystal radius r_c of cations used in ion-replacement experiments. r_c of Li^+ from Gourary and Adrian (1960); r_c of other metals from Wyckoff (1968). The integrity of *Halobacterium* sp. cells is maintained in concentrated solutions of cation chlorides with cation crystal radius lower than the value shown by arrows.

effect becomes noticeable at concentrations above 1–2 M. LiCl is unique in that it is more effective at 40°C than at 20°C.

Figure 16 shows the results of replacing a portion of the NaCl by KCl, RbCl, or CsCl. At the concentrations shown in the figure there were no losses of bacterial protein. There were, however, losses of cell K in CsCl and RbCl. Since other work (Ginzburg and Ginzburg, unpublished) has shown that Rb and Cs can be bound in place of K, the results of these experiments are not completely comparable with those of Fig. 13, in which no ion other than K was bound.

A preliminary report on the effect of replacing Cl^- by some other anion on amount of cell K has appeared already (Ginzburg *et al.*, 1971b). The relative effectiveness of the halide anions and of a few organic anions is shown in Fig. 17. This shows that NaCl and NaBr are equally effective in maintaining cell K; NaI is indistinguishable from the two former salts at concentrations up to 2 M, but no further increase occurs at higher concentrations. No organic anion was found to be as effective as NaCl or NaBr; cell K content was considerable, however, with 2 M sodium perchlorate. Smaller amounts were found with the sodium salts of thiocyanate and salicylate.

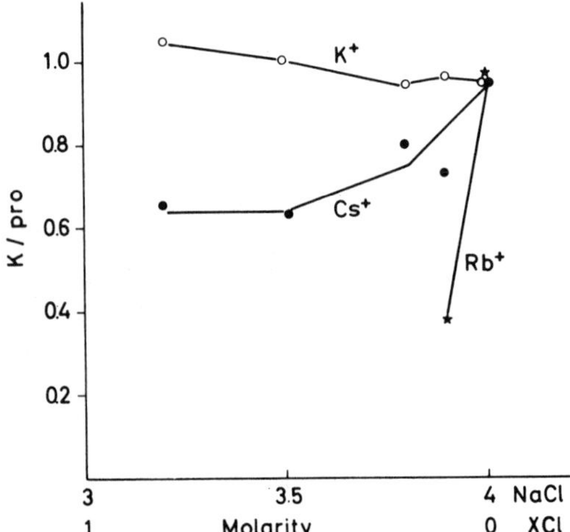

Fig. 16. Effect of mixtures of NaCl and KCl, CsCl, or RbCl on cell K content of starved *Halobacterium* sp. K/pro: μequiv cell K per 100 μg protein. Total salt concentration in medium maintained at 4 M. Progression from right to left represents progressive replacement of Na^+ by X^+, where X^+ is K^+, Cs^+, or Rb^+.

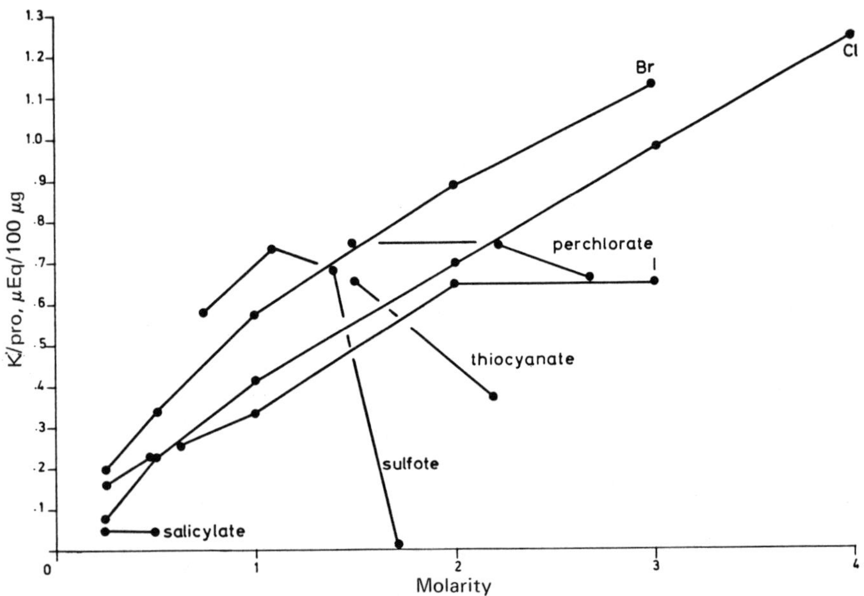

Fig. 17. Effect of different Na salt solutions on cell K content of starved *Halobacterium* sp. K/pro: μequiv cell K per 100 μg protein. Temperature of experiments, 37°C.

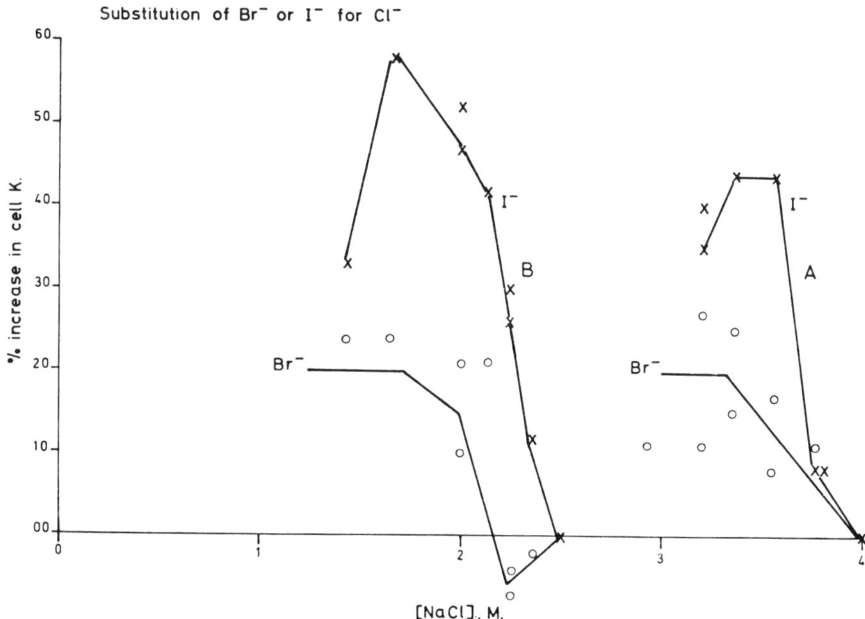

Fig. 18. Effect of mixtures of NaCl and NaBr or NaI on cell K content of starved *Halobacterium* sp. Cell K as percentage of amount found in NaCl. In groups A and B the total Na^+ concentration was maintained at 4 and 2.5 M, respectively. Progression from right to left represents progressive replacement of Cl^- by Br^- or I^-. Temperature of experiment: 40°C.

A detailed comparison of the chloride, bromide, and iodide ions was made by substituting small amounts of Br^- or I^- for Cl^- while keeping the overall salt concentration constant. The amounts substituted were not enough to cause protein breakdown. The results of these experiments are shown in Fig. 18. Group A refers to bacteria kept at a constant $[Na^+]$ of 4 M, up to 0.8 M Cl^- being substituted by Br^- or I^-. One notices that at 40°C the order of anion effectiveness is $I^- > Br^- > Cl^-$ when 0.4–0.6 M of the alternative ion was substituted for Cl^-. At 20°C, I^- was significantly more effective than Cl^-. It should be noted that in these experiments there was a net, nonmetabolic uptake of cell K in presence of NaI or NaBr. In group B the total salt concentration was maintained at 2.5 M and up to 1 M NaCl was substituted by NaBr or NaI. In these bacteria it was only at 40°C that NaI and NaBr brought about increases in the amount of bound cell K.

The major conclusion to be drawn from these experiments is that NaCl and NaBr are equally effective in maintaining cell K, while NaI at low concentrations is significantly more effective, and becomes even more so as the temperature is increased from 20 to 40°C.

D. Discussion

The experimental work described in this section has shown that the integrity of the bacterial cell and the amount of cell potassium are determined by the *p*H of the medium on the one hand and by the nature and concentration of the major salt in the medium on the other. This conclusion applies both to metabolizing and to starving bacteria: Cell K fell with decrease in NaCl or *p*H in the medium in very similar fashion in both types of cell (Figs. 5 and 9). The cell K content is thus determined by equilibrium properties of the system, rather than by the metabolic activity of the bacteria. In starved cells K is lost when the NaCl in the medium is decreased, when an unsuitable ion is substituted for Na in the medium, or when the temperature is changed from 20°C. Conversely, there is a gain of cell K against the concentration gradient when at least two out of the above three conditions are reversed. The cell K is therefore present in a state of equilibrium with the cell surroundings.

The cell K consists of two fractions, one balanced by Cl^-, K_{Cl}, and the other by organic anions, K_x. These two fractions reactions reacted differently to changes in the outside medium; thus only K_{Cl} was affected by decreases in NaCl concentration of the outside medium (Fig. 11). The difference between K_{total} and K_{Cl} gives the amount of K_x; this latter stayed almost constant throughout the whole range of NaCl concentrations tried. In Fig. 11 the average amount of K_x was 0.24 μequiv per 100 μg protein. Now it is known from protein analyses of other *Halobacterium* species (Bayley, 1966; Reistad, 1966) that about 30% of the amino acid residues are either aspartate or glutamate. Since there are about 900 amino acid residues per 10^5 g of halophilic protein, there must be 270 carboxylate groups per 10^5 g, or 0.27 μequiv COO^- per 100 μg protein. On the assumption that the amino acid composition of our *Halobacterium* is similar to that of the species analyzed, there is good agreement between the amount of K_x and that of COO^- in the protein. On a rough estimate, the amounts of other inorganic anions seem to be equivalent with those of the basic amino acid residues and can be neglected.

When the *p*H was reduced from 7.0 to 5.5, about half of the K_x was lost, presumably in exchange for H^+ since changes in Na balance those of Cl. A *p*H of 5.5 is in the range of titration of carboxylate groups on proteins, the exact *pK* being modified by electrostatic interactions and other factors (Edsall and Wyman, 1958). Thus the experiments shown in Fig. 8 support the hypothesis that K_x is balanced by carboxylate groups.

The experimental evidence supports the conclusion that K_{Cl} is affected primarily by the NaCl concentration of the medium, while K_x is affected by *p*H. However, when the *p*H was lowered below 5.0, there was a total loss of potassium, suggesting that the *p*H affects both the state of ionization of the fixed COO^- groups and K_{Cl} in some other, unknown way. It should be stressed that the electrostatic balancing of K by these COO^- groups can explain neither the K^+/Na^+ specificity nor the phenomenon of K binding; additional factors must be involved.

Some insight into the effect of low *p*H is seen in Fig. 6, which shows rates of K loss from bacteria held at *p*H 4. At each temperature the kinetics of K loss can be described by two kinetic constants; these cannot be related to the two fractions of K in the cell. From an Arrhenius plot of log [K] versus $1/T$ (Fig. 7) it can be seen that the enthalpy of activation of the initial loss is around −7000 cal/mol and that of the second phase is about −18,000 cal/mol. Neither value fits the diffusion of H^+ or K^+ in aqueous solution, or the ionization of the carboxylate group. The high enthalpies of activation may indicate the presence of a microenvironment other than the usual aqueous one. One should contrast the rate of K loss when the *p*H is lowered with the rate of loss when the NaCl concentration is lowered (Table II); this latter is very much faster than the former.

Let us now discuss effects of the concentration of the major salt in the medium. Figure 10 demonstrates these effects on cell K at different levels of KCl in the medium. It has been shown in Section I that the desorption isotherm of K^+ at 4 M NaCl has an apparent dissociation constant of 0.1–0.2 mM. As a first approximation all the lines in Fig. 10 can be taken to be coincident. Thus in the range 2–20 mM KCl the increase of cell K with NaCl concentration in the medium must be due to a change in the binding capacity of the system, and not to a change in dissociation constant. Of course it cannot be ruled out that at KCl concentrations below 2 mM the amount of cell K might be affected by differences in the dissociation constant.

The effect of salt concentration on amount of cell K was found to be affected by the temperature of the cells (Fig. 13). At higher concentrations of NaCl, $MgCl_2$, and $CaCl_2$ there was more cell K at 20° than at 40°; in NaCl there was more cell K at 20° than at 0° or 10°. This dual effect of temperature indicates that the cell K content is controlled by at least two reactions.

The nature of the salt solution in the medium affects the integrity of the bacteria as well as the amount of cell K; these effects are summarized in Fig. 15, which shows that the bacteria lose K and disintegrate in solutions of salts with cations with a crystal radius above 1.1 Å. When the crystal radius of the cation is smaller, the cells remain intact and the cell K may be higher even than in solutions of NaCl. We are thus confronted with the paradox that these bacteria cannot withstand immersion in solutions of, e.g., 4 M KCl although the apparent internal concentration of K is of the same order, and although the cell membrane is freely permeable to K^+. Since the same ion has different effects according as to whether it is inside or outside the cell, it appears that the cell proteins must be involved in two ionic interactions, one short range (e.g., with cell K) and the other long range—unfavorable with K, favorable with cations with a small crystal radius. These small cations are known as "structure-making" when present in aqueous solution. In Samoilov's terms they are positively hydrated, in contrast to the larger cations, which are negatively hydrated (Samoilov, 1965). Figure 15 resembles qualitatively Fig. 5 of Samoilov (1972); in both figures the transition between the two groups of cations occurs between Na and K at a crystal radius of 1.1 Å.

The influence of cations on water structure has been measured by macroscopic viscosity (Stokes and Mills, 1965), NMR spectroscopy (Giese *et al.*, 1970), and self-diffusion of water (Wang, 1954). Samoilov has stressed that the cation–water molecule interactions take place over a very short range. Within the bacteria, on the other hand, the interaction in which the structure-making cation is involved appears to be relatively long range. At present there is no way of relating the effects which structure-making ions have on water in aqueous solutions with the effects they have within the halophilic cell, although it is reasonable to suggest that these cations affect the cell proteins through their effects on the cell water.

Berendsen has discussed the subject of water of hydration of protein (Berendsen, 1974). He distinguishes between specific water of hydration, amounting to 0.2–0.5 mol of water per 100 g of protein, and nonspecific water of hydration, amounting to about 3 mol per 100 g of protein. Nonspecific water of hydration does not freeze at −80°C; the molecules rotate more slowly than molecules in the ordinary liquid state although they are not strongly immobilized.

Berendsen's discussion referred to proteins in water or in dilute solutions. In contrast, the proteins in intact halophilic bacteria are dissolved in concentrated salt solution and it is of interest to see how his estimations may be applied to more salty proteins. In the halophilic cell, it can be estimated that there are about 11 mol of water per 100 g of protein (Ginzburg *et al.*, 1970). Since 70% of the cell water does not freeze at −18°C[1]—the temperature at which the medium freezes—11 × 0.7 mol of cell-water may bear some resemblance to Berendsen's nonspecific water of hydration. The cell-water of *Halobacterium* could be regarded as a two-phase system, one phase consisting of a modified form of nonspecific water of hydration (*sw*) and the other with properties close to that of the bulk phase (*w*). It is suggested that the dielectric constant of the *sw* phase is lower than that of the *w* phase. In all media of low dielectric constant ion-pair formation is favored (Robinson and Stokes, 1959), and in *sw*, KCl and potassium carboxylate would tend to be in the associated form. The model suggested is shown in Fig. 19(c).

[1] Unpublished observations of H. Edzes, B. Z. Ginzburg, M. Ginzburg, and H. J. C. Berendsen.

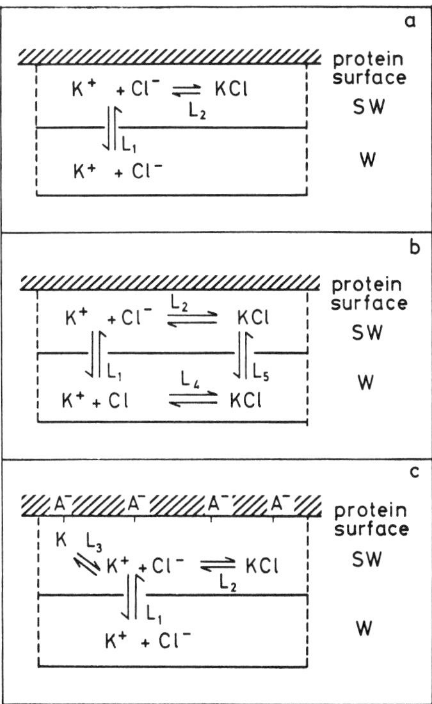

Fig. 19. Model of potassium binding in *Halobacterium* sp. *w*, water phase. *sw*, water influenced by protein surface. A^-, fixed organic anions. For the definition of the L parameters, see Eq. (1). (a) Simplified model showing association of K^+ and Cl^- in *sw* phase. (b) Association of K^+ and Cl^- in both phases. (c) Complete model showing association of K^+ and Cl^- in *sw* phase and electrostatic interaction of K^+ and A^-.

It is suggested that cations of the structure-making group do not enter the *sw* phase, due to an unfavorable ΔG of hydration. An analogous situation is found in the inner Helmholtz layer at electrified interfaces of metal electrodes (Bockris and Reddy, 1970; Gierst *et al.*, 1960).

The concentration ratio of K in the cell relative to K outside is

$$\frac{C_{sw}^{\mathrm{K}}}{C_w^{\mathrm{K}}} = L_1\left(1 + \frac{L_1}{L_2}\, C_w^{\mathrm{K}} + \frac{A_T}{L_3 + C_w^{\mathrm{K}} L_1}\right) \tag{1}$$

where C_w^{K} is the concentration of K in the water (w) phase, C_{sw}^{K} is the concentration of K in the surface-water (sw) phase, L_1 is the coefficient of distribution of K between the w and sw phases, L_2 is the equilibrium constant between ion-pair KCl and dissociated form, L_3 is the association constant of K to carboxylate on protein, and A_T is the total amount of fixed anionic charges.

A derivation of Eq. (1) is given in the appendix. The equation is a formal description of the model. It expresses explicitly the separation of the cell into two fractions, K_{Cl}, represented by the second term, and K_x, represented by the third term within the parentheses. Since K_{Cl} is the major form of K in the cell, the model predicts that most of the cell K will be in the form of

KCl, i.e., will exist as an ion-pair with Cl. In principle, measurements of the conductivity of the cell solution should check this prediction. L_3, the association constant of K to carboxylate groups on the protein, should be affected by the value of the dielectric constant. The size of the third term in Eq. (1) is limited by A_T, the total amount of fixed anionic charges; it can be expected to be at its maximum when L_3 is smallest for a given L_1 and outside KCl concentration. It was shown before that there is a good correspondence between K_x and A_T, the number of fixed carboxylate groups.

The model, as it stands, can explain several of the phenomena described in this section:

1. It obviates the need for a special binding material.

2. It can lead to an explanation of cell Na and Cl concentrations. This subject is dealt with more fully elsewhere.

3. It can explain the temperature effects described in this section; if the cell K is a function of several different reactions, one would expect disjunct temperature effects of the type shown in Fig. 6.

4. It would explain the release of KCl when the cells are ruptured, since the K is not "bound," but is retained within a special phase.

5. It explains the specificity between K and Na as due to the elimination of Na from the *sw* phase on thermodynamic grounds. Whether the mechanism responsible for the maintenance of the *sw* phase is related to the structure-making properties of cations with small crystal radii remains an open question.

E. Summary

The organism used for this work was a species of *Halobacterium* which grows optimally at NaCl concentrations of not less than 3.5 M. The cells of this organism contain 3–4 mol potassium, which is retained when the rate of metabolism is reduced to its lowest level, even though the outer membrane is permeable to K^+; cell K is therefore apparently bound. Measurements were made of the variation in cell K with change in external *p*H, NaCl concentration, and nature of the major salt in the ambient medium. Both starving and growing bacteria were tested, as far as possible. Cell K consisted of two portions, one balanced by Cl^- (K_{Cl}) and the other apparently by fixed organic anions (K_x). Maximum amounts of cell K were found at *p*H 7; as the *p*H was lowered, K_x was lost, and below *p*H 5.5, K_{Cl}. The amount of cell K was dependent upon the [NaCl] of the medium, starving bacteria tolerating [NaCl] as low as 0.5 M, while growing bacteria required at least 2 M NaCl. Starving bacteria retained their integrity and potassium only in solutions of salts with cations with a crystal radius equal to or less than

1.1 Å (mono- or divalent cations) and with small anions. It is postulated that interactions between the protein surface and major salt solution cause ordering of the water in this interphase, K^+ rather than Na^+ being favored there due to the difference in hydration energy of the two cations. Then K^+ would serve to balance the COO^- groups belonging to acidic amino acid residues and known to be abundant in halophilic proteins; this portion of the cell K may be equivalent to K_x. The association of $K^+ + Cl^-$ to KCl would be favored by the low dielectric constant of the interphase "bound" water and would account for the K_{Cl} fraction of the cell K.

APPENDIX

The model postulates that water near the surface of the halophilic protein is influenced by interactions between the salt and the protein, and accordingly acquires properties different from those of the bulk water. There are therefore two phases, *sw* and *w*. The following assumptions are made concerning the *sw* phase:

1. The dielectric constant of the phase is low and the dissociation of salts is lower than in the bulk phase.
2. Ions such as Na and other order-making ions do not enter this phase due to an unfavorable ΔG of hydration.
3. It is not assumed that $\mu^0_{KCl}(PT)$ in the *sw* and the *w* phases are equal; thus it cannot be said whether

$$L_1 = \exp[(\mu^0_{K(sw)} - \mu^0_{K(w)})/RT] = 1 \tag{A1}$$

We will first test the distribution of K between the *sw* and *w* phases. For this step the simplified model (Fig. 19a) is used:

$$L_1 = [K^+]_{sw}/[K^+]_w \tag{A2}$$

$$L_2 = [K^+]_{sw}[Cl^-]_{sw}/[KCl]_{sw} \tag{A3}$$

It is assumed that

$$[K^+]_{sw} = [Cl^-]_{sw} \tag{A4}$$

Thus

$$L_2 = [K]^2_{sw}/[KCl]_{sw} \tag{A2}$$

and we have

$$\text{total K in } w \text{ phase} \equiv C_w{}^K = [K^+]_w \tag{A6}$$

$$\text{total K in } sw \text{ phase} \equiv C^K_{sw} = [K^+]_{sw} + [KCl]_{sw} \tag{A7}$$

The ratio of the total K concentration in *sw* to that in *w* is

$$\frac{C_{sw}^{\mathrm{K}}}{\mathrm{C}_w{}^{\mathrm{K}}} = \frac{[\mathrm{K}^+]_{sw} + [\mathrm{KCl}]_{sw}}{[\mathrm{K}^+]_w} = \frac{[\mathrm{K}^+]_{sw}}{[\mathrm{K}^+]_w} + \frac{[\mathrm{KCl}]_{sw}}{[\mathrm{K}^+]_w} \tag{A8}$$

Since

$$[\mathrm{KCl}]_{sw} = [\mathrm{K}]_{sw}^2/L_2 \tag{A9}$$

we have

$$C_{sw}^{\mathrm{K}}/C_w{}^{\mathrm{K}} = L_1 + ([\mathrm{K}]_{sw}^2/L_2[\mathrm{K}^+]_w) \tag{A10}$$

On multiplying the second term by $[\mathrm{K}]_w/[\mathrm{K}]_w$, we obtain

$$C_{sw}^{\mathrm{K}}/C_w{}^{\mathrm{K}} = L_1 + (L_1{}^2/L_2)\, C_w{}^{\mathrm{K}} \tag{A11}$$

So far, the association of $\mathrm{K}^+ + \mathrm{Cl}^- \rightarrow \mathrm{KCl}$ in the bulk water phase has been neglected. For the sake of completeness, this will now be checked explicitly (Fif. 19b). By definition,

$$L_4 \equiv [\mathrm{K}]_w{}^2/[\mathrm{KCl}]_w \tag{A12}$$

From (A6) and (A7)

$$\frac{C_{sw}^{\mathrm{K}}}{C_w{}^{\mathrm{K}}} = \frac{[\mathrm{K}^+]_{sw} + [\mathrm{KCl}]_{sw}}{[\mathrm{K}^+]_w + [\mathrm{KCl}]_w} \tag{A13}$$

$$= \frac{[\mathrm{K}^+]_{sw}}{[\mathrm{K}^+]_w + [\mathrm{KCl}]_w} + \frac{[\mathrm{KCl}]_{sw}}{[\mathrm{K}^+]_w + [\mathrm{KCl}]_w} \tag{A14}$$

From (A12),

$$[\mathrm{KCl}]_w = [\mathrm{K}]_w{}^2/L_4\,.$$

By substitution in (A14)

$$\frac{[\mathrm{K}^+]_{sw}}{[\mathrm{K}^+]_w + [\mathrm{KCl}]_w} = \frac{[\mathrm{K}^+]_{sw}}{[\mathrm{K}^+]_w + ([\mathrm{K}]_w{}^2/L_4)}$$

$$= \frac{[\mathrm{K}^+]_{sw}}{[\mathrm{K}]_w{}^2\{(1/[\mathrm{K}^+]_w) + (1/L_4)\}}$$

$$\frac{[\mathrm{K}^+]_{sw}}{[\mathrm{K}^+]_w + [\mathrm{KCl}]_w} = \frac{[\mathrm{K}^+]_{sw}}{[\mathrm{K}]_w{}^2}\,\frac{[\mathrm{K}^+]_w\, L_4}{[\mathrm{K}^+]_w + L_4}$$

$$= \frac{L_1 L_4}{[\mathrm{K}^+]_w + L_4} \tag{A15}$$

It is assumed that in water KCl is mostly in the dissociated form, i.e., $[K^+]_w \gg [KCl]_w$. It follows that $[K^+]_w/[KCl]_w \gg 1$ and $L_4 \gg [K^+]_w$. We can neglect $[K^+]_w$ relative to L_4, and from (A15)

$$\frac{[K^+]_{sw}}{[K^+]_w + [KCl]_w} = L_1 \tag{A16}$$

On substituting from (A5) in the numerator of (A16) we obtain

$$\frac{[K^+]^2_{sw}/L_2}{[K^+]_w\{(1/[K^+]_w) + (1/L_4)\}} = \frac{[K]^2_{sw}}{[K]_w{}^2} \frac{1/L_2}{(1/[K^+]_w) + (1/L_4)} \tag{A17}$$

$$= \frac{L_1{}^2}{L_2} \frac{[K^+]_w \, L_4}{[K^+]_w + L_4} \tag{A18}$$

Since $[K^+]_w \ll L_4$, we obtain

$$[K^+]_{sw}/([K^+]_w + [KCl]_w) = L_1{}^2[K^+]_w/L_2 \tag{A19}$$

or [Eq. (A11)]

$$C^K_{sw}/C_w{}^K = L_1 + (L_1{}^2/L_2)\, C_w{}^K$$

Thus, on the assumption that KCl is largely dissociated in the *w* phase, the model of Fig. 19(b) yields the same result as that of Fig. 19(a).

We now add the remaining parameter needed to complete the model (Fig. 19c), i.e., the reaction between K^+ and the fixed anions A^-:

$$L_3 = [K^+]_{sw}[A^-]_{sw}/[KA]_{sw} \tag{A20}$$

where A^- is the ionized fixed organic anion and KA denotes bound K; we have

$$\frac{C^K_{sw}}{C_w{}^K} = \frac{[K^+]_{sw} + [KCl]_{sw} + [KA]_{sw}}{[K^+]_w} \tag{A21}$$

On substituting $[KA]_{sw}$ from Eq. (A20), we obtain

$$[KA]_{sw}/[K^+]_w = [K^+]_{sw}[A^-]_{sw}/L_3[K^+]_w \tag{A22}$$

Since $[K^+]_{sw}/[K^+]_w = L_1$, we have

$$[KA]_{sw}/[K^+]_w = L_1[A^-]_{sw}/L_3 \tag{A23}$$

We introduce A_T:

$$[A_T]_{sw} = [KA]_{sw} + [A^-]_{sw} \tag{A24}$$

From Eq. (A10)

$$[A_T] = (L_3[\mathrm{KA}]_{sw}/[\mathrm{K}]_{sw}) + [\mathrm{KA}]_{sw} \tag{A25}$$

$$= [\mathrm{KA}]_{sw}\{1 + (L_3/[\mathrm{K}^+]_{sw})\} \tag{A26}$$

$$= [\mathrm{KA}]_{sw}\{([\mathrm{K}^+]_{sw} + L_3)/[\mathrm{K}^+]_{sw}\} \tag{A27}$$

we have

$$[\mathrm{A}^-] = [A_T] - [\mathrm{KA}]$$

$$= [A_T] - \{[A_T][\mathrm{K}^+]_{sw}/([\mathrm{K}^+]_{sw} + L_3)\} \tag{A28}$$

$$= [A_T]([\mathrm{K}^+]_{sw} + L_3 - [\mathrm{K}^+]_{sw})/([\mathrm{K}^+]_{sw} + L_3) \tag{A29}$$

$$= [A_T]\, L_3/(L_3 + [\mathrm{K}^+]_{sw}) \tag{A30}$$

Since $[\mathrm{K}^+]_{sw} = L_1[\mathrm{K}^+]_w$, we have

$$[\mathrm{A}^-] = A_T\, L_3/(L_3 + L_1[\mathrm{K}^+]_w) \tag{A31}$$

From (A24)

$$\frac{[\mathrm{KA}]_{sw}}{[\mathrm{K}^+]_w} = \frac{L_1}{L_3}\frac{L_3}{L_3 + L_1[\mathrm{K}^+]_w} \tag{A32}$$

$$= \frac{A_T L_1}{L_3 + L_1[\mathrm{K}^+]_w} \tag{A33}$$

Thus

$$\frac{C_{sw}^{\mathrm{K}}}{C_w^{\mathrm{K}}} = L_1\left(1 + \frac{L_1}{L_2}\, C_w^{\mathrm{K}} + \frac{A_T}{L_3 + C_w^{\mathrm{K}} L_1}\right)$$

Acknowledgments

This work was performed with the help of a grant-in-aid from the Israel National Academy of Sciences and Humanities to M.G. The technical assistance of Mrs. Liliana Richman is gratefully acknowledged.

The idea of studying ion transport in halophilic bacteria was orginally suggested by Prof. Aharon Katchalsky, who gave unstinting guidance and criticism until his death.

REFERENCES

Adamson, A. W., 1967, "Physical Chemistry of Surfaces," 2nd ed., pp. 568–569, Interscience, New York.

Bayley, S. T., 1966, Composition of ribosomes of an extremely halophilic bacterium, *J. Mol. Biol.* **15**:420.

Berendsen, H. J. C., 1974, Specific interactions of water with biopolymers, *in* "Water, A Comprehensive Treatise" (F. Franks, ed.), Vol. 5, Chapter 6, p. 293, Plenum Press, New York.

Bockris, J. M., and Reddy, A. K. N., 1970, "Modern Electrochemistry," Vol. 2, p. 750.

Christian, J. H. B., and Waltho, J. A., 1962, Solute concentrations within cells of halophilic and non-halophilic bacteria, *Biochim. Biophys. Acta* **65**:506.

Edsall, J. T., and Wyman, J., 1958, "Biophysical Chemistry," pp. 522–540.

Gierst, L., Nicholas, E., and Tytgat-Vandenberghen, L., 1960, Double-layer studies using depolarizers as probe—A reassessment, *Croatica Chem. Acta* **42**:117.

Giese, K., Kaatze, U., and Pottel, R., 1970, Permittivity and dielectric and proton magnetic relaxation of aqueous solutions of the alkali halides, *J. Phys. Chem.* **74**:3718.

Ginzburg, M., Sachs, L., and Ginzburg, B. Z., 1970, Ion metabolism in a Halobacterium. I. Influence of age on intracellular concentrations, *J. Gen. Physiol.* **55**:187.

Ginzburg, M., Sachs, L., and Ginzburg, B. Z., 1971a, Ion metabolism in a Halobacterium. II. Ion concentrations in cells at different levels of metabolism, *J. Membrane Biol.* **5**:78.

Ginzburg, M., Ginzburg, B. Z., and Tosteson, D. C., 1971b, The effects of anions on K^+ binding in a Halobacterium species, *J. Membrane Biol.* **6**:259.

Gourary, B. S., and Adrian, F. J., 1960, *in* "Solid State Physics," Vol. 10, p. 127.

Katchalsky, A., Oplatka, A., and Litan, A., 1964, The dynamics of macromolecular systems, *in* "Molecular Architecture in Cell Physiology" (T. Hayashi and A. G. Szent-Gyorgyi, eds.), Prentice-Hall, Englewood Cliffs, New Jersey.

Larsen, H., 1967, Biochemical aspects of extreme halophilism, *Advan. Microbiol. Physiol.* **1**:97.

Reistad, S. T., 1966, On the composition and nature of the bulk protein of extremely halophilic bacteria, *J. Mol. Biol.* **15**:420.

Robinson, R. A., and Stokes, R. H., 1959, "Electrolyte Solutions," Butterworths, London.

Samoilov, O. Ya., 1965, "Structure of Aqueous Electrolyte Solutions and the Hydration of Ions," Consultants Bureau, New York.

Samoilov, O. Ya., 1972, Residence times of ionic hydration, *in* "Water and Aqueous Solutions: Structure, Thermodynamics and Transport Processes" (R. A. Horne, ed.), Chapter 14, Wiley–Interscience, New York.

Steinhardt, J., and Reynolds, J. A., 1969, "Multiple Equilibria in Proteins." Academic Press, N.Y.

Stokes, R. H., and Mills, R., 1965, *in* "The International Encyclopedia of Physical Chemistry and Chemical Physics" (E. A. Guggenheim, J. E. Meyer, and F. C. Tomkins, eds.), Vol. 3, Pergamon Press, London.

Wang, J. H., 1954, Effect of ions on the self-diffusion and structure of water in aqueous electrolytic solutions, *J. Phys. Chem.* **58**:686.

Wyckoff, R. G., 1968, "Crystal Structures," Interscience, New York.

Index